Extremophilic Enzymatic Processing of Lignocellulosic Feedstocks to Bioenergy

Rajesh K. Sani • R. Navanietha Krishnaraj
Editors

Extremophilic Enzymatic Processing of Lignocellulosic Feedstocks to Bioenergy

 Springer

Editors
Rajesh K. Sani
Department of Chemical and Biological
  Engineering
South Dakota School of Mines and
  Technology
Rapid City, South Dakota
USA

R. Navanietha Krishnaraj
Department of Chemical and Biological
  Engineering
South Dakota School of Mines and
  Technology
Rapid City, South Dakota
USA

ISBN 978-3-319-54683-4   ISBN 978-3-319-54684-1   (eBook)
DOI 10.1007/978-3-319-54684-1

Library of Congress Control Number: 2017942740

Printed on acid-free paper

This Springer imprint is published by Springer Nature
The registered company is Springer International Publishing AG
The registered company address is: Gewerbestrasse 11, 6330 Cham, Switzerland

Preface

Biochemical processes have been realized as the ideal option for replacing physicochemical processes in an efficient, eco-friendly, and economical manner. The understanding of the enzymes, their catalysis, and their applications are mandate for the engineers working in the industry. Today, most industries which were making use of chemical processes have replaced several of their processes with bioprocesses because of the several advantages. Hence, it becomes equally important for the engineers to understand the bioprocess on a par with chemical processes. Enzymes are widely used for several industrial processes these days. Different enzymes have been explored for real-time applications in various industries such as biofuel, detergent, brewing, culinary, dairy, paper industry, food processing, starch, molecular biology research, as well as biosensor development. Enzymes have been thought to be less advantageous than microbial whole cell catalysts as they are fragile and get denatured easily. However, with the findings of a new way for exploiting enzymes that can operate in severe operating conditions from extremophilic organisms, the scope of using enzymes for industrial applications has improved tremendously.

The extremophilic enzymes can operate at lower or higher pH conditions, different range of temperatures, and different pressures, etc. The idea of exploring the enzymes from extremophiles is not new. For example, Taq polymerase, a thermostable enzyme with a half-life of greater than 2 h at 92.5 °C and which can function at around 70–80 °C, was isolated from a thermophilic bacteria *Thermus aquaticus* in 1976. The Taq polymerase is being used for amplification of DNA in polymerase chain reaction for over four decades. Thermophilic enzymes will likely have major applications in the selective synthesis of economically valuable compounds in a large-scale setup. These enzymes are promising candidates for the development of amperometric biosensors for the detection of analyte for diagnosis, biomedical, food industry, etc. Extremophilic enzymes are promising candidates for carrying out operations in adverse conditions such as space, mining (biomining/bioleaching), deep sea, etc. All these motivated us to write the current text focused on industrially important extremophilic enzymes. The knowledge of extremophilic enzymes is essential for chemists, biochemists, chemical/biochemical/bioprocess engineers, biotechnologists, molecular biologists, genetic engineers, as well as computational biologists. The research activities are going on at rapid speed in identifying the sources and applications of extremophilic enzymes; however, we are still in infancy in terms of taking extremophilic enzyme technologies for real-time application. There are a few extremophilic enzymes which have been taken up for real-time applications, while hundreds of extremophilic enzymes will be used for industrial applications in the near future.

It has been established that extremophilic enzymes play important roles in many kinds of bioprocessing, e.g., in conversion of biomass into biofuels. Existing enzymatic technologies (e.g., hydrolysis of lignocellulose into sugars) have several limitations including very slow enzymatic hydrolysis rates, low yields of products (often incomplete hydrolysis), high dosages of enzymes, and sensitive to microbial contamination problems. These limitations can be overcome using extremophilic enzymes. This book introduces the fundamentals of enzymatic processes, various renewable energy resources, and their pretreatment processes. This book presents in-depth review of extremophilic enzymes which can be used in several biotechnological processes. In addition, the book provides the knowledge on how to engineer enzymes for enhanced conversion of lignocellulosic feedstocks to biofuels. This book will support the readers to get a clear understanding on this upcoming field of science and engineering of extremophilic enzymes in such a way besides understanding the concept that they will be in position to design the bioprocesses for production of the suitable/desired enzyme from the ideal source for their desired application. This book can be used for academia, research, and industry. Utmost care has been taken to address the basic concepts in extremophilic enzymatic processing so that it would be useful for the beginners. The activities and key questions are also included at the end of every chapter to improve the reasoning ability of the reader in a specific topic.

Chapter 1 is the introduction to the book. It begins with the growing demand for the enzymatic processes and the advantages of the extremophilic enzymes over others. It covers the various sources of extremozymes such as thermophiles, psychrophiles, barophiles, acidophiles, alkaliphiles, desiccation-resistant microorganisms, etc. It emphasizes the need for knowledge, understanding, and skill on working with extremophilic enzymes. By the end of the chapter, the beginner will get a clear essence of identifying the ideal source for the extremophilic enzyme, identifying the suitable enzyme for the desired bioprocess operation, and application of the extremozymes.

Chapter 2 deals with the basic concepts in enzymatic bioprocesses. It is important for the readers to first understand the microbial catalysis and only then they will be able to recognize the advantages of the enzymatic processes or extremophilic enzymatic systems. Hence, Chap. 2 is planned to cover the basic concepts of enzymes are introduced and finally the concepts about extremophiles, extremophilic enzymes, and their advantages over others are described.

Chapter 3 deals with the different approaches for the pretreatment of lignocellulosic feedstocks. Lignocellulosic biomass is the most abundantly available feedstock that comes from agricultural, forestry, municipal, and domestic sources. The use of lignocellulosic biomass for microbial/enzymatic process can greatly help in lowering down the costs of operation, but its recalcitrant nature is its major limitation. The chapter begins with the purpose of pretreatment and gives a detailed description and comparison of various pretreatment methods of lignocellulosic biomass including physical, chemical, physiochemical, and biological. Physical processes such as mechanical comminution and extrusion, chemical pretreatment processes such as acid-based pretreatments, alkali-based pretreatment, and organosolv are elaborately described. Physicochemical processes such as steam explosion pretreatment, hydrothermal pretreatment, and ammonia fiber explosion (AFEX) are also addressed. The chapter provides information about the effect of pretreatment of lignocellulosic biomass on the process as well as product yield. The chapter also gives a clear idea about the economic and environmental evaluation of pretreatment processes for treating lignocellulosic biomass.

Chapters 4–13 describe various extremophilic enzymes that are used for different applications including lignocellulosics hydrolysis and saccharification. Each chapter discusses about an extremophilic enzyme, its source and molecular structure, catalysis, and its applications. These chapters also provide relevant literature on those selected extremophilic enzymes. Chapter 4 deals with extremophilic cellulases which have an elevated industrial demand especially from paper industry and biofuel sector. The chapter describes glycosyl hydrolases, which are involved in the hydrolysis of lignocellulosics and their classification and structural features. The chapter discusses the metagenomic approaches for isolating novel cellulases genes. The chapter also covers important aspects such as methods for isolation of cellulase producers and cellulase activity assays as well as approaches for strain improvement which are very important for the industry personnel or researchers.

Chapter 5 covers extremophilic xylanases, their applications, their production, and properties. The chapter discusses about the structure and occurrence of xylan, a substrate for xylanase. The chapter also addresses the approaches for improving microbial xylanases. Chapter 6 describes in detail about the lytic polysaccharide monooxygenases, their occurrence, classification, structure, types, mechanism of reaction catalysis, as well as their applications.

Chapter 7 covers the various concepts about extremophilic amylases and their occurrences in detail from various sources such as Thermophiles/Thermoacidophiles, Psychrophiles/Psychrohalophiles, Alkaliphiles, Halophiles, and Archaea. The chapter also discusses about the genetic engineering approaches that are used for enhancing the amylase activity. Finally, this chapter covers the various applications of amylases in food industry, detergents, fermentation industry, etc.

Chapter 8 deals with another demanding extremophilic lignolytic enzymes. Lignin acts as a cement and hinders the hydrolysis of cellulose present in plant biomass. This chapter covers the major lignolytic enzymes, namely, manganese peroxidase, lignin peroxidase, and laccase. The molecular structure, catalytic cycle, mode of action, common substrate, microbial source, and effect of various operating conditions for these enzymes are discussed in detail.

Chapter 9 covers with the extremophilic pectinases. The chapter covers two aspects: pectin and pectin-degrading enzymes. The first section covers the occurrence and distribution of pectic substances, structure, and their classification. The second part of the chapter covers the nomenclature of the pectinase enzymes, their classification, microbial source, in vitro assays, as well as their applications. The chapter also covers different extremophilic pectinases such as acidic, alkaline, thermostable, and cold-active pectinases. The chapter ends with the state of the art in an industrial/commercial perspective and the future prospects of extremophilic pectinases.

Chapter 10 deals with extremophilic esterases and discusses about its major types such as thermophilic, psychrophilic, halophilic, piezophilic, and polyextremophilic esterases in detail. The chapter discusses about the stability of the esterases against temperatures, chemicals, their characteristics, and immobilization strategies. Chapter 11 makes a special discussion about the relevance of esterases to "Lignocellulosic Feedstocks." It deals with extremophilic esterases for bioprocessing of lignocellulosic feedstocks. It discusses about the structure and mode of action of the esterases. It covers different types of esterases, namely, acetyl xylan esterases, acetyl mannan esterases, feruloyl esterases, glucuronoyl esterases, and complexed hemicellulases.

Chapter 12 discusses about the chitinases from different sources such as bacteria, fungi, plants, and insects. The chapter also describes about chitin and its derivatives. The chapter provides clear insights about catalysis mechanism of chitinases, chitinase production, applications of chitinases, and molecular biology/genetic engineering approaches for improving the extremophilic chitinases. Chapter 13 deals with extremophilic lipases, their structures, and catalytic mechanisms. It covers the different types of extremophilic lipases, namely, thermophilic, psychrophilic, alkaliphilic, acidophilic, and halophilic lipases. It also discusses the structural characteristics of extremophilic lipases with a special emphasis on structural features that contribute to stability. Two major sources of extremophilic lipases, namely, lipase P1 from *Bacillus stearothermophilus* and lipase from *Archeoglobus fulgidus*, are discussed.

The final Chap. 14 deals with bioprospection of extremozymes for conversion of lignocellulosic feedstocks to bioethanol and other biochemicals. It covers the various interesting topics such as different approaches, e.g., microbial, enzymatic, and metagenomic, in search of extremozymes. It also discusses in detail about the protein engineering strategies such as rational design and directed evolution of extremophilic glycoside hydrolases and semi-rational protein engineering and design for improving the catalytic rates of the enzymes.

Contents

List of Contributors

Rodrigo Volcan Almeida Departamento de Bioquímica, Instituto de Química, Laboratório de Microbiologia Molecular e Proteínas, Programa de Pós-graduação em Bioquímica, Universidade Federal do Rio de Janeiro, Rio de Janeiro, RJ, Brazil

Cristiane Dinis AnoBom Departamento de Bioquímica, Instituto de Química, Laboratório de Biologia Estrutural de Proteínas, Programa de Pós-graduação em Bioquímica, Universidade Federal do Rio de Janeiro, Rio de Janeiro, RJ, Brazil

Manuel Becerra Facultade de Ciencias, Departamento de Bioloxía Celular e Molecular, Grupo EXPRELA, Centro de Investigacións Científicas Avanzadas (CICA), Universidade da Coruña, A Coruña, Spain

Mohit Bibra Department of Chemical and Biological Engineering, South Dakota School of Mines and Technology, Rapid City, SD, USA

Jenny M. Blamey Swissaustral USA, Athens, GA, USA

Fundación Científica y Cultural Biociencia, Ñuñoa, Santiago, Chile

Faculty of Chemistry and Biology, University of Santiago, Santiago, Chile

Freddy Boehmwald Fundación Científica y Cultural Biociencia, Ñuñoa, Santiago, Chile

Gabriela Coelho Brêda Departamento de Bioquímica, Instituto de Química, Laboratório de Microbiologia Molecular e Proteínas, Programa de Pós-graduação em Bioquímica, Universidade Federal do Rio de Janeiro, Rio de Janeiro, RJ, Brazil

María Esperanza Cerdán Facultade de Ciencias, Departamento de Bioloxía Celular e Molecular, Grupo EXPRELA, Centro de Investigacións Científicas Avanzadas (CICA), Universidade da Coruña, A Coruña, Spain

Bhupinder Singh Chadha Department of Microbiology, Guru Nanak Dev University, Amritsar, India

Ram Chandra Environmental Microbiology Division, Indian Institute of Toxicology Research, Lucknow, UP, India

Paramageetham Chinthala Department of Microbiology, Sri Venkateswara University, Tirupati, India

Paul Christakopoulos Biochemical and Chemical Process Engineering, Division of Chemical Engineering, Department of Civil, Environmental and Natural Resources Engineering, Luleå University of Technology, Luleå, Sweden

Aditi David Department of Chemical and Biological Engineering, South Dakota School of Mines and Technology, Rapid City, SD, USA

Rafael Alves de Andrade Departamento de Bioquímica, Instituto de Química, Laboratório de Microbiologia Molecular e Proteínas, Programa de Pós-graduação em Bioquímica, Universidade Federal do Rio de Janeiro, Rio de Janeiro, RJ, Brazil

Departamento de Bioquímica, Instituto de Química, Laboratório de Biologia Estrutural de Proteínas, Programa de Pós-graduação em Bioquímica, Universidade Federal do Rio de Janeiro, Rio de Janeiro, RJ, Brazil

Karina de Godoy Daiha Departamento de Bioquímica, Instituto de Química, Laboratório de Microbiologia Molecular e Proteínas, Programa de Pós-graduação em Bioquímica, Universidade Federal do Rio de Janeiro, Rio de Janeiro, RJ, Brazil

Leticia Dobler Departamento de Bioquímica, Instituto de Química, Laboratório de Microbiologia Molecular e Proteínas, Programa de Pós-graduação em Bioquímica, Universidade Federal do Rio de Janeiro, Rio de Janeiro, RJ, Brazil

Juan José Escuder Facultade de Ciencias, Departamento de Bioloxía Celular e Molecular, Grupo EXPRELA, Centro de Investigacións Científicas Avanzadas (CICA), Universidade da Coruña, A Coruña, Spain

Giannina Espina Fundación Científica y Cultural Biociencia, Ñuñoa, Santiago, Chile

Pablo Fuciños International Iberian Nanotechnology Laboratory (INL), Braga, Portugal

Roberto González-González Department of Food and Analytical Chemistry, University of Vigo, Ourense, Spain

María-Isabel González-Siso Facultade de Ciencias, Departamento de Bioloxía Celular e Molecular, Grupo EXPRELA, Centro de Investigacións Científicas Avanzadas (CICA), Universidade da Coruña, A Coruña, Spain

Prasada Babu Gundala Department of Botany, Sri Venkateswara University, Tirupati, India

Margarita Kambourova Institute of Microbiology, Bulgarian Academy of Sciences, Sofia, Bulgaria

Naveen Kango Department of Applied Microbiology, Dr. HariSingh Gour Vishwavidyalaya (A Central University), Sagar, MP, India

Baljit Kaur Department of Microbiology, Guru Nanak Dev University, Amritsar, India

Vineet Kumar Department of Environmental Microbiology, Babasaheb Bhima Rao Ambedkar Central University, Lucknow, UP, India

Olalla López-López Facultade de Ciencias, Departamento de Bioloxía Celular e Molecular, Grupo EXPRELA, Centro de Investigacións Científicas Avanzadas (CICA), Universidade da Coruña, A Coruña, Spain

Antonio D. Moreno Department of Biology and Biological Engineering, Industrial Biotechnology, Chalmers University of Technology, Gothenburg, Sweden

Department of Energy, Biofuels Unit, Ciemat, Madrid, Spain

Marcelo Victor Holanda Moura Departamento de Bioquímica, Instituto de Química, Laboratório de Microbiologia Molecular e Proteínas, Programa de Pós-graduação em Bioquímica, Universidade Federal do Rio de Janeiro, Rio de Janeiro, RJ, Brazil

Madhu Nair Muraleedharan Biochemical and Chemical Process Engineering, Division of Chemical Engineering, Department of Civil, Environmental and Natural Resources Engineering, Luleå University of Technology, Luleå, Sweden

R. Navanietha Krishnaraj Department of Chemical and Biological Engineering, South Dakota School of Mines and Technology, Rapid City, SD, USA

Lisbeth Olsson Department of Biology and Biological Engineering, Industrial Biotechnology, Chalmers University of Technology, Gothenburg, Sweden

Rocío Peralta Fundación Científica y Cultural Biociencia, Ñuñoa, Santiago, Chile

María Luisa Rúa Department of Food and Analytical Chemistry, University of Vigo, Ourense, Spain

Ulrika Rova Biochemical and Chemical Process Engineering, Division of Chemical Engineering, Department of Civil, Environmental and Natural Resources Engineering, Luleå University of Technology, Luleå, Sweden

Rajesh K. Sani Department of Chemical and Biological Engineering, South Dakota School of Mines and Technology, Rapid City, SD, USA

Felipe Sarmiento Swissaustral USA, Athens, GA, USA

Hemant Soni Department of Applied Microbiology, Dr. HariSingh Gour Vishwavidyalaya (A Central University), Sagar, MP, India

Sheelu Yadav Department of Environmental Microbiology, Babasaheb Bhima Rao Ambedkar Central University, Lucknow, UP, India

About the Editors

Rajesh K. Sani is an Associate Professor in the Department of Chemical and Biological Engineering and Chemistry and Applied Biological Sciences at the South Dakota School of Mines and Technology, South Dakota. He joined the South Dakota School of Mines and Technology as an Assistant Professor in 2006. Prior to this, he worked as a Postdoctoral Researcher and Research Assistant Professor at the Washington State University, Pullman, WA, and focused his research on Waste Bioprocessing. He also served as an Associate Director of NSF Center for Multiphase Environmental Research at the Washington State University. He received his BS in Mathematics from the Meerut University in India, his MS in Enzyme Biotechnology from Devi Ahilya University in India, and his PhD in Environmental Biotechnology from the Institute of Microbial Technology in India.

Due to his interdisciplinary background, Sani has been integrating engineering with biological sciences in his teaching as well as research endeavors. For over 12 years, Sani has engaged in a constant endeavor to improve his teaching skills to become an effective instructor and communicator. In Washington State University's School of Chemical and Bioengineering and Center for Multiphase Environmental Research, he taught a variety of engineering courses including Integrated Environmental Engineering for Chemical Engineers, Bioprocess Engineering, and Current Topics in Multiphase Environmental Research—a team taught interdisciplinary course to undergraduate and graduate students. Over the last 9 years at the South Dakota School of Mines and Technology, he has been teaching various science and engineering courses including Microbiology for Engineers, Biochemistry Laboratories, Bioinformatics, Molecular Biology for Engineers, Microbial Genetics, and Microbial and Enzymatic Processing to students of various disciplines of Chemical Engineering, Environmental Engineering, Applied Biological Sciences, Chemistry, Interdisciplinary Studies, Biology, Medical, and Paleontology. Sani has received several awards including the outstanding student research (India), Department of Biotechnology Scholarship (India), the Council of Scientific and Industrial Research (India), and Science and Technology Agency (Japan).

Sani group's research includes extremophilic bioprocessing of lignocellulose-based renewables for biofuels and bioproducts and bioprospecting of extremophilic microorganisms for developing more efficient and cost-effective biofuel (bioenergy) production technologies. Over the past 11 years, he has been the PI or co-PI on over $12 million in funded research. Several of his accomplishments in research and advising include (i) postdocs supervised (7); (ii) graduate students supervised (MS students, 10; and PhD, 6), and (iii) undergraduate students and K12 teachers supervised (over 35). He has one patent and five invention disclosures, and he has published over 55 peer-reviewed articles in high impact factor journals and contributed in several book chapters. He is currently acting as editor and coeditor for three textbooks which will be published by Springer International Publishing AG. In addition, he has been a proposal reviewer and panelist for the Federal Agencies: (i) National Science Foundation, (ii) U.S. Army Research Office, (iii) Department of Energy, (iv) U.S. Geological Survey, and (v) User Facility—Environmental Molecular Sciences Laboratory. He also serves the Industrial Microbiology profession as "Biocatalysis Program Committee Member" of the Society for Industrial Microbiology and Biotechnology (SIMB) and technical session chair at the Annual American Institute of Chemical Engineers (AIChE) and SIMB and is an associate editor.

R. Navanietha Krishnaraj is currently a B-ACER fellow and Research Professor at the Department of Chemical and Biological Engineering, South Dakota School of Mines and Technology, USA. Prior to this, he worked at the Department of Biotechnology, National Institute of Technology Durgapur, India. He received his B.Tech in Biotechnology and PhD in Chemical Engineering in the field of microbial fuel cells from the CSIR—Central Electrochemical Research Institute, Karaikudi, India. He recently received the prestigious Bioenergy Award for Cutting Edge Research (B-ACER) from the Department of Biotechnology, Government of India, and the Indo-U.S. Science and Technology Forum. His areas of research interest include bioelectrocatalysis and bioenergy. He has taught bioinformatics and computational biology courses to undergraduate students. He is a life member of several renowned professional societies. He is the faculty advisor for the Indian Society for Technical Education, Durgapur Chapter.

Chapter 1
Introduction to Extremozymes

R. Navanietha Krishnaraj and Rajesh K. Sani

What Will You Learn from This Chapter?
This chapter introduces the basic concepts of enzymes, applications of enzymes and advantages of microbial bioprocesses over the enzymatic bioprocesses. The chapter gives an introduction about the extremozymes, its sources, and advantages of extremozymes over other enzymatic processes. This chapter also explains the different types of extremozymes from thermophilic, hyperthermophilic, psychrophilic, barophilic, acidophilic, alkaliphilic, xenophilic, halophilic as well as metal-resistant microorganisms. This chapter gives a broad outline about extremophilic enzymatic processes which is a prerequisite for the readers to understand the following chapters.

Biotechnology and bioprocess engineering are a boon to mankind. Biochemical engineers make use of the microorganisms as catalysts for the wide range of applications including food processing, water treatment, solid waste disposal, and production of organic acids, vitamins, antibiotics, and therapeutic molecules. Microorganisms utilizes the substrates as the source of energy and produce primary and secondary metabolites. They convert the substrate to product either in a single reaction or a linear/complex series of reactions. Each of these reactions are carried out by a single or a set of enzymes.

Biotechnology research and bioprocess industry have grown at rapid pace to incredible heights that they have developed bioprocesses for almost all traditional chemical processes. The bioprocesses, which are ecofriendly and economical, are green alternatives to the chemical processes. They can operate at ambient physical conditions such as temperature, pH and pressure unlike chemical processes which demands very high temperature, pressure, or a specific pH. The biological processes

R. Navanietha Krishnaraj • R.K. Sani (✉)
Department of Chemical and Biological Engineering, South Dakota School of Mines and Technology, 501 East St. Joseph Street, Rapid City, SD 57701-3995, USA
e-mail: Rajesh.Sani@sdsmt.edu

© Springer International Publishing AG 2017
R.K. Sani, R.N. Krishnaraj (eds.), *Extremophilic Enzymatic Processing of Lignocellulosic Feedstocks to Bioenergy*, DOI 10.1007/978-3-319-54684-1_1

also do not require any special apparatus or sophisticated processes for the production of desired product as in the case of chemical process. Besides these, the microbial processes can make use of the waste organic materials such as effluent or solid waste from agri-food or any other industry as the substrate. This helps greatly in reducing the costs of operation besides bioremediation/disposal of wastes from the environment. The major issue with the microbial processes is that their metabolic pathways are very complex leading to several undesired products. Therefore purification of the products especially in the case of therapeutic molecules/foods/single cell proteins is very difficult. In some cases, microorganisms can also release some toxins into the reaction system. These major limitations can be circumvent by the use of specific enzymes which can confer sensitivity and selectivity to the reaction.

Enzymes, also known as biocatalysts, produced by the microorganisms that can catalyze a particular reaction or a set of reactions. Enzymes are generally proteins, however, ribozyme is an exception. Some enzymes need cofactors or coenzymes for their catalytic activity. The use of enzymes for different applications have been explored well. Technologies have improved in such a way that there is innumerable number of products based on enzymes. Enzymes are indispensable to research as well as modern life. However, these enzymes have also certain limitations. They are so fragile they get denatured easily because they are mainly composed of proteins. Few enzymes can only be operated under very narrow operating conditions. In addition, purification of enzymes is a tedious job. The use of extremophilic enzymes can help to overcome some of the limitations (Rothschild and Mancinelli 2001).

Extremophilic enzymes can be operated under adverse conditions such as a high or low temperature, pressure, extreme radiations and pH conditions. The operating conditions of most enzymes depend on the microorganisms from which they are isolated. These enzymes can be isolated from thermophiles, hyperthermophiles, psychrophiles, barophiles, acidophiles, alkaliphiles, xenophiles, halophiles as well as metal-resistant microorganisms. Figure 1.1 shows the various sites in USA (South Dakota, Wyoming, and Washington) and India (Himachal Pradesh and Haryana) where extremophiles are present. The extremophilic enzymes have several advantages over the mesophilic enzymes. These extremophilic enzymes can operate a much broader range of conditions besides being stable, and have much longer shelf life. These enzymes also possess higher activity and high rate the catalysis when compared with the normal enzymes. They are more resistant to proteolysis and are more robust to organic solvents. They can be overexpressed to very high levels using heterologous host-vector systems. The high structural stability of the extremozyme also helps in engineering the enzymes by genetic engineering or site directed mutagenesis/protein engineering approaches. It also helps in improving the immobilization processes onto the wide range of carriers either by surface immobilization by functionalization/covalent bonding/adsorption by weak Vander walls forces or entrapment/encapsulation avoiding the mass transfer limitations leading to improvement in effectiveness. The extremozymes can be produced from extremophilic microorganisms including bacteria, algae, fungi or even from plants growing in adverse conditions (Anitori 2012; Atomi 2005).

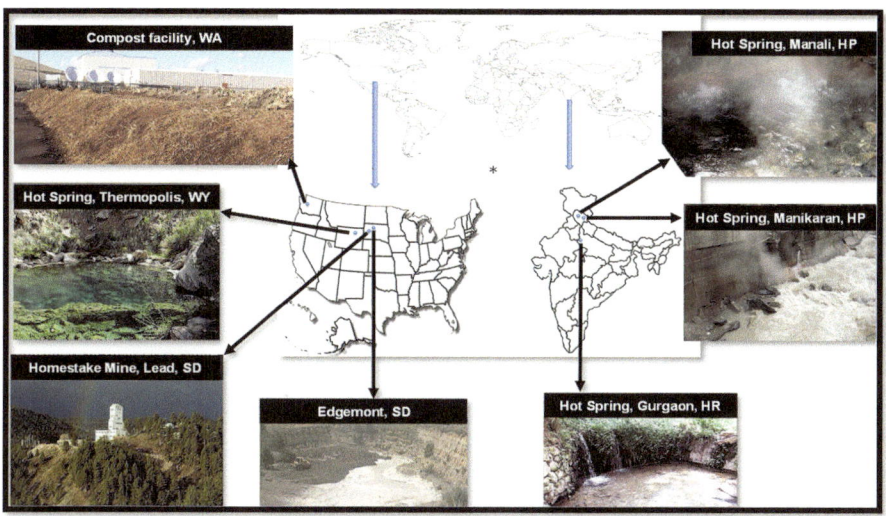

Fig. 1.1 Presence of extremophiles at various sites in USA (South Dakota, Wyoming, and Washington) and India (Himachal Pradesh and Haryana)

The extremophilic enzymes have wide range of applications when compared with the normal enzymes. For example, thermostable enzymes have several potential advantages e.g., higher specific activity and greater half-lives. Carrying out hydrolysis at higher temperature can ultimately lead to improved performance through decreased enzyme dosage and reduced hydrolysis time, thus, resulting in decreased hydrolysis costs. Besides these, high temperature can also help in avoiding the mesophilic contamination and thus prevent from undesired reactions. Thermophilic proteases find its applications in hydrolysis in food, feed, brewing, and baking. Thermophilic glycosyl hydrolases namely amylases, pullulanase, glucoamylases, glucosidases, cellulases and xylanases are shown to have applications in processing the polysaccharides such as starch, cellulose, chitin, pectin, and textiles. Thermophilic lipases, proteases and esterases have been widely used in detergent industry. Thermophilic xylanases are used for bleaching paper. Thermophilic DNA polymerases has been used in molecular biology for PCR. Like thermophilic enzymes, the psychrophilic proteases, amylases and cellulases have also been used in detergent industry. Extremophilic oxidoreductases are widely used for the real time development of electrochemical biosensors. Halophilic peptidases have been used for peptide synthesis. Acidophilic proteases have been used for detergent, food, and feed industry.

Several investigations have been carried out to understand the structural features that confer better stability to thermophilic enzymes when compared with mesophilic enzymes. literature suggest that the increased surface charge, increased protein core hydrophobicity, and replacement of exposed 'thermolabile' amino acids together can lead to the increase in the stability of the thermophilic enzymes. Halophilic enzymes can exhibit catalysis at a very high salt concentration (e.g. KCl concentrations of 4 M and NaCl concentrations of >5 M). The halophilic enzymes have a relatively large number of negatively charged amino acid residues on their surfaces which helps them to adapt to this environmental pressure without getting precipitated. However, halophilic enzymes have several limitations that they are not soluble in surroundings with lower salt concentrations which hinders the use of halophiles in these environment (Egorova and Antranikian 2005; Demirjian et al. 2001).

Extremozymes have immense potential for applications in industries including agricultural, chemical, and pharmaceutical. So far, a very small percentage of the extremozymes have been explored for industrial applications. With research advancements in highly stable extremozymes from different organisms, the number of biotechnology products may also increase. In summary, there is no doubt that extremozymes will significantly improve the scope of biotechnology towards real time applications.

Take Home Message

- The bioprocesses have several advantages over chemical processes. The bioprocess operations can operate at ambient physical conditions such as temperature, pH and pressure whereas chemical processes require very high temperature, pressure, or a specific pH. The bioprocesses are also ecofriendly and economical.
- The bioprocess operations can be mediated by enzymes or microorganisms. Enzymatic processes have advantages such as high rate of catalysis, specificity and selectivity; however, suffers from limitations such as high cost and narrow range of operating conditions. Enzymatic processes are prone to denature at high temperature. The use of extremozymes will help to circumvent these shortcomings.
- Extremozymes are those enzymes which can be operated under adverse conditions such as a high or low temperature and pressure and extreme radiations and pH conditions. These enzymes can be isolated from thermophiles, hyperthermophiles, psychrophiles, barophiles, acidophiles, alkaliphiles, xenophiles, halophiles as well as metal-resistant microorganisms. The extremozymes can be used for wide range of applications including agricultural, chemical, and pharmaceutical sectors.

References

Anitori RP (ed) (2012) Extremophiles: microbiology and biotechnology. Caister Academic Press, Norfolk. isbn:978-1-904455-98-1

Atomi H (2005) Recent progress towards the application of hyperthermophiles and their enzymes. Curr Opi Chem Biol 9(2):166–173

Demirjian DC, Morís-Varas F, Cassidy CS (2001) Enzymes from extremophiles. Curr Opin Chem Biol 5(2):144–151

Egorova K, Antranikian G (2005) Industrial relevance of thermophilic Archaea. Curr Opin Microbiol 8(6):649–655

Rothschild LJ, Mancinelli RL (2001) Life in extreme environments. Nature 409(6823):1092–1101

Chapter 2
Fundamentals of Enzymatic Processes

R. Navanietha Krishnaraj, Aditi David, and Rajesh K. Sani

What Will You Learn from This Chapter?

A basic and clear understanding about enzymes is essential for the readers before they begin to learn about the extremozymes (enzymes in extreme conditions). The chapter begins with the basic concepts of enzymes, roles of enzymes in biological systems, components of enzymes, detailed list of applications of the enzymes and the history of enzymology. Specificity is the key characteristic of the enzyme and has crucial role in terms of selectivity and catalytic activity of the enzyme. The section on specificity of enzymes explains five different types of specificity namely Absolute Substrate specificity, Broad specificity (Group specificity), Bond specificity (Relative specificity), Stereochemical specificity and Reaction specificity. The chapter covers the different methods of classification of enzymes and enzyme nomenclature. The chapter gives a clear explanation about the mechanisms of enzyme- substrate interactions with special emphasis on Lock and Key Theory and Induced Fit Hypothesis. Different units of enzyme activity (Katal, IU, Turnover number), different models of enzyme kinetics, types of enzyme inhibition and different strategies for immobilization of are also addressed in this chapter. Finally, the chapter describes the various applications of extremozymes.

R. Navanietha Krishnaraj • A. David • R.K. Sani (✉)
Department of Chemical and Biological Engineering, South Dakota School of Mines and Technology, 501 East St. Joseph Street, Rapid City, SD 57701-3995, USA
e-mail: Rajesh.Sani@sdsmt.edu

© Springer International Publishing AG 2017
R.K. Sani, R.N. Krishnaraj (eds.), *Extremophilic Enzymatic Processing of Lignocellulosic Feedstocks to Bioenergy*, DOI 10.1007/978-3-319-54684-1_2

2.1 Introduction

Enzymes are biocatalysts produced by all living organisms such as plants, animals, human beings including microorganisms. Enzymes help them to carry out their metabolic reactions. They are generally proteins in nature. All enzymes are not proteins. Ribozyme is an enzyme that is made up of nucleic acids. Ribozymes are involved in the cleavage of phosphodiester bond in the hydrolysis of hnRNA to mRNA. There are over 20,000 genes that are coding for the proteins in the human genomes. Most of these genes code for enzymes which are involved in various metabolic reactions. For instance, the saliva contains several enzymes such as amylase, protease, lipase, DNase, and RNase which are helpful in the digestion process. Enzyme like lysozyme confers natural immunity to our body.

The enzymes help in accelerating the biochemical reaction which converts the substrate into product. The substance on which the enzyme acts is called a substrate. The region of the enzyme on which the substrate binds is called as the active site of the enzyme. The enzyme has two components. Prosthetic group is the non-protein component of the enzyme and the apoenzyme is the protein part of the enzyme. The apoenzyme and the prosthetic group are together known as the holoenzyme. If prosthetic group is inorganic, then it is called cofactor and if the prosthetic group is organic, it is called coenzyme. The enzymes help in decreasing the activation energy required for the catalytic reaction. Lower the activation energy, higher is the reaction catalysis. The enzyme–substrate interaction causes the redistribution of electrons in the chemical bonds of the substrate. Generally, the enzymes are larger than their substrate. However, there are some exceptions like DNA polymerase. The enzymes have several advantages over the chemical catalysts. They have very high catalytic rates and high specificity when compared with the chemical catalysts.

Carbonic anhydrase hydrolyses carbon dioxide to carbonic acid and it can catalyze the hydration of 10^5 carbon dioxide molecules per second. Most of the chemical catalysts are not very specific and they catalyze related compounds as well and end up in producing undesired products. Both enzymes and chemical catalysts help in lowering the activation energy but there are some differences. The chemical catalysts may be simple organic or inorganic molecules/compounds/materials and have low molecular weights. Majority of enzymes are high molecular weight globular proteins with a few exceptions. The chemical catalysts require a very high operating conditions such as high temperature, high pressure etc. However, enzymatic processes can occur in normal mild operating conditions of temperature, pressure, pH etc. Certain enzymes are produced in inactive forms such as trypsinogen, chymotrypsinogen, pepsinogen etc. and they are called zymogens. These enzymes after getting activated, are termed as trypsin, chymotrypsin, pepsin, respectively. If these enzymes are not produced in inactive forms, then these enzymes may damage the proteins of the host cells/tissues.

Enzymes are indispensable to all biological systems. They play a key role in all anabolic and catabolic reactions in the biological systems. Enzymes operate the central dogma of life. The upregulation and downregulation of enzymes in the biological systems lead to genetic diseases. For instance, deficiency of glucocerebrosidase, an enzyme which breaks down of a glucocerebroside leads to Gaucher's disease. This disease can be treated using recombinant imiglucerase enzyme or glucosylceremide synthetase inhibitor. Similarly, deficiency of the enzyme alpha-galactosidase A leads to Fabrys disease which leads to progressive accumulation of lipids in kidney, heart, and other organs. Another inherited disease state namely type 1 mucopolysaccharidosis is caused by the deficiency of the enzyme alpha-L-iduronidase.

Enzymes are used for the treatment of wide range of diseases. For example, asparaginase is used for the treatment of leukemia. Leukemic cells are devoid of the essential amino acid, asparagine. Generally, the cancer cells receive asparagine from the normal cells. When the asparaginase is provided to the cancer patient, the enzyme utilizes asparagine in normal cells and the tumor cells do not receive asparagine. Similarly, collagenase is used for the treatment of skin ulcers. Rhodonase is used for the treatment of cyanide, and can be used for cyanide poisoning. Glutaminase is used for the treatment of leukemia. Urease hydrolyses urea and is used for hyperureamia. α-amylase can be used to hydrolyze starch and treatment of digestive disorders. Bilirubin oxidase is used for the treatment of hyoerbilirubinemia. The enzyme uricase helps in the oxidation of uric acid to 5-hydroxyisourate, and can be used for the treatment of gout. Streptokinase and urokinase comes under the group of fibrinolytics and can be used for thrombolysis and treatment of myocardial infarction. Enzymes are also used for the diagnosis of different diseases. The enzyme levels are used as the markers for several diseases. Enzymes are also used for the development of biosensors. For instance, liver cirrhosis can be diagnosed with the levels of liver enzymes such as alanine aminotransferase (ALT), alkanine protease (ALP) and bilirubin oxidase.

Enzymes are also used for the development of electrochemical sensors such as amperometric sensor, impedance sensor etc. In amperometry sensor, the enzymes are used for the oxidation or reduction of the substrates. The analyte concentration is estimated from the correlation between the current and the concentration of the analyte at the specific oxidation/reduction potential. Reports are also available for the detection and destruction of explosives (2,4-Dinitroanisole) using immobilized enzymes (e.g., DNAN demethylase encapsulated in biogenic silica). Enzymes are playing an important role in the research areas. Without enzymes it would be impossible to understand and elucidate the molecular mechanisms in biological systems. A thermostable enzyme, Taq polymerase plays a key role in the polymerase chain reactions. Molecular scissors are yet another wonder molecules which are restriction enzymes that can help in excising the DNA in the specific position of the nucleotides. Scientists from NASA have developed a thermostable cellulolytic enzyme and a synthetic cellulosome (rosettazyme) by genetic engineering approaches for the hydrolysis of cellulose in space.

2.2   History of Enzymology

Enzymes have been using in biological processes since 400 BC. Yeast is well known for the fermentation for the production of alcoholic beverages. Payen and Persoz, for the first time in 1883, showed that the extracts of germinating barley hydrolyzed starch to dextrin and sugar which provided the clues about the enzymatic catalysis. They showed that very small amount of the extracts of germinating barley was able to liquefy large amount of starch indicating the very high rates of catalytic activity of the extract (Payen and Persoz 1833). Further, the extract was shown to be thermos-labile and the active catalytic substance could be precipitated out from aqueous solution using alcohol. This active catalytic agent obtained from the extract of germinating barley was called as diastase. Later, it was realized that the extract dextrin was composed of a mixture of amylases. The diastase was then used to produce dextrin which is used for the production of alcoholic beverages from fruits and bread. Similarly, malt extract (amylases, amyloglucosidases) was used for the hydrolytic processes. Until, Berzilius made his hypothesis, the catalytic agents were termed as ferments. Berzelius was the first to hypothesize that ferments were catalysts. Further, Wagner classified ferments as organized and unorganized ferments in 1857. Finally, in 1878, Kuhne coined the term enzymes for the catalytic agent which were previously called as ferments. The term "Enzyme" was derived from "In-Zymase". Zymase was the enzyme produced by yeast which converts sugar to alcohol and carbon dioxide. However, the scientific community was not convinced about the concept of enzymatic catalysis for long time. The enigma in the theory of fermentation existed for a long time as some vital factor which is different from the chemical forces is present in the extracts of the living organisms and this mediates the reactions. In fact, Liebig and group believed fermentation is simply a decay process.

The first company for the production of enzyme for cheese making was started by Christian Hansen's Laboratory in 1874 (Buchholz and Poulson 2000). Two decades later, Emil Fischer started his investigations on the specificity of the enzymes in 1894. He carried out a series of experiments to assess the specificity of different enzymes using several glycosides and oligosaccharides. He showed that invertin extracted from yeast can hydrolyze the α-methyl-D-glucoside but not β-methyl-D-glucoside. Whereas, emulsion (a commercial product of Merck) was able to hydrolyze the β-methyl-D-glucoside and not α-methyl-D-glucoside. With these results, Fischer proposed the "lock and key mechanism" on the interactions of the enzyme on the substrate. Further, in 1894 it was Fisher who first proposed that the enzymes were proteins in nature. It took more than 20 years for the scientific community to get convinced with this concept. Like Fisher, Buchner also made significant contributions in the field of fermentation and enzymology. He used the cell free extracts of yeast cells for the fermentation of sugar into alcohol and carbon dioxide. The concept of Buchner was contradictory to the theory of Louis Pasteur which states that the alcoholic fermentation was mediated by the presence and action of living cells and the required vital force which they termed as "avis vitalis".

However, Buchner showed that the fermentation can be carried out using enzymatic catalysis without the use of live cells or other vital forces.

The first biochemical process for the production of isomaltose from yeast extract (α-glycosidase) was demonstrated by Croft-Hill in 1898 (Sumner and Somers 1953). The work of Sumner laid the foundation for the use of enzymes for biochemical processes. Sumner has made significant contribution for enzymology and he is called as the father of enzymology. Sumner confirmed that the enzymes are proteins in nature. He isolated urease from jack beans and crystallized urease in 1926. He also developed a general crystallization method for the enzymes. With the crystallization method developed by Sumner, Northrop from Rockfeller institute crystallized pepsin. In 1946 Sumner and Northrop got Nobel Prize for their discovery that "enzymes can be crystallized". In 1948, Sumner's contributions were recognized and he was elected to the National Academy of Sciences, USA.

In summary, enzymes are key players in biological systems. Nobel prizes have been awarded to several scientists working in different areas directly or indirectly related to enzymes. Some of their contributions are briefly discussed here. In 1965, François Jacob, André Lwoff, and Jacques Monod jointly received the Nobel Prize in Physiology or Medicine for their discoveries concerning genetic control of enzyme and virus synthesis. Restriction enzymes have key roles in molecular genetics, and for this discovery Werner Arber, Daniel Nathans and Hamilton O. Smith jointly received the Nobel Prize in Physiology or Medicine in 1978. Elizabeth H. Blackburn, Carol W. Greider and Jack W. Szostak jointly received the Nobel Prize in Physiology or Medicine in 2009 "for the discovery of how chromosomes are protected by telomeres and the enzyme telomerase".

2.3 Specificity of Enzymes

Specificity is the inherent property of enzymes, and is one of the crucial factors that make enzymes advantageous over the chemical catalysis and microbial processes. Specificity of the enzyme to the substrate is a very interesting molecular recognition mechanism. It is based on the structural and configurational complementarity between the enzyme and the substrate. The ratio of k_{cat}/K_m provides the information about enzyme specificity where k_{cat} refers to the turnover number and Km refers to the Michaelis Menten constant. Km is the substrate concentration required by the enzyme to operate at half its maximum velocity. The details about turnover number and Michaelis Menten constant are discussed in detail in the later part of the chapter under sections "Enzyme Activity and Enzyme Kinetics". These properties make the enzymes useful for diagnostic and research applications. Enzymes are highly specific for the substrate (also called reactant) and the reaction they catalyze. Different enzymes exhibit different degrees of specificity to the substrate. Enzyme specificity is due to the way an enzyme interacts with the substrate molecule to form an enzyme–substrate complex (also called transition-state complex). In the enzyme–substrate complex, the substrate binds to a specific site on the enzyme called the active site through weak, non-covalent interactions (hydrogen bonding and Van der Waals interactions). The difference in the energy of the free substrates and the enzyme–substrate complex sites of enzyme are very selective for a particular substrate molecule. This is the reason for the high specificity of enzymes. After the formation of the enzyme substrate complex, the reaction proceeds and the products are formed.

Enzymes have been widely used for the development of biosensors for the selective detection of different analytes at very low concentration. Specificity of the enzyme also helps in preventing the unwanted reactions in the bioprocess operations/fermentations. The microorganisms mediate a wide range of reactions in the system and produce undesired products and even toxins which necessitates the need for a challenging downstream process. Enzymes that are used for therapeutic applications should be highly specific to the desired molecule. However, it is not always that highly specific enzymes have greater industrial importance. In some cases, like microbial fuel cell or treatment of wastewater, an enzyme which has a broad range of specificity is preferred. Specificity of the enzymes can be classified into five different types as described in the Manual of Clinical Enzyme Measurements published in 1972. The specificities of enzymes can be categorized as follows:

Absolute Substrate Specificity The enzyme with absolute substrate specificity can act on only a specific substrate and will mediate only a specific reaction. For example, the enzyme urease can mediate hydrolysis of urea, but it cannot act on thiourea. The enzyme urease is very specific for the substrate and fails to hydrolyze if the methyl and alkyl groups are replaced by NH_2 groups or oxygen is replaced by sulfur molecules. For example, lactase can act on lactose, maltase can act on maltase, and sucrase can act on sucrose. Carbonic anhydrase can act only on carbonic acid. Uricase can act only on uric acid. Arginase can act only on arginine.

Broad Specificity or Group Specificity Group specific enzyme can act on a group of substrates that have specific functional groups, such as amino, phosphate, and methyl groups. Certain enzymes can act not only on the specific bond, but on the structure surrounding it. For example, hexokinase will not only act on glucose but also on other hexoses. Peptidases act on peptide bonds, but differ in their specificity depending on the amino acids making these bonds. Pepsin is an endopeptidase that acts on the peptide bonds formed by the aromatic amino acids such as phenylalanine, tyrosine and tryptophan. Trypsin is another endopeptidase that acts on the peptide bonds in which amino groups contributed by basic amino acids such as histidine, arginine and lysine. In the same way, chymotrypsin acts on the peptide bonds having carboxyl group of the aromatic amino acids. Aminopeptidase is an exopeptidase that specifically hydrolyses the peripheral bond on the amino terminal of the polypeptide chain. Carboxypeptidase specifically hydrolyses the peripheral bond on the carboxyl terminal of the polypeptide chain.

Relative Specificity or Bond Specificity Bond specificity of the enzyme refers to the activity of the enzyme on specific bonds. For example, proteases act on peptide bonds formed by any amino acid in the protein. Amylase can act on α-1,4-glycosidic bonds in dextrin, starch and glycogen. Lipase can mediate the hydrolysis of different ester bonds in triglycerides.

Stereochemical Specificity The stereochemical specificity is the most interesting and important characteristic of enzymes. An enzyme can act not only on a particular substrate but also on a specific optical configuration. For example, the enzyme α-glycosidase can act only on α-glycosidic bonds of glycogen, starch and dextrin. The enzyme, β-glycosidase can act only on β-glycosidic bonds of the cellulose. Similarly, L-amino acid oxidase can act on L-amino acids but not on D-amino acids. D-amino acid oxidase can act on D-amino acids but not on L-amino acids.

Reaction Specificity A substrate is acted upon by different enzymes and each one gives rise to different products. Different enzymes act on a single substrate and gives rise to different products. This kind of specificity is called reaction specificity.

There are few other categories of enzyme specificity namely geometric specificity and cofactor specificity which are covered in the Manual of Clinical Enzyme Measurements. In geometrical specificity, different substrates having similar molecular geometry can be acted upon by a single enzyme. For example, alcohol dehydrogenase can oxidize ethanol and methanol as ethanol and methanol have similar molecular geometry.

Co-factors specificity is not between the enzyme and the substrate and it is between the enzymes and the cofactors. Certain enzymes require cofactors for their activity. However, each enzyme requires the specific cofactor for their activity. The appropriate combination of enzyme and co-factor is required to mediate the catalysis of substrate in the enzymatic reaction.

2.4 Classification of Enzymes

The terms classification and nomenclature are often used synonymously in enzymology. The classification refers to the grouping of enzymes based on certain common properties in relatively lesser number of groups. The term Nomenclature refers to systematic and scientific method of classification to identify the specific enzymes based on its detailed biocatalytic reactions.

Enzymes can be classified by different ways such as depending on the constituents of enzyme, role of metal ions, substrate, reaction, reaction and substrate and so on. One simple way of classifying enzymes is as simple enzymes and conjugated enzymes. Simple enzymes are composed of only proteins. The hydrolysis of simple enzymes gives amino acids. The conjugated enzymes are made up of protein part called apoenzyme and the non-protein part called prosthetic group. The cofactor may be organic or inorganic. If the cofactor is organic, then it is called as the coenzyme and if the cofactor of the conjugated enzyme is inorganic, it is termed as cofactor. Coenzymes are mainly composed of vitamins or vitamin derivatives.

Some of the examples of coenzymes are NAD^+, $NADP^+$ and FAD^+. Inorganic metal ions such as iron, magnesium or zinc act as cofactors for the enzymes.

Many enzymes have metal ions as the cofactor and it is required for the activity of the enzyme. The enzymes having metal ion as cofactor may be classified as metalloenzymes and metal activated enzymes. Metals helps in mediating the biocatalytic activity of the enzymes by different ways. Actually the metals are not involved in the catalytic activity of the enzyme directly but they activate the enzyme by changing its shape. Metal activated enzymes have metals ions that are loosely bound onto the enzyme and this enzyme are more prone to lose the metal ion during purification and in turn may lose enzymatic activity. These enzymes always require the higher concentration of metal ions than the concentration of the enzyme. Loss of metal ion will lead to decrease in the enzymatic activity but not the loss of activity. In contrast, the metalloenzymes loses the activity when the cofactor metal ion is lost. The metal ion is tightly bound to the apoenzyme in the metalloenzymes. Some metalloenzymes, require one or more metal ion for its activity. For example, the enzyme superoxide dismutase require $Cu2^+$ and $Zn2^+$ for its activity. Fe, Zn, Cu and Mn are some of the cofactors of metalloenzymes.

Certain enzymes are produced by the cells of the organisms at all times and are called as normal enzymes. These enzymes help in the normal metabolic processes. For example, amylase in the saliva helps in the digestion of starch and lysozyme helps in the innate immune response of our body. However, certain enzymes are produced only when they are exposed to some drugs and are called drug metabolizing enzyme. Cytochrome P450, cytochrome b5, and NADPH-cytochrome P450 reductase are the examples of drug metabolizing enzymes.

The other simple way of classifying the enzyme is based on the site of release of the enzyme. The enzymes that are produced extracellularly are called exozymes. Certain enzymes are produced within the cells and are called endozymes. These terms should not be confused with isozymes. If a specific reaction can be catalyzed by two different enzymes, then these enzymes are called isozymes or isoenzymes. The isozymes are homologous enzymes and they have different amino acid sequences. They have different kinetic and regulatory features. They also differ in their K_M and V_{max} values. The isoenzymes of Lactate dehydrogenase are typical examples. The Lactate dehydrogenase produced by different organs differ in their amino acid sequence and their levels of expression.

As in the case of classification of enzymes, there are different ways for naming the enzymes. Generally the names of the enzymes ends with the suffix "ase". But, it is not always the case. The name of the enzymes namely pepsin, trypsin, rennin, papain etc. do not end with the suffix "ase". Depending on the substrate on which the enzyme acts, the enzyme can be classified as amylase (if it acts of starch), cellulase (if enzyme acts on cellulose), protease (if it acts on protein), lipase (if enzyme acts on lipid) and so on.

The enzymes can also be named depending on the type of reaction that it mediates. For example, the enzymes mediating oxidase reaction are termed oxidases, the enzymes mediating reduction reactions are reductases, the enzymes mediating dehydrogenation are dehydrogenases, the enzymes mediating transamination reaction are transaminases etc. Sometimes, the enzymes are named based on the substrate and the reaction it catalyzes. Pyruvate dehydrogenase is an example for this type.

The different types of conventions used for naming enzymes created lot of confusions among the researchers. Sometime, a single enzyme is named by two different names by two different conventions. Hence there was a need for a rational classification system for naming the enzymes. This was initiated by Enzyme Commission. Since many enzymes have more than one common name (based on different conventions), EC numbers were introduced and each enzyme was given with the specific EC number. Every enzyme code consists of the letters "EC" followed by four numbers separated by periods. The EC nomenclature classified enzymes based on the type of reaction it mediates. Each category covers the group of enzymes that catalyses a similar group of reactions. In an EC number code, the first digit indicates the general type of reaction mediated by the enzyme. The first digit ranges from 1 to 6 indicating the six different types of reactions.

The six different categories of enzymes are as follows:

1. **Oxidoreductases**: It mediates oxidation or reduction reactions. Dehydrogenase and oxidase are the examples for this class.
2. **Transferases**: It mediates the transfer of a functional group from one molecule to another. These functional groups may be phosphate, methyl and glycosyl groups. Example for this category includes transaminases and kinases.
3. **Hydrolases**: It helps in the hydrolysis reaction where the molecule is split into two or more smaller molecules by the addition of water. Proteases, nucleases and phosphatases are the examples. Proteases and Nucleases mediate hydrolysis reactions and they hydrolyse proteins and nucleic acids respectively.
4. **Lyases**: It mediates the lysis reaction by the cleavage of C–C, C–O, C–S and C–N bonds other than oxidation or hydrolysis reactions. This is a type of elimination reaction. Decarboxylase is an example.
5. **Isomerases**: It mediates the isomerisation reaction wherein the substrate in one isomeric form is converted to its other isomeric forms. There is the atomic rearrangement within the molecule leading to change in structural formula but not molecular formula. Isomerases are of different classes namely geometric, structural, enantiomers and stereoisomers. Rotamase, isomerase, mutase, epimerase and racemase are some of the examples.
6. **Ligases**: It mediates the catalysis of the reaction which joins two molecules. The chemical potential energy is required for this reaction and it is coupled with the hydrolysis of a disphosphate bond in a nucleotide triphosphate such as ATP. The enzymes peptide synthase, aminoacyl-tRNA synthetase, DNA ligase and RNA ligase are some of the examples for this category.

The series of next three numbers in the EC code defines and narrow the details of the reaction type specifically. The second and third numbers of the code indicates the enzyme's sub-class and sub-sub-class respectively. These numbers explain the reaction with respect to the compound, group, bond or product involved in the reaction. The final digit of the EC code is called as the serial identifier and it provides insights about specific metabolites and cofactors involved.

2.5   Mechanism of Enzymes

The binding of the substrate to the active site of the enzyme and its interaction is the basic mechanism of enzymatic catalysis. The active site is the catalytic region of the enzyme on which the substrate binds. Once the substrate binds to the active site of the enzyme, it causes the redistribution of electrons in the chemical bonds of the substrate. This redistribution of electrons in the substrate causes the biochemical transformation of the substrate to form products. Once the substrate is converted into product in the active site of the enzyme, the products are released from the enzyme and the next substrate molecule binds on the active site of the enzyme again and this cycle continues. The substrate can interact with the active site of the enzyme by different ways based on opposite charges, hydrogen bonding, hydrophobic non-polar interaction, and coordinate covalent bonding.

The unique geometric shape of the active site provides clue for the specificity of the enzyme to the specific substrate. There is a correlation between the geometric shapes of substrate to the geometric shape of active site of the enzyme. Two major hypothesis namely lock and key theory and Induced fit hypothesis have been proposed to explain the mechanism of interaction of the enzyme and substrate depending on the geometric shape.

Lock and Key Theory   The specific interaction between the enzyme and its substrate was first postulated by Emil Fisher in 1894 using Lock and Key analogy. In this hypothesis, Email Fisher postulated that the enzyme has a rigid structure and its active site has a defined geometric shape. Only the substrate whose geometric shape is complementary to the active site will be able to bind to the active site of the enzyme and gets catalysed. It is similar mechanism of the lock and the key. The enzyme acts as the lock and the substrate acts as the key. If the key exactly suits the lock, only then the key can used to open the lock. Even a very small changes in the shape/size of the key or if the key is not positioned properly, then the lock cannot be opened.

Induced Fit Theory   However, the Fisher's lock and key theory which explained enzyme as the rigid structure had limitations and failed to support all the experimental evidences on enzyme–substrate interactions. To circumvent this issue, a new theory called the induced-fit hypothesis has been developed. In contrast the lock and key mechanism, the induced fit hypothesis proposed that the enzyme structure is flexible and not rigid.

It is proposed that the substrate has a crucial role in determining the structure of the enzyme and the shape of the active site. The structure of the enzyme is partially flexible. On binding of the substrate to the enzyme, the enzyme structure changes accordingly and mediates the catalysis of the substrate. This theory also describes the reason behind the irreversible inhibition of enzyme wherein inhibitors can bind to the enzyme and causes the distortion of the enzyme. The substrate molecules which has a smaller geometric size when compared with the geometric size of the active site of the enzyme, cannot induce the structural change of the enzyme and they could not react/get catalysed. The specific substrate can only induce the change in the structure of the specific enzyme and get catalysed.

The enzyme activities are regulated in a highly systematic manner in the living systems. Certain enzymes are produced in inactive forms and they get activated whenever their catalytic activity is required. Enzymes such as pepsin, trypsin and chymotrypsin are produced in in active forms namely pepsinogen, trypsinogen and chymotrypsinogen respectively. These are proenzymes and it is inappropriate to term them as inactive enzymes. The term "inactive enzymes" refers to those enzymes which have lost their activity due to physical/chemical/metabolic factors or any other reasons. But zymogens are molecules that needs to be activated to make it an active enzyme. It is apt to term zymogens as the inactive precursor of enzymes. Some of the digestive enzymes and coagulation factors are synthesized as zymogens. The synthesis of digestive enzymes as zymogens is a safe mechanism as most of these digestive enzymes are proteolytic and if the enzymes are synthesized in active form, they have greater chance of hydrolyzing the proteins in the cells synthesizing them. If the zymogens are synthesized in actively forms, it leads to certain diseases in biological systems. Acute pancreatitis is one such example wherein the pancreatic enzymes e.g. trypsin, phospholipase A2, and elastase are activated in premature state.

Allosteric enzymes have a different interesting regulatory mechanism when compared with the normal enzymes. Allosteric enzymes have two different sites namely catalytic site and the regulatory site. The molecules which binds to these regulatory sites are called modulators or effectors. The positive modulators will mediate catalysis and negative modulators will inhibit the catalysis. Unlike the normal enzymes, the allosteric enzymes do not follow the Michaelis–Menten Kinetics as they have multiple active sites. Binding of the substrate to one active site of allosteric enzyme will influence the binding of the substrate to its next active site and this phenomenon is called cooperativity. Cooperativity is an interesting feature of allosteric enzyme. If the binding of first substrate onto the one active site facilitates binding of subsequent substrate molecule onto the next active site of the substrate molecule, then it is called "positive cooperativity". On the contrary, if the binding of the first substrate to its active site decelerates the binding of the next substrate to other active site, it is termed as "negative cooperativity". In some cases, various enzymes combine to form supramolecular complexes that allows the direct transfer of metabolites from one enzyme to the other without entering the bulk solution and this process is termed as metabolic channelling. The multi-enzyme complex systems may be static or dynamic.

2.6 Enzyme Activity

To assess the activity of the enzyme, it is more important to understand the units of the enzyme. Generally, the enzymes are quantified based on the unit of activity rather than in terms of the amount that is physically present (weight). This is mainly because of two reasons. It is difficult to purify the enzyme and the enzyme is easily prone to denaturation or loss of activity. The activity of the enzyme is generally represented as International Unit (IU) which is widely used. The unit is defined as the amount of enzyme which catalyzes the transformation of 1 micromole of the substrate per minute under standard conditions. The other unit of enzyme activity is Katal (referred as to kat). It is the amount which catalyzes the transformation of one mole of substance per second (1 kat = 60,000,000 U). There are few non-standard units for enzyme activity namely Soxhlet, Anson, and Kilo Novo units. However, these units are not generally used. The activity of the enzyme is a tool to assess the quality of the enzyme in terms of catalytic activity and it is of use in industrial applications. The specific activity is another important parameter for enzyme activity. Specific activity is the number of enzyme units per mg of enzyme protein. It can be denoted as units/mg (U mg^{-1}). Generally the definitions of activity are described with reference to the term "standard conditions". The term "standard conditions" refer to optimal conditions of properties such as pH, ionic strength, temperature, substrate concentration, and the concentration of cofactors/coenzymes. However, these conditions depends on the type of application and experimental conditions.

The enzyme activity depends on several factors such as temperature, pH, pressure, cofactor like metal ions etc.

Temperature The enzymatic reactions can operate at a narrow range of temperature. With increase in temperature up to the optimum temperature, the enzymatic reaction rate increases but after optimal temperature is reached, the reaction rate decreases with increasing temperature. The increase in temperature causes the increase in collision between the molecules with higher energy (increase in kinetic energy) leading to enhanced enzymatic catalytic rates. Like other catalytic reactions, enzymatic reactions also obey Arrhenius equation which states that rate of reaction will exponentially increase with increase in temperature. The Arrhenius equation is as follows:

$$k = Ae^{-\Delta G^*/\mathrm{RT}}$$

where k is the kinetic rate constant for the reaction, A is the Arrhenius constant or frequency factor, G* is the standard free energy of activation (kJ M^{-1}) which depends on entropic and enthalpic factors, R is the gas law constant and T is the absolute temperature. Different enzymes have different optimal temperatures and even a single enzyme from different organisms might differ in their optimal temperature. Actually the enzymes cannot decrease the activation energy of the same barrier. The enzyme directs the reaction in an alternate chemical pathway which has a lower activation energy. Most enzymes are stable for months if they are refrigerated at 0–4 °C. However, storing the enzymes below 0 °C is not advisable. Storing the enzymes below 0 °C requires the additives such as glycerol and such very low temperature often causes denaturation due to the stress and pH variation caused by ice-crystal formation.

Effect of pH on Enzyme Activity The enzymes can remain active and accelerate catalytic reactions only in a narrow range of pH. The drastic changes in pH from its optimal levels will change the stability of the enzymes. With increase or decrease in the pH, the various acid and amine groups on side chains of the protein is disturbed which in return affects the structure of the enzyme and its catalysis. The pH of the solution affects the state of ionization of acidic or basic amino acids. The change in the state of ionization of amino acids in a protein leads to changes in the ionic bonds which in turn affects the structure of the enzyme. Different enzymes differ in their optimal pH values. For example, lipase in pancreas has the optimal pH of 8.0 and lipase from stomach has the optimal pH of 4.0 to 5.0. Pepsin has an optimal pH value of 1.5–1.6 and trypsin has an optimal pH value of 7.8–8.7. The changes in ionic strength in the enzymatic system also affect the activity of the enzyme and its catalytic rates.

The changes in pressure also affects the enzymes in two different ways. The high pressure applied to the system will affect the conformation of the enzyme leading to loss of catalytic activity. On the other hand, if the enzymatic reaction systems (such as in oxygenases and decarboxylases enzyme systems) involve dissolved gases, the increase in pressure will lead to increase in gas solubility. This causes the disturbance in the reaction system leading to shift in the equilibrium position of the reaction due to any difference in molar volumes between the reactants and products.

2.7 Enzyme Kinetics

The enzyme catalyzed reaction can be of zero, first or second order. For an elementary reaction occurring in one direction, the order of reaction is equal to the molecularity. For non-elementary reaction, the rate order is experimentally determined to analyze what happens to the rate of reaction as the concentration of one of the reactants change. For first order reaction, if you double the concentration of the reactant A, the rate also doubles (Rate α [A]). For second order reaction, if you double the concentration of A the reaction rate will increase by 4 times (Rate α $k[A]^2$). For zero order reaction, the rate is apparently independent of the reactant concentration.

Enzyme kinetics involves the study of mechanisms and rates of enzyme catalyzed reactions at different reactions conditions such as varying the substrate and enzyme concentrations. Various environmental factors influence the Enzyme activity, including substrate concentration, pH, temperature and presence of inhibitors

Fig. 2.1 Relationship
between reaction velocity
and substrate concentration

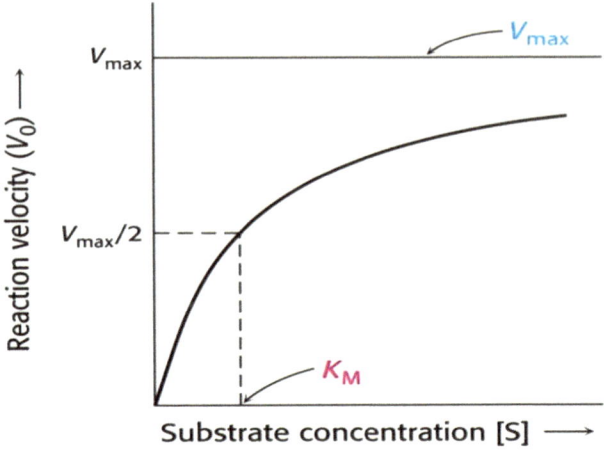

and activators. Substrate concentration is one of the most important influencing factor. The rate of enzyme catalyzed reaction increases with increase in substrate concentration up to the optimal substrate concentration. This is because more the amount of substrate molecules present, an enzyme binds substrate more often, and the rate of product formation (reaction velocity) is greater than at lower substrate concentration. After reaching an optimal substrate concentration, the further increase in substrate concentration causes no further increase in reaction velocity because no more active sites are available on the enzyme molecules, that is, the enzyme is saturated with the substrate and is converting substrate to product as rapidly as possible. This means enzyme is working at maximum velocity (V_{max}). Relationship between reaction velocity and substrate concentration is shown in Fig. 2.1.

Michaelis–Menten Kinetics: The relation between the substrate concentration and reaction rate/velocity is described by the following Michaelis–Menten equation

$$v = V_{max} \cdot [S]/(Km + [S])$$

The Michaelis constant (Km) is that substrate concentration at which the reaction rate (velocity) is half its maximum. Its significance is that it acts as indicator for the affinity of the enzyme for its substrate. The lower the Km value, more is the affinity of the enzyme to that substrate. In other words, a low Km means that the enzyme achieves its maximum catalytic efficiency at low substrate concentrations and therefore is more efficient as a catalyst or has more affinity for that particular substrate.

Lineweaver–Burk plot is obtained by taking the reciprocal of the Michaelis–Menton equation. It was widely used to calculate the Km and V_{max} before non-linear regression software were developed. Figure 2.2 depicts the Lineweaver–Burk plot.

Fig. 2.2 Lineweaver–Burk plot

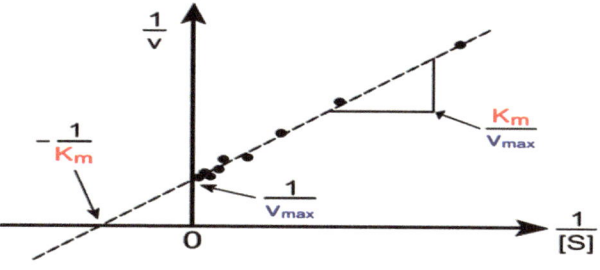

Fig. 2.3 Hanes–Woolf plot

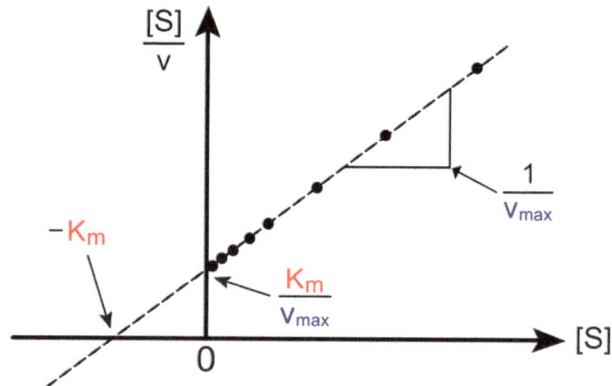

Although it is still used for representation of kinetic data, non-linear regression or alternative linear forms of the Michaelis–Menten equation such as the Hanes–Woolf plot or Eadie–Hofstee plot are generally used for the calculation of parameters. The drawback for Lineweaver–Burk plot is that it is error prone as small errors in measurement gets magnified as y-axis takes the reciprocal of the rate of reaction

Hanes–Woolf plot (shown in Fig. 2.3) is also obtained on rearranging the Michaelis–Menton equation to get the ratio of the initial substrate concentration [S] to the reaction velocity v. It linearizes the Michaelis–Menten equation but has a drawback that neither ordinate nor abscissa represent independent variables: both are dependent on substrate concentration.

Enzymes can have more than one substrate binding site. When such enzymes catalyze reactions, the rate increases in a sigmoidal manner as the substrate concentration increases instead of the hyperbola curve in case of enzymes with just one binding site. The Hill equation is commonly used to study the kinetics of reactions that exhibit a sigmoidal behavior. The rate of many transporter-mediated processes can be analyzed by the Hill equation. Hyperbola curve demonstrates saturation of the enzyme or transporter at high substrate concentrations. Saturation is caused by the fact that there is a fixed number of enzyme or transporter molecules, each with a fixed number of substrate binding sites. At high substrate concentrations, all of the binding sites have substrate bound and each enzyme or transporter molecule is working as fast as its intrinsic rate to catalyze the reaction.

Reactions that exhibit a sigmoidal curve also exhibit saturation at high substrate concentration. However, at low substrate concentrations, a very different behavior is observed (compared to a hyperbolic relationship). At low substrate concentrations, the rate increases only incrementally with increases in the substrate concentration. As the substrate concentration increases further, small increases in the substrate concentration lead to large increases in the reaction rate. At very high substrate concentrations, the rate exhibits saturation, where additional increases in the substrate concentration no longer increase the reaction velocity. This type of saturation kinetics is adequately described by the Hill equation.

2.8 Enzyme Inhibitors and Types

Enzyme inhibitors are molecules that prevents the enzyme catalysis by interacting with the enzyme molecule. Enzyme inhibitors have several applications in the development of drugs, pesticides etc. However, some of the inhibitors inhibit the key enzymes of our body and causes diseases and even death.

The enzyme's catalytic activity can be altered by binding various small molecules to it, sometimes at its active site, and sometimes at a site distant from the active site. Usually these alterations involve a reduction in the enzyme's ability to accelerate the reaction; less commonly, they give rise to an increase in the enzyme's ability to accelerate a reaction. Enzyme inhibitors are molecules which interact in some way with the enzyme and consequently slow down, or in some cases stop catalysis. Enzyme inhibitors can be specific or non-specific. Non-specific methods of inhibition include any physical or chemical changes which ultimately denatures the enzyme. The specific inhibitors act on a single enzyme or closely related enzymes and are further characterized into three main types: competitive, uncompetitive and non-competitive. These types of enzyme inhibition can be explained on the basis of their effect on the formation of enzyme–substrate complex and is explained in the following paragraphs.

Competitive Inhibitors These inhibitor molecules are similar, in chemical structure and molecular geometry, to the substrate of the enzyme. Therefore, as the name suggests, they compete with the substrate for binding to the active side of the enzyme. The inhibitor binds to the active site of the enzyme and thus prevent the formation of the Enzyme–Substrate complex. In other words, enzyme–inhibitor (E-I) complex is formed in instead of enzyme–substrate complex. Therefore, binding of substrate to enzyme in inhibited and the product formation is decreased. However, Competitive inhibition is reversible and levels of inhibition depends upon the relative concentrations of the substrate and the inhibitor molecules as the inhibitors compete with the substrate to bind the active site of the enzyme.

Drug disulfuran (Antabuse) which is used to help people to get rid of alcoholism, acts as a competitive inhibitor for the aldehyde oxidase enzyme which causes the accumulation of acetaldehyde resulting in unpleasant side effects such as nausea and vomiting. Another example of a useful application of competitive inhibition is in the case of methanol poisoning. Ethanol is given as an antidote in these cases as it acts as a competitive inhibitor to methanol, thus preventing its oxidation to form-aldehyde and formic acid. Formic acid is the actual toxic compound that effects the optic nerve causing blindness and its production is stopped by administration of ethanol.

It is important to note that the competitive inhibitor affects the binding of the substrate but not the reaction velocity i.e. it affects the Km of the enzyme but not the V_{max}. As discussed earlier, Km value is a measure of the amount of the substrate required to reach half the maximum reaction velocity (V_{max}). It is also known that the presence of the competitive inhibitor, forces to increase the substrate concentration in order to achieve that half maximal velocity. It therefore, increases the Km.

Non-competitive Inhibitors A non-competitive inhibitor interacts with the enzyme at a site other than the active site called the allosteric site. This site could be very close to the active site or far from it. Non-competitive inhibitors are usually reversible, but are not influenced by concentrations of the substrate since they do not compete with substrate molecules as in the case for a reversible competitive inhibitor. Therefore, Km is unchanged, but V_{max} is reduced. A non-competitive inhibitor reacts with the enzyme–substrate complex, and slows the rate of reaction to form the enzyme–product complex. This means that increasing the concentration of substrate will not relieve the inhibition, since the inhibitor reacts with the enzyme–substrate complex. An example of non-competitive inhibition is the inhibition of cytochrome enzymes by cyanide.

Uncompetitive Inhibitors Uncompetitive inhibition is a very rare class of inhibition. An uncompetitive inhibitor binds to the enzyme and enhances the binding of substrate (so reducing Km), but the resultant enzyme–inhibitor-substrate complex only undergoes reaction to form the product slowly, so that V_{max} is also reduced. While uncompetitive inhibition requires that an enzyme–substrate complex must be formed, non-competitive inhibition can occur with or without the substrate present. In addition, uncompetitive inhibition works best when substrate concentration is high. Lithium in the phosphoinositide cycle is an example of uncompetitive inhibition which ameliorates manic-depressive psychosis.

2.9 Immobilization of Enzymes

Enzyme immobilisation is a technique for entrapment of the enzyme onto a distinct support or matrix. The support or matrix on which the enzymes are immobilized is called a carrier and it allows the substrate or effector or inhibitor molecules to interact with the immobilised enzyme. The enzyme amino acylase of *Aspergillus oryzae* was the first enzyme to be immobilized for the production of L-amino acids. The enzyme can be immobilised onto the carrier either permanently or temporarily for a certain period of time. Three different polymers such as natural polymers, synthetic polymers and inorganic materials can be used as carriers for the immobilisation of enzymes. Alginate, chitosan, collagen, carrageenan, gelatin, cellulose, starch, and pectin are some of the natural polymers that are used for immobilisation. Alginate is an inert natural polymer and has good water retaining capacity. Calcium or magnesium alginate are the commonly used polymers for the preparation of natural matrices. Chitosan and chitin are the structural polysaccharides occurring naturally in the cell wall of fungi that are widely used polymers for immobilisation. Collagen is the proteinaceous support with good porosity and water retaining capacity. The side chains of the amino acids in the collagen and that of enzyme can form covalent bonds to permanently hold the enzyme to the support. Gelatin is the partially hydrolyzed collagen with good water holding capacity and is used for immobilisation of enzymes. Carrageenan, is a sulphated polysaccharide extracted from red algae, is a gelling agent that can hold proteins.

Polyvinyl chloride (PVC), Diethylaminoethyl cellulose (DEAE cellulose), and UV activated Polyethylene glycol (PEG) are some of the synthetic polymers that are used for the enzyme immobilisation. Zeolites, ceramics, diatomaceous earth, glass, silica, activated carbon, and charcoal are some of the examples for the inorganic materials.

There are a different types of carriers or supports that are used for immobilization of enzymes. The carrier used should be cheap and easily available. The ideal carrier should also have ease of functionalisation and should not inhibit the catalytic activity of the enzyme. The good carrier should offer good mass transfer characteristics and better diffusion rates leading to enhanced effectiveness. The effectiveness can be calculated by the ratio of rate of the reaction to the rate of diffusion. The immobilization of enzymes onto the carrier has several advantages such as enhanced functional efficiency of enzyme, reuse of enzyme, decreasing the reaction time, decreasing the cost of operation, high enzyme substrate ratio, and enhanced reproducibility of the process.

The immobilisation of enzymes has certain disadvantages. In some cases, the immobilisation of enzymes causes the loss of catalytic activity. The immobilisation of enzyme involves high cost and technical difficulties in the recovery of active enzyme and purification. There is a high chance that enzyme becoming unstable and losing its catalytic activity after it is recovered from the carrier. The immobilisation strategy does not suit for most industrial operations.

The immobilisation of enzyme can be carried out by five different methods namely adsorption, covalent bonding, entrapment, copolymerization and encapsulation. Adsorption is the simplest technique for the immobilisation of enzyme onto carrier. It is based on formation of weak (low energy) bonds to stabilise the enzymes onto the external surface of the carrier. Ionic interaction, hydrogen bonds and Van der Waal forces are the weak bonds involved in the adsorption process. Adsorption is the most traditional method for enzyme immobilisation.

Charcoal was used as the carrier for immobilisation of invertase in 1916. Materials such as aluminium oxide, clay, starch, and ion exchange resins are used as adsorbants/carriers. The smaller the size of the carrier, the larger is the surface area and higher is the rate of immobilisation and catalysis. The adsorption process also have the advantage that it will not suffer from pore diffusion limitations since the enzymes are immobilised on the external surface of the carrier. The adsorption of enzyme onto the carrier has several advantages such as simplicity, low cost for immobilisation, no need for reagents, minimum activation steps and no loss of enzyme structure/catalytic activity than other immobilisation methods.

The adsorption of enzyme onto the carrier can be carried out by four different methods namely static loading process, dynamic process, reactor loading process and electrode position process. Adsorption of enzyme onto the carrier by static process involves the carrier to be placed in the solution containing the enzyme without stirring. In dynamic process, the carrier is placed in the solution and mixed by agitation/stirring. Reactor loading process involves the transfer of enzyme solution from the carrier to the reactor by continuous agitation process. Electrode position process makes use of the electric field for the migration of the enzyme to the surface of the carrier. The enzyme immobilisation by adsorption technique has certain disadvantages such as the enzyme is prone to desorb from the carrier and the efficiency of this immobilization technique is poor.

The second important technique for immobilisation of the enzyme on the matrix by forming covalent bond between the chemical groups in enzyme and to the chemical groups on the support. Chemical bonding is an important and stable method for enzyme immobilization. The choice of the carrier is very important for immobilisation using this covalent bonding technique. The carrier should have chemical groups such as amino groups, imino groups, hydroxyl groups, carboxyl groups, thiol groups, methylthiol groups, guanidyl groups, imidazole groups or phenol ring which can help in forming the covalent group with the enzyme. Similarly the enzyme should have functional groups such as alpha carboxyl group at 'C' terminal of enzyme, alpha amino group at 'N' terminal of enzyme, phenol ring of Tyrosine, thiol group of Cysteine, hydroxyl groups of serine and threonine, imidazole group of histidine and indole ring of tryptophan. Different matrices such as cellulose, agarose, DEAE cellulose, polyacrylamide, collagen, gelatin, amino benzyl cellulose, porous glass, cyanogen bromide etc can be used as support for enzyme immobilization. The covalent bonding can be mediated by diazotion, peptide bond formation and using poly functional reagents. The covalent bonding technique is a simple method that provides a strong linkage of enzyme to the carrier/matrix and avoids leakage/desorption of enzymes. The covalent bonding technique for enzyme immobilisation is widely used for several industrial applications. But, in some cases, covalent bonding leads to enzyme inactivation/chemical modification of the enzyme which in turn affects the conformation of the enzyme and its catalytic activity.

Entrapment is another method for enzyme immobilisation wherein the enzymes are physically entrapped within a porous carrier. Generally, water soluble polymer is used for the preparation of carriers for entrapment of enzymes. The enzymes are immobilised onto the matrix either covalently or non-covalently. The nature of the carrier and pore size are some of the criteria for choice of matrix. Pore size of the carrier can be modified based on the concentration of the polymer used. Entrapment technique is not suitable for the low molecular weight enzymes from the matrix as it causes leakage. Polyacrylamide gels, agar, alginate, gelatin, and carrageenan are some of the matrices that are used as matrices for entrapment. Entrapment of enzymes can be made in the gels, fibers or in microcapsules. This is a simple, quick, and economical method of immobilisation. It requires mild operating conditions for the entrapment and has lesser change of conformational changes in structure of the enzyme. Entrapment techniques has certain limitations such as leakage/loss of the enzyme, loss by diffusion, and is easily prone to microbial contamination.

The next method of immobilisation is cross linking method in which functional groups between the enzymes are cross linked directly by covalent bonds. Polyfunctional reagents such as glutaraldehyde and diazonium salt are generally used as crosslinking agents for the immobilisation by crosslinking or copolymerisation technique. It is a simple and an economical method for immobilisation where pure enzymes are not required. It is widely used for several industrial and commercial applications. This technique has certain limitations—the crosslinking agents that are used for co-polymerisation of enzymes are prone to denature the structure or activity of the enzyme.

The fifth method for immobilisation is the encapsulation. It is done by immobilising the enzymes within membrane capsule. Semi permeable membranes such as nitro cellulose or nylon are generally used as capsules. The mass transfer shortcoming is the major shortcoming of this technique. This technique allows immobilisation of a large quantity of enzyme within the capsule. It is a simple and inexpensive method. The membrane capsule allows only the substrate of smaller size to enter the capsule and this technique has pore size limitation.

Applications of Immobilized Enzymes Immobilized enzymes are widely used for several industrial biotechnology applications such as production of alcohols, organic acids, amino acids, drugs, antibiotics etc. The use of immobilized enzymes greatly helps to cut down the costs of bioprocesses. Immobilized enzymes are also used widely in the food industry for the production of corn syrup, jams, jellies etc. Immobilised pectinases and cellulases are widely used in food industries. Biomedical sector is one major area where the immobilized enzymes are widely used in the diagnosis of ailments and therapy. Immobilized enzymes based biosensors are widely used for detection of different analytes such as glucose, urea, nitric oxide, dopamine, phenol etc. Enzyme Immobilization strategies are also very useful for the targeted drug delivery and controlled drug release at the site of infection. Immobilised enzymes are also widely used in research especially in molecular biology. The other areas where the immobilised enzyme can be used are for the production of bio-diesel from vegetable oils, treatment of wastewater, textile industry and detergent industry.

2.10 Application of Extremozymes

Extremophiles are microorganisms that survive under harsh environmental conditions that can include atypical temperature, pH, salinity, pressure, nutrient, oxygen, water, and radiation levels. Thus, extremophiles are a robust group of organisms producing stable enzymes, which are often capable of tolerating changes in environmental conditions such as high/low pH and temperature. These organisms and their enzymes have wide range of potential applications in biotechnology. One application is that extremophilic enzymes, also called extremozymes can be exploited for conversion of biomass to biofuel. Biomass is biological material derived from living, or recently living organisms. It most often refers to plants or plant-based materials which are specifically called lignocellulosic biomass. As an energy source, biomass can either be used directly via combustion to produce heat, or indirectly after converting it to various forms of biofuel. Biomass can be converted to other usable forms of energy like methane gas or transportation fuels like ethanol and biodiesel.

The five most basic type of extremozymes and their applications are as follows:

Thermophilic Enzymes High temperatures for chemical reaction is favorable due to the increase in substrate solubility, better mixing, high reaction rates, and low viscosity. Another important advantage of carrying out reactions at high temperatures is pathogen removal. Thus, thermophilic enzymes which are active at high temperatures can be used to catalyze such high temperature chemical processes. An enzyme or protein is considered thermostable when they have a high defined unfolding (transition) temperature (Tm), or a long half-life at a selected high temperature. The use of such enzymes in maximizing reactions accomplished in the food and paper industry, detergents, drugs, toxic wastes removal, and drilling for oil is being studied extensively. Thermophilic enzymes specifically hemicellulose degrading is employed in treatment of wood to obtain pulp as it is carried out at high temperatures. Thermophilic xylanases (xylan degrading) also reduce the consumption of Chlorine in the bleaching process. Thus, decreasing the release of hazardous organic halogens to the environment. Thermostable enzymes have also been used in the pharmaceutical industry along with conventional chemical synthesis for production of drugs. L-aminoacylase from *Thermococcus litoralis* developed at Exeter is used for the resolution of amino acids and analogues (Toogood et al. 2002). The gamma lactamase from *Sulfolobus solfataricus* is used for the production of optically pure gamma lactam—the building block for anti-viral carbocyclic nucleotides (Toogood et al. 2004). The alcohol dehydrogenase from *Aeropyrum pernix* for the production of optically pure alcohols (Guy et al. 2003). Another major applications for thermostable enzymes are starch liquefaction using amylases from thermophilic *Bacillus* sp. and proteases for food processing and detergents. Thermostable DNA polymerases are greatly useful in the PCR process.

Psychrophilic Enzymes These enzymes are produced by psychrophilic (organisms that thrive at very low temperatures, usually below 5 °C). Psychrophilic enzymes are therefore cold active and heat labile. They have a number of advantages in the field of biotechnology, majorly because of their high k_{cat} at low to moderate temperatures. Also, they can be used at lower concentration as they have high activities, and thus reduce the cost of producing large amount of the enzyme. So additional heating costs for their optimum activity is not required. Due to this, these enzymes, typically belonging to the lipase and protease class have been used as additives to detergents for washing at room temperature. Further, due to their heat lability, they can be selectively deactivated easily by slight increase in temperatures. Their industrial applications have been widely explored.

Psychrophilic enzymes finds immense applications in food industry for improving the digestibility and removing hemicellulose from feed, meat tenderizing, ripening of cheese, dough fermenting, stabilizing wine and beverages (Bialkowska et al. 2009; Collins et al. 2006; Tutino et al. 2009; Wang et al. 2010). They also find applications in detergent industries, (Tutino et al. 2009; Wang et al. 2010); Biofuels and energy production (Dahiya et al. 2006; Hildebrandt et al. 2009; Ueda et al. 2010); pharmaceutical, (Dahiya et al. 2006; Joseph et al. 2008); Textile industries (Collins et al. 2006; Ueda et al. 2010); environmental biotechnology for bioremediation; biobleaching of pulp and paper and tanning (Joseph et al. 2008; Wang et al. 2010) and chemical synthesis (of peptides, epoxides, oligosaccharides and other organic compounds) by reverse hydrolysis in organic solvents (Aurilia et al. 2008; Joseph et al. 2008).

Acidophilic Enzymes Acidophilic enzymes derived from acidophiles are adapted to work under low pH. In other words the optimum pH for their activity lies is the acidic range. An acidophilic β-galactosidase enzyme purified from *Teratosphaeria acidotherma* AIU BGA-1 has been found to show high activity at extremely low pH of 1 (Chiba et al. 2015). Another acidophilic enzyme, β-mannanase from *Gloeophyllum trabeum* CBS900.73 with significant transglycosylation activity have been found recently and posseses feed digesting ability (Wang et al. 2016). Xylanase produced by an acidophile *Penicillium oxalicum* GZ-2 has great potential to be used in biofuels, animal feed, and food industry applications (Liao et al. 2012).

Alkaline Enzymes Alkaliphiles are microorganisms that can grow in alkaline environments, i.e. pH > 9.0 Alkaline enzymes obtained from organisms living in these environments are able to function under high alkaline pH values because of their stability/activity under these conditions. Alkaline enzymes often show activities in a broad pH range, thermostability, and tolerance to oxidants compared to neutral enzymes (Fujinami and Fujisawa 2010).

From many decades, many alkaline protease enzymes have been tested for their compatibility with commercial detergents (Devi et al. 2016; Gupta et al. 2002; Haddar et al. 2009; Phadatare et al. 1993). They also have wide applications in tannery and food industries, medicinal formulations, and processes like waste treatment, silver recovery and resolution of amino acid mixtures (Agrawal et al. 2004; Horikoshi 1999; Sinha et al. 2014). Alkaline proteases have been used in the preparation of protein hydrolysates of high nutritional value. The protein hydrolysates play an important role in blood pressure regulation and are used in infant food formulations. The possible utilization of alkaline protease secreted by *Penicillium* sp. for hydrolysis of soy protein, a byproduct of soybean industries has also been done (Agrawal et al. 2004; Fujinami and Fujisawa 2010). Other alkaline enzymes, e.g. alkaline cellulases, alkaline amylases, and alkaline lipases, are also adjuncts to detergents for improving cleaning efficiency.

Barophilic Enzymes These enzymes are produced by deep sea Piezophilic/barophilic microorganisms and are stable at high pressures. They can be used for enzymatic processes at high pressure such as in food processing industry for sterilization of food, deep sea waste disposal, production of novel natural products and catabolic activities. These enzymes can also be used for other high pressure bioreactors. The application of these enzymes in various industrial processes is still in its budding stage.

Take Home Message

- Enzymes are biological catalysts that can mediate the biological reactions. They are produced by all living organisms. Most enzymes are proteins except ribozymes. All enzymes are not proteins and all proteins are not enzymes.
- Enzyme Commission has classified enzymes into six classes: Oxidoreductases, Transferase, Hydrolase, Lyase, Isomerase and Ligase.
- Specificity is the inherent charecetristic of the enzyme. The enzyme specificity can be categorized into five groups: Absolute Substrate specificity, Broad specificity (Group specificity), Bond specificity (Relative specificity), Sterochemical specificity, and Reaction specificity.
- Active site is the region of the enzyme where the substrate binds. Lock and Key Theory states both the structure of enzyme and the substrate are rigid whereas Induced Fit Theory describes that the structure of enzyme is partially flexible. The binding of the substrate onto enzyme causes structural changes in the enzyme and mediates the catalysis of the substrate.
- The activity of the enzyme can be represented as International Unit (IU), Katal, and Turnover number.
- The relation between the substrate concentration and reaction rate/velocity is described by Michaelis–Menten equation.
- Enzyme inhibition can be reversible or irreversible. The inhibitors of enzymes can be classified as competitive, non- competitive, and un-competitive.
- The enzymes can be immobilized onto carriers by different methods such as adsorption, covalent bonding, cross linking, entrapment, and encapsulation.

References

Agrawal D, Patidar P, Banerjee T, Patil S (2004) Production of alkaline protease by Penicillium sp. under SSF conditions and its application to soy protein hydrolysis. Process Biochem 39:977–981

Aurilia V, Parracino A, D'Auria S (2008) Microbial carbohydrate esterases in cold adapted environments. Gene 410:234–240

Buchholz K, Poulson PB (2000) Overview of history of applied biocatalysis. In: Straathof AJJ, Adlercreutz P (eds) Applied biocatalysis. Harwood Academic Publishers, Amsterdam

Bialkowska AM, Cieslinski H, Nowakowska KM, Kur J, Turkiewicz M (2009) A new beta-galactosidase with a low temperature optimum isolated from the Antarctic arthrobacter sp. 20B: gene cloning, purification and characterization. Arch Microbiol 191:825–835

Chiba S, Yamada M, Isobe K (2015) Novel acidophilic beta-galactosidase with high activity at extremely acidic pH region from Teratosphaeria acidotherma AIU BGA-1. J Biosci Bioeng 120:263–267

Collins T, Hoyoux A, Dutron A, Georis J, Genot B, Dauvrin T, Arnaut F, Gerday C, Feller G (2006) Use of glycoside hydrolase family 8 xylanases in baking. J Cereal Sci 43:79–84

Dahiya N, Tewari R, Hoondal GS (2006) Biotechnological aspects of chitinolytic enzymes: a review. Appl Microbiol Biotechnol 71:773–782

Devi SG, Fathima AA, Sanitha M, Iyappan S, Curtis WR, Ramya M (2016) Expression and characterization of alkaline protease from the metagenomic library of tannery activated sludge. J Biosci Bioeng 122:694–700

Fujinami S, Fujisawa M (2010) Industrial applications of alkaliphiles and their enzymes – past, present and future. Environ Technol 31:845–856

Gupta R, Beg QK, Lorenz P (2002) Bacterial alkaline proteases: molecular approaches and industrial applications. Appl Microbiol Biotechnol 59:15–32

Guy JE, Isupov MN, Littlechild JA (2003) The structure of an alcohol dehydrogenase from the hyperthermophilic Archaeon aeropyrum pernix. J Mol Biol 331:1041–1051

Haddar A, Agrebi R, Bougatef A, Hmidet N, Sellami-Kamoun A, Nasri M (2009) Two detergent stable alkaline serine-proteases from Bacillus mojavensis A21: purification, characterization and potential application as a laundry detergent additive. Bioresour Technol 100:3366–3373

Hildebrandt P, Wanarska M, Kur J (2009) A new cold-adapted beta-D-galactosidase from the Antarctic arthrobacter sp. 32c – gene cloning, overexpression, purification and properties. BMC Microbiol 9:151

Horikoshi K (1999) Alkaliphiles: some applications of their products for biotechnology. Microbiol Mol Biol Rev 63:735–750

Joseph B, Ramteke PW, Thomas G (2008) Cold active microbial lipases: some hot issues and recent developments. Biotechnol Adv 26:457–470

Liao H, Xu C, Tan S, Wei Z, Ling N, Yu G, Raza W, Zhang R, Shen Q, Xu Y (2012) Production and characterization of acidophilic xylanolytic enzymes from Penicillium oxalicum GZ-2. Bioresour Technol 123:117–124

Payen A, Persoz JF (1833) Mémoire su la diastase, les principaux produits de ses réactions, et leurs applications aux arts industriels. Ann Chim Phys 53:73–92

Phadatare SU, Deshpande VV, Srinivasan MC (1993) High activity alkaline protease from Conidiobolus coronatus (NCL 86.8.20): enzyme production and compatibility with commercial detergents. Enzyme Microb Technol 15:72–76

Sinha R, Srivastava AK, Khare SK (2014) Efficient proteolysis and application of an alkaline protease from halophilic Bacillus sp. EMB9. Prep Biochem Biotechnol 44:680–696

Sumner JB, Somers GF (1953) Chemistry and methods of enzymes, third edition. Soil Sci 76 (2):166

Toogood HS, Hollingsworth EJ, Brown RC, Taylor IN, Taylor SJ, McCague R, Littlechild JA (2002) A thermostable L-aminoacylase from Thermococcus litoralis: cloning, overexpression, characterization, and applications in biotransformations. Extremophiles 6:111–122

Toogood HS, Brown RC, Line K, Keene PA, Taylor SJC, McCague R, Littlechild JA (2004) The use of a thermostable signature amidase in the resolution of the bicyclic synthon (rac)–γlactam. Tetrahedron 60:711–716

Tutino ML, di Prisco G, Marino G, de Pascale D (2009) Cold-adapted esterases and lipases: from fundamentals to application. Protein Pept Lett 16:1172–1180

Ueda M, Goto T, Nakazawa M, Miyatake K, Sakaguchi M, Inouye K (2010) A novel cold-adapted cellulase complex from Eisenia foetida: characterization of a multienzyme complex with carboxymethylcellulase, beta-glucosidase, beta-1,3 glucanase, and beta-xylosidase. Comp Biochem Physiol B Biochem Mol Biol 157:26–32

Wang F, Hao J, Yang C, Sun M (2010) Cloning, expression, and identification of a novel extracellular cold-adapted alkaline protease gene of the marine bacterium strain YS-80-122. Appl Biochem Biotechnol 162:1497–1505

Wang C, Zhang J, Wang Y, Niu C, Ma R, Wang Y, Bai Y, Luo H, Yao B (2016) Biochemical characterization of an acidophilic beta-mannanase from Gloeophyllum trabeum CBS900.73 with significant transglycosylation activity and feed digesting ability. Food Chem 197:474–481

Chapter 3
Pretreatment of Lignocellulosic Feedstocks

Antonio D. Moreno and Lisbeth Olsson

Abbreviations

AFEX	ammonia fiber explosion
ARP	ammonia recycled percolation
CELF	co-solvent enhanced lignocellulosic fractionation
COSLIF	cellulose and organic solvent-based lignocellulosic fractionation
CrI	cellulose crystallinity index
CSF	combined severity factor
Cu-AHP	copper-catalyzed alkaline hydrogen peroxide
DP	degree of polymerization of cellulose
EA	extractive ammonia
GHG	greenhouse gas
ILs	ionic liquids
LCA	life-cycle assessment
LCCs	lignin-carbohydrate complexes
NMMO	N-methylmorpholine-N-oxide
SAA	soaking in aqueous ammonia
SF	severity factor
SPORL	sulfite pretreatment to overcome recalcitrance of lignocellulose
WRV	water retention value

A.D. Moreno (✉)
Department of Biology and Biological Engineering, Industrial Biotechnology,
Chalmers University of Technology, 412 96 Gothenburg, Sweden

Department of Energy, Biofuels Unit, Ciemat, Avda. Complutense 40, 28040 Madrid, Spain
e-mail: davidmo@chalmers.se; david.moreno@ciemat.es

L. Olsson
Department of Biology and Biological Engineering, Industrial Biotechnology,
Chalmers University of Technology, 412 96 Gothenburg, Sweden

© Springer International Publishing AG 2017   31
R.K. Sani, R.N. Krishnaraj (eds.), *Extremophilic Enzymatic Processing of
Lignocellulosic Feedstocks to Bioenergy*, DOI 10.1007/978-3-319-54684-1_3

3.1 Introduction

After the first oil crisis in the mid-late twentieth century, the need for a more sustainable and renewable energy system became clear. Today, 40 years later, a continuous increase in the energy demand and the urgency in reducing carbon emissions are still good reasons for us to give up our dependence on non-renewable fossil resources. Consequently, the scientific community is in the process of helping to develop and implement novel technologies and lay the basis of a new bio-based economy with relevant products for the energy, chemical, material, and food sectors. In this context, lignocellulosic biomass represents a great source of renewable raw material for the implementation of this bio-based economy since it is widely available, relatively inexpensive and do not compete with food production.

Lignocellulose is the most abundant renewable organic matter in nature, with an estimated annual production of more than 10^{10} MT worldwide (Sánchez and Cardona 2008). It includes agricultural wastes, forest products, and energy crops, and its chemical composition varies depending on the raw material (Sun and Cheng 2002). The main components of lignocellulosic biomass are cellulose, hemicellulose, and lignin. From the biochemical point of view, cellulose and hemicellulose incorporate sugars as monomers (glucose, xylose, mannose, arabinose, etc.), while lignin is built up of three primary phenylpropane units (p-coumaryl, coniferyl, and sinapyl alcohol). Due to their different properties, each polymer has a specific role in lignocellulosic materials. Cellulose fibers interact to each other by hydrogen bonds, forming a compact, crystal structure that confers rigidity and stability. Hemicellulose, on the other hand, acts as a link between cellulose and lignin, while lignin provides a recalcitrant matrix together with cellulose and hemicellulose and protects these two components from chemical and biological degradation. Lignin is also able to form ether- and ester-type covalent bonds with hemicellulose, forming lignin-carbohydrate complexes (LCCs) (Jørgensen et al. 2007). These intrinsic properties of lignocellulose render a three-dimensional structure that is difficult to disrupt. For an optimal use of lignocellulosic feedstocks as an energy source, a pretreatment step to improve its digestibility is therefore essential.

In a sugar biorefinery platform, lignocellulose is converted to several value-added compounds such as ethanol, organic acids, or lipids, by enzymatic hydrolysis and microbial/chemical catalysis processes (Fig. 3.1). Enzymatic hydrolysis of cellulose is carried out by cellulases, which major components are endo-1,4-β-D-glucanases (EC 3.2.1.4., attack cellulose regions with low crystallinity, creating new free chain ends), exo-1,4-β-D-glucanases (EC 3.2.1.91., degrade the cellulose molecule by releasing cellobiose units from the free chain ends), and 1,4-β-D-glucosidases (EC 3.2.1.21., hydrolyze cellobiose to produce glucose) (Jørgensen et al. 2007). Similarly, hemicellulases [endo-1,4-β-D-xylanases (EC 3.2.1.8), endo-1,4-β-D-mannanases (EC 3.2.1.78), α-L-arabinofuranosidases (EC 3.2.1.55), etc.] are responsible for breaking down hemicellulose polymer (Jørgensen et al. 2007). However, the recalcitrant structure of lignocellulosic materials limits the accessibility of hydrolytic enzymes to cellulose and hemicellulose and leaves a demand of

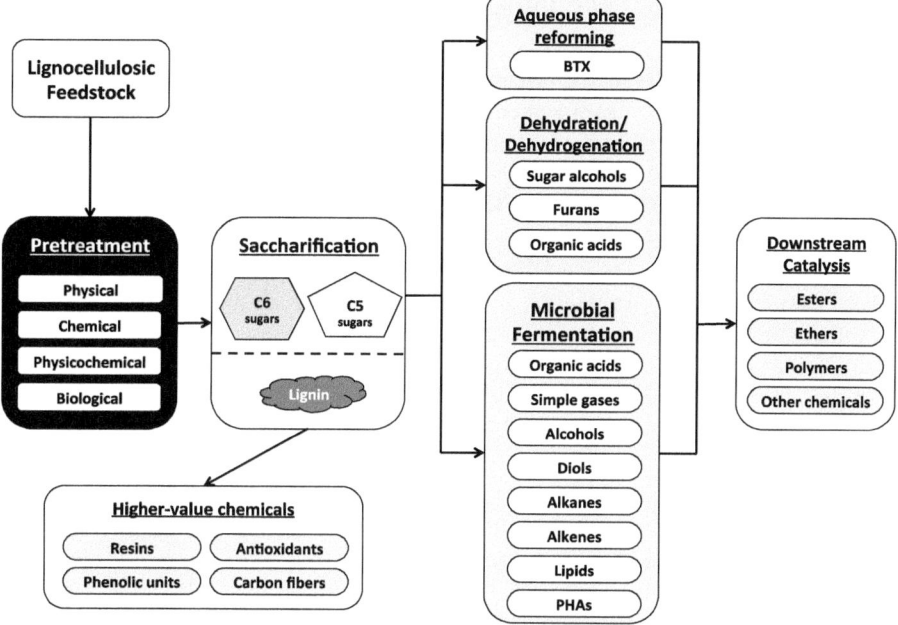

Fig. 3.1 Possible pathways and chemical compounds in a sugar biorefinery platform. BTX: benzene, toluene, xylene; C6: hexoses; C5: pentoses; PHAs: polyhydroxyalkanoates

a pretreatment step prior to enzyme addition. The aim of the pretreatment process is to alter the structural characteristics of lignocellulose and increase the accessibility of cellulose and hemicellulose to hydrolytic enzymes. The pretreatment is most certainly a crucial step during lignocellulosic biomass processing, since it not only has a great impact on final yields, but also makes an important contribution to overall costs.

The enzymatic hydrolysis of lignocellulose is limited by many factors, including the crystallinity and the degree of polymerization (DP) of cellulose, the water swelling capacity, moisture content, surface area, and lignin content. Therefore, to improve the hydrolisability of lignocellulosic feedstocks, one or more of these parameters should be modified during the pretreatment process (Karimi and Taherzadeh 2016). The mechanisms responsible for the effectiveness of pretreatment vary according to the nature of the pretreatment itself. Of the factors mentioned, an increase in the enzyme-available surface area and the pore size, together with the disruption of LCC linkages and/or lignin and hemicellulose removal, are the main considerations to take into account for an effective pretreatment process (Jørgensen et al. 2007). Nevertheless, it is important to note that several factors change during the pretreatment process and no factor can be single out as being especially important for improving the enzymatic hydrolysability of lignocellulosic feedstocks.

3.2 Pretreatment Technologies

A large number of diverse pretreatment technologies have been suggested to overcome the complex physicochemical, structural, and compositional barriers that hinder the digestibility of lignocellulosic biomass (Hendriks and Zeeman 2009; Alvira et al. 2010; Tomás-Pejó et al. 2011; Bensah and Mensah 2013; Moreno et al. 2015; Akhtar et al. 2016). Still, it is not possible to identify the best pretreatment method, and the choice of an adequate pretreatment technology will depend on factors such as type of lignocellulosic biomass, the final targeted product (s), economics, and environmental impact. For instance, in the case of bioethanol production, pretreatment methods offering lignin recovery at the end of the process are preferred since it can potentially offer an added value (Fig. 3.1).

Pretreatment technologies can be classified into physical, chemical, and biological methods. The combination of different pretreatment technologies has been also proposed in several cases, due to the added benefits of the synergistic effects and the potential cost reductions (Moreno et al. 2015; Bensah and Mensah 2013). Thus, combined physicochemical pretreatments reduce the amounts of chemicals or solvents required during the process, use milder process conditions, and decrease enzyme loadings in the subsequent saccharification process. Biological pretreatments can be also combined with chemical/physicochemical pretreatments in order to increase delignification yields. Table 3.1 summarizes all pretreatment technologies developed up to date, listing the advantages and disadvantages of each one.

3.2.1 Physical Pretreatments

3.2.1.1 Mechanical Comminution and Refining

During mechanical pretreatment, the structure of the cell wall is disrupted by three main stress factors: cutting, shearing and compression. Basically, there is a reduction in particle size and the outer walls of fibers are pulled out and primary walls are removed. These effects increase the enzyme-available surface area and reduce the cellulose crystallinity index (CrI) and DP, improving the accessibility of carbohydrates to hydrolytic enzymes (Gharehkhani et al. 2015). The final particle size of the material will depend on the technique used. For instance, 10- to 30-mm particles are obtained after chipping processes while a particle size of 0.2–2 mm is usually observed after milling or grinding (Sun and Cheng 2002). Methods such as stirred ball milling, hammer milling, disk refining, roller (Szego) milling, high-pressure homogenization, and mechanical jet smash are the main techniques used for these pretreatment processes (Alvira et al. 2010; Kim et al. 2016). Among them, ball mills, disk mills and roller (Zsego) mills are the major scalable methods and can be adapted for dry and wet samples. The industrial use of mechanical pretreatments is however limited due to the long milling times and the high energy demands, which

Table 3.1 Comparison of different technologies developed for pretreatment of lignocellulosic feedstocks

Pretreatment	Mode of action	Saccharification yields[a]	Biomass degradation[b]	Biomass spectra/effectiveness[c]	Additional observations	Reference(s)
Physical pretreatments						
Mechanical	Reduce CrI, DP, and particle size	Low	−	++	High operational costs. Effective when combined with hydrothermal/chemical technologies	Gharehkhani et al. (2015), Kim et al. (2016)
Extrusion	Increase enzyme-available surface area and reduce CrI and DP	Medium/very high	−	++; HB	Higher saccharification yields when combined with chemical catalysts	Zheng and Rhemann (2014), Duque et al. (2014)
Chemical pretreatments						
Acid-based pretreatment	Hemicellulose solubilization; lignin modification and cellulose hydrolysis (dependent on acid concentration)	Medium/very high	+++	+++; HB/HWB	Dilute-acid pretreatments are more advantageous than strong acid pretreatments	Kootstra et al. (2009), Tomás-Pejó et al. (2011)
Alkali-based pretreatment	Delignification; cellulose swelling and partial hemicellulose modification	High	−/+	+; HB	Higher saccharification yields when combined with other pretreatment technologies	Zhao et al. (2008), Karimi and Taherzadeh (2016)
Organosolv	Lignin removal and partial hydrolysis of hemicellulose	Medium	+/++	+++; HB/SWB/HWB	High-quality lignin	Pan et al. (2005), Mesa et al. (2011), Wildschut et al. (2013)
Ionic liquids (ILs)	Biomass fractionation/lignin removal and hydrolysis of hemicellulose	High/very high	+	++; HB/SWB/HWB	Possibility of lignin revalorization. An efficient recovery system is needed	Brandt et al. (2013)

(continued)

Table 3.1 (continued)

Pretreatment	Mode of action	Saccharification yields[a]	Biomass degradation[b]	Biomass spectra/effectiveness[c]	Additional observations	Reference(s)
Soaking in aqueous ammonia (SAA)	Lignin removal	High	–	+; HB	Very long pretreatment times	Kim and Lee (2005)
Cellulose and organic solvents-based lignocellulosic fractionation (COSLIF)	Biomass fractionation	Very high	–	+++; HB/HWB	Mild reaction conditions	Zhang et al. (2007)
Co-solvent enhanced lignocellulosic fractionation (CELF)	Lignin removal and hydrolysis of hemicellulose	High	–	++; HB	Higher yields obtained in simultaneous saccharification and fermentation processes	Nguyen et al. (2016)
Ozonolysis	Lignin removal	Medium/very high	–/+	+/++; HB/HWB	High cost of large amount of ozone needed	Travaini et al. (2015)
Aqueous N-methylmorpholine-N-oxide	Biomass fractionation; cellulose swelling and increased porosity	High/very high	–/+	++; HB/HWB	Possibility of simultaneous pretreatment and saccharification	Bensah and Mensah (2013)
Combined physicochemical pretreatments						
Steam explosion	Hemicellulose hydrolysis and lignin redistribution	Medium/very high	+/+++	+++; HB/HWB	Low environmental impact	Wang et al. (2015)
Hydrothermal pretreatment	Hemicellulose hydrolysis and lignin redistribution	Medium/very high	++/+++	++; HB	Lower biomass degradation at pH 4-7	Alvira et al. (2010)
Ammonia fiber explosion (AFEX)	Cellulose swelling and deacetylation of hemicellulose	Very high	–	++; HB	Small particle size needed	Akhtar et al. (2016)

Pretreatment	Mechanism	Yield[a]	Biomass degradation[b]	Applicability[c]	Advantages	References
Extractive ammonia (EA)	Lignin solubilization and cellulose decrystallization	Very high	−	++; HB	Lignin with high purity recovered and intact, native lignin functionality	da Costa Sousa et al. (2016)
Wet oxidation	Lignin oxidation and hemi-cellulose hydrolysis	Medium	−/+	++; HB	Low energy demand due to an exothermic reaction	Klinke et al. (2002), Olsson et al. (2005)
Microwave pretreatment	Delignification and/or hemicellulose hydrolysis	Medium/high	+/++	+++	Need for addition of chemicals	Xu (2015)
Ultrasound pretreatment	Lignin removal	Medium	+	++	Possibility of integrating ultrasonication and enzymatic hydrolysis	Bussemaker and Zhang (2013)
Ammonia recycle percolation (ARP)	Lignin removal and partial hydrolysis of hemicellulose	Very high	−	+; HB/HWB		Bussemaker and Zhang (2013)
Supercritical fluids	Delignification and hydrolysis of hemicellulose	Medium	+	++	Higher delignification when combined with organic solvents	Schacht et al. (2008)
Sulfite pretreatment to overcome recalcitrance of lignocellulose (SPORL)	Lignin sulfonation and hydrolysis of hemicellulose	Medium/high	+	++		Bensah and Mensah (2013)
Biological pretreatments						
Microbial pretreatment	Delignification	Low/high	−	+	Reduced cellulose content	Moreno et al. (2015)
Enzymatic pretreatment	Delignification	Low/high	−	+/++; HB/HWB	Short reaction times	Moreno et al. (2015)

[a] Low: yields lower than 30%; Medium: yields of 30–60%; High: yields of 60–80%; Very high: yields higher than 80%.
[b] −: no biomass degradation; +: low biomass degradation; ++: medium biomass degradation; +++: high biomass degradation.
[c] +: low applicability to different types of biomass; ++: it can be applied to some extent to different types of biomass; +++: high applicability to different types of biomass; HB: effective on herbaceous biomass; SWB: effective on softwood biomass; HWB: effective on hardwood biomass

implies in general an increase in the overall costs (Hendriks and Zeeman 2009; Kim et al. 2016). Nevertheless, mechanical pretreatments are commonly used to reduce particle size in all type of lignocellulosic feedstocks before subjecting raw materials to other pretreatment technologies.

3.2.1.2   Extrusion

Extrusion is a thermo-mechanical pretreatment process. It is based on the effect exerted by the tight rotation of a single or a twin screw at a certain temperature. The equipment configuration can vary by using different types of screw elements. Also, it can be coupled to the addition of chemical and biological catalysts (Zheng and Rhemann 2014; Duque et al. 2014). Depending on the screw speed and barrel temperature, the structure of lignocellulose is modified by promoting defibrillation and shortening of fibers, which—in a similar way to milling processes—increases the surface area available to hydrolytic enzymes and reduces CrI and DP. The integration of chemical catalysts (acid or alkali) and/or biological catalysts (hydrolytic enzymes) results in a water-swollen effect—that increases water retention values (WRV) of fibers—and a more efficient biomass fractionation, leading to higher sugar yields in the subsequent saccharification step (Zheng and Rhemann 2014; Duque et al. 2014). Extrusion process has been applied to a wide range of lignocellulosic feedstocks (eucalyptus, pine sawdust, switchgrass, wheat straw, municipal waste, sugarcane bagasse, etc.) with and without acid or alkali as catalysts, increasing sugar yields during the saccharification steps by 40–95% (Zheng and Rehmann 2014). Also, when combining alkali-extrusion and bioextrusion, 73% sugar yields where obtained from barley straw (Duque et al. 2014).

In contrast to the high energy demanding mechanical methods, extrusion is considered a promising pretreatment technology for biomass use because of its adaptability to process modifications (alkali or enzyme addition) and its versatility regarding the use of different raw materials.

3.2.2   Chemical Pretreatments

3.2.2.1   Acid-Based Pretreatments

Acid-based pretreatments can use either concentrated or dilute acids as catalysts. The main objective of both acid pretreatment methods is to solubilize hemicellulose, increasing the porosity and making cellulose more accessible to enzymes. In general, concentrated-acid pretreatments (acid concentration higher than 30%) are less popular for industrial applications than dilute-acid pretreatments (acid concentration between 0.5% and 5%), due to equipment corrosion problems, difficulties in acid recovery, and considerable degradation products formation (Tomás-Pejó et al.

2011). Dilute-acid pretreatment is usually performed at temperatures above 100 °C, and it also promotes biomass degradation.

The most common acid catalyst is H_2SO_4. Still, HCl, H_3PO_4, HNO_3 and organic acids such as acetic acid, maleic acid, and acetic acid have been tested (Kootstra et al. 2009). The use of organic acids has shown advantages for reducing the formation of degradation products. Nevertheless, these pretreated materials usually show lower hydrolysability, and high temperatures (>170 °C) are required to reach glucose yields comparable to those obtained with strong acids (Kootstra et al. 2009). Although dilute-acid pretreatment is especially suitable for biomass with low lignin content, this technology has been assessed with a wide range of ligno-cellulosic feedstocks (wheat straw, olive tree, pine, municipal solid waste, switch-grass, aspen, corn cob), showing about 70–90% solubilization of hemicellulose (in both xylans and mannans), and about 60–80% of glucose yield from the enzymatic hydrolysis of pretreated materials (Cara et al. 2008; Satimanont et al. 2012; Arora and Carrier 2015).

3.2.2.2 Alkali-Based Pretreatments

Alkaline pretreatments improve the accessibility of lignocellulosic biomass by mainly removing lignin polymer. Also, partial hemicellulose modification, higher WRV and lower CrI values are usually observed (Karimi and Taherzadeh 2016). The effectiveness of these pretreatment processes depends on the lignin content, and they are more effective on agricultural residues (such as wheat straw, corn stover, or sugarcane bagasse) than on woody materials due to the lower lignin content. Although some inhibitory compounds are generated during the process, alkaline pretreatments cause less sugar degradation than acid pretreatments.

The most common alkali compounds used for biomass pretreatment are NaOH, KOH, lime [$Ca(OH)_2$], NH_3 and NH_4OH. These processes are preferably performed at room temperature or temperatures below 60 °C, with residence times varying from seconds to days. Due to the lower costs and less stringent safety requirements, lime—which can be easily recovered at the end of the process with CO_2—and ammonia are the preferred alkali compounds (Zhao et al. 2008; da Costa Sousa et al. 2016).

Alkali-based pretreatments have shown to remove 20–60% lignin content (Zhao et al. 2008; Akhtar et al. 2016). Alkali-pretreated feedstocks usually show sacchar-ification yields of about 50–70%. However, higher saccharification yields (up to 95%) can be obtained by combining alkali-based processes with other pretreatment methods (e.g. mechanical or biological pretreatments) or with oxidant agents such as H_2O_2 or copper-catalyzed alkaline hydrogen peroxide (Cu-AHP) (Moreno et al. 2015; Akhtar et al. 2016; Bhalla et al. 2016). Due to the lower lignin content, the non-specific binding of cellulases is reduced and lower enzyme loadings can be used in comparison to dilute-acid pretreatments.

Several ammonia-based pretreatments, including soaking in aqueous ammonia (SAA), ammonia fiber explosion (AFEX; described in Sect. 3.2.3.3), extractive ammonia (EA, described in Sect. 3.2.3.4), or ammonia-recycle percolation (ARP; described in Sect. 3.2.3.5), have been developed during last years, showing very promising results (Akhtar et al. 2015; da Costa Sousa 2016). SAA is an interesting chemical-based technology with mild reaction conditions. During SAA pretreatment, biomass is submerged in aqueous ammonia solution at temperatures below 75 °C for an extended period of time (up to several weeks). Optimal ammonia concentrations and solid:liquid ratios will depend on the feedstock used, ranging from 15% to 30% and from 1:2 to 1:15, respectively (Kim and Lee 2005). This pretreatment can efficiently remove lignin polymer (up to 75%) with low formation of degradation products, and keeping both cellulose (100%) and hemicellulose (85%) fractions in the solid residue. The presence of hemicellulose in the pretreated material makes the addition of hemicellulases imperative for improving saccharification yields.

3.2.2.3   Organosolv

Organosolv pretreatment uses organic or aqueous solvents to remove lignin polymer and hydrolyze hemicellulosic sugars. This method is performed at temperatures of about 100–250 °C. Acids (HCl, H_2SO_4, oxalic acid, or salicylic acid) bases (mainly NaOH) or salts [$MgCl_2$, $Fe_2(SO_4)_3$] can also be added as catalysts (Bensah and Mensah 2013). Organosolv-pretreated materials can be split into three different fractions: (1) cellulose fibers, (2) solid lignin (which is obtained after removing the solvent from the liquid phase), and (3) a liquid phase containing hemicellulosic sugars (Tomás-Pejó et al. 2011). Ethanol, methanol, acetone, ethylene glycol, and other low-molecular-weight alcohols are the main solvents used in organosolv pretreatment. Solvents must be recovered after each pretreatment process for its reutilization and to avoid the inhibitory effect that these compounds exert on hydrolytic enzymes and fermentative microorganisms. Although organosolv pretreatment can be potentially used for any feedstock (e.g. softwoods, sugarcane bagasse, or wheat straw) with high lignin removal (60–90%, with high purity) and saccharification yields (up to 90%), it implies high capital costs and promotes the formation of degradation compounds (Pan et al. 2005; Mesa et al. 2011; Wildschut et al. 2013).

3.2.2.4   Other Chemical Pretreatments

In addition to acid, alkali, and organosolv pretreatment methods, ionic liquids (ILs), cellulose and organic solvents-based lignocellulosic fractionation (COSLIF), co-solvent enhanced lignocellulosic fractionation (CELF), ozonolysis, or aqueous N-methylmorpholine-N-oxide (NMMO) are also considered as chemical pretreatment technologies:

ILs ILs are organic or inorganic cations and anions that remain in liquid form below 100 °C. They are considered green solvents because they are non-flammable, non-volatile, and recyclable, and can be used for either partial or complete biomass deconstruction (Brandt et al. 2013). When complete biomass dissolution is desired, acidic or acidified ILs can be used for the hydrolysis of carbohydrates. On the other hand, ILs promote lignin removal (up to 70%) and partial hemicellulose hydrolysis (up to 85%) during partial deconstruction of biomass. Similar to organosolv pretreatment, ILs offer the possibility of lignin revalorization, with the advantage of handling non-volatile, non-odorous, and relatively safe liquors. In spite of producing low amounts of degradation products, ILs can act as inhibitory compounds themselves. Certain ILs have shown to inhibit cellulase activity by 70–85% with a concentration of 10% (v/v) (Brandt et al. 2013). Also, the economic feasibility of these pretreatment technologies is another drawback, and an energy-efficient recycling method is therefore required to compensate the high costs of ILs.

Imidazolium salts are very commom ILs (Brandt et al. 2013). Recently, tertiary amines such as $[FurEt_2NH][H_2PO_4]$ and $[p\text{-}AnisEt_2NH][H_2PO_4]$ (also called bioionic liquids), have been synthesized by reductive amination of aldehydes derived from lignin and hemicellulose (Socha et al. 2014). In presence of these lignin-derived ILs, enzymatic hydrolysis of switchgrass released 90–96% of potential glucose and 70–76% of potential xylose. However, they showed to be less effective than [C2mim][OAc] toward lignin removal.

COSLIF The COSLIF technology is a pretreatment method that can effectively fractionate lignocellulosic biomass into amorphous cellulose, lignin, hemicellulose, and acetic acid (Zhang et al. 2007). It is based on the sequential use of a non-volatile cellulose solvent [concentrated (80–90%) phosphoric acid], a highly volatile organic solvent (acetone or ethanol), and water. In a similar way than SAA, COSLIF pretreatment requires mild reaction conditions (50 °C and atmospheric pressure), but with the great advantage that the overall process takes place in few hours. COSLIF processes also give other benefits, including less degradation products formation, the possibility of obtaining high sugar yields with fast hydrolysis rates (up to 97% yield within 24 h of saccharification at low enzyme loadings), and the possibility of isolating high-value lignin, acetic acid, and hemicellulose. Moreover, it can be applied to a wide range of lignocellulosic feedstocks with only minor modifications of the operation conditions. All these favorable considerations—quite apart from the need of an efficient solvent recycling process—make COSLIF technology an attractive pretreatment process for the biorefinery concept.

CELF Similar to COSLIF technology, CELF pretreatment combines an aqueous solution of a biomass-sourced green solvent, THF (compound regenerated catalytically from the biomass degradation product furfural), with dilute-acid pretreatment, promoting delignification and solubilization of biomass with minimal sugar degradation (Nguyen et al. 2016). In comparison to dilute acid pretreatment alone [0.5% (w/w) H_2SO_4], the present of THF [1:1 (v:v) THF:water ratio] in CELF pretreatment enhanced final ethanol titers and yields by 25%, during a simultaneous saccharification and fermentation process of pretreated corn stover (Nguyen et al.

2016). Due to extensive biomass deconstruction, low enzyme loadings are required in the subsequent saccharification process. Still, further research is needed to fully understand the beneficial mechanisms promoted by CELF in altering the physico-chemical properties of lignocellulose.

Ozonolysis Ozone can be used as oxidizing agent to hydrolyze and remove lignin polymer from lignocellulosic biomass with slight modifications of cellulose and hemicellulose. Among other feedstocks, ozonolysis has been successfully applied to wheat straw, sugarcane bagasse and poplar, reducing the lignin content by up to 60% (Travaini et al. 2015). The process is in general performed at room temperature and low amounts of degradation products are generated. The main disadvantage of ozonolysis is the high amount of ozone required for the process, which limits its industrial application.

Aqueous NMMO NMMO is an environmentally friendly solvent that in a similar way to ILs can fractionate biomass at moderate temperatures (80–130 °C). This solvent disrupts hydrogen bonds in cellulose fibers, increasing porosity and WRV, and reduces CrI and DP by partial cellulose solubilization. In addition, NMMO is easy to recover and generates low amount of degradation products. During the enzymatic hydrolysis of NMMO-pretreated biomass, >90% sugar yields have been observed (Bensah and Mensah 2013). At concentrations of about 15–20% (w/w), NMMO does not affect cellulase activity, offering the possibility of a simultaneous pretreatment and saccharification process.

3.2.3   Combined Physicochemical Pretreatments

Physicochemical pretreatments are technologies that combines both physical and chemical effects to improve the hydrolysability of raw materials. These technologies include, among others, steam explosion, hydrothermal methods, AFEX, EA, microwave, ultrasound, or wet oxidation.

3.2.3.1   Steam Explosion Pretreatment

Steam explosion pretreatment combines the physical effect of an explosive decompression with a chemical autohydrolysis promoted by the solubilization of acetyl groups. The process is performed at high temperatures and pressures (160–250 °C, 6–50 bar), reached by injecting saturated steam into the reactor. Steam explosion promotes hydrolysis of hemicellulosic sugars and redistribution and partial solubilization of lignin polymer. Residence time (t) and temperature (T) are the main factors influencing the effectiveness of this pretreatment method. The reaction ordinate (R_o) correlates these two parameters in a single Eqn. ((3.1)) and it is used to estimate the severity factor (SF) ((3.2)) for the evaluation and comparison between pretreatment processes.

$$R_o = t \times e^{(T-100/14.75)} \tag{3.1}$$

$$SF = Log\ (R_o) \tag{3.2}$$

Relatively low capital investment, low environmental impact, and complete sugar recovery have positioned steam explosion technology as one of the most widely used for lignocellulose pretreatment. As major drawback, the harsh conditions applied during the process cause severe biomass degradation and generate abundant degradation products (Wang et al. 2015). In order to compromise the better biomass accessibility and the lower biomass degradation, SF values between 3 and 4.5 have been considered to be optimal. Steam explosion has been used to pretreat almost the full range of lignocellulosic feedstocks, showing sugar yields of about 90% in a subsequent enzymatic hydrolysis (Tomás-Pejó et al. 2011; Wang et al. 2015). It is remarkable that the process can be performed without a previous reduction in chip size and without the addition of chemical catalysts, offering a substantial benefit for cost savings. Notwithstanding, an acid catalyst is often used, especially in softwoods, to reduce pretreatment temperatures and residence times, and decrease the amount of degradation products formed during the process. When a catalyst is used, a combined severity factor (*CSF*) ((3.3)), which includes the pH parameter in the previous Eqn. ((3.2)), is used to compare acid-catalyzed steam explosion pretreatments.

$$CSF = Log\ (R_0) - pH \tag{3.3}$$

3.2.3.2 Hydrothermal Pretreatment

Hydrothermal pretreatment (also known as liquid hot water) uses high pressures to keep water in the liquid phase at temperatures ranging from 160 °C to 240 °C. Similar to steam explosion, these conditions favors hemicellulose autohydrolysis and redistribution and partial solubilization of lignin, enhancing the accessibility of the cellulose polymer. Hydrothermal pretreatment also promotes the formation of high amounts of degradation products. As an attempt to control biomass degradation, the pH can be maintained between 4 and 7 to guarantee a mild autohydrolysis process.

Using the hydrothermal technology, about 80% of the hemicellulosic fraction has been removed from materials such as corn stover, sugarcane bagasse, and wheat straw, improving the digestibility of these lignocellulosic materials (Alvira et al. 2010).

3.2.3.3 Ammonia Fiber Explosion (AFEX)

The AFEX pretreatment is another physicochemical process that uses anhydrous ammonia at high pressures and temperatures between 60 °C and 100 °C (Akhtar et al. 2016). The temperature range can be further extended to 140 °C without the need of external heat, due to exothermal reactions between ammonia and water (da Costa Sousa et al. 2016). Equivalent to steam explosion pretreatment, a fast depressurization of the reactor leads to a rapid expansion of the ammonia gas, which causes the physical disruption of lignocellulosic fibers. In consequence, biomass digestibility is increased due to higher WRVs and lower CrI and DP. Also, deacetylation of hemicellulose polymer is usually observed. Recovered pretreated materials consist of a solid fraction (since ammonia is completely evaporated) with slight modifications in hemicellulose and lignin content. Hemicellulases are therefore required to increase enzymatic saccharification yields when using AFEX as pretreatment method.

AFEX technology has a greater effect on agricultural residues, in contrast to woody biomass, where it has shown to be less effective. One major advantage of AFEX technology is that no degradation products are generated during the process. In spite of minor lignin or hemicellulose removal, enzymatic hydrolysis can result in high sugar yields (above 90% in agricultural residues), even at low enzyme loadings (da Costa Sousa et al. 2016). This result can be explained by the fact that ammonia affects lignin in a way that reduces its ability to non-specifically adsorb hydrolytic enzymes.

3.2.3.4 Extractive Ammonia (EA)

EA represents a step forward in the use of ammonia for lignocellulose pretreatment. This pretreatment uses liquid ammonia at elevated temperatures to solubilize lignin polymer and to modify crystalline cellulose, making it more accessible to hydro-lytic enzymes. During EA pretreatment, biomass with low moisture levels [up to 10% (w/w)] are mixed with ammonia up to 6:1 ammonia:biomass ratio and heated to about 120 °C for 30 min (da Costa Sousa et al. 2016). A nitrogen overpressure is used to maintain ammonia in the liquid phase. In contrast to AFEX pretreatment, EA requires external heat to reach reaction temperatures due to the lower moisture levels. After pretreatment, ammonia can be recovered and recycle. However, 0.022 g of ammonia per 100 g biomass is lost during each cycle due to ammonia-biomass interactions. EA pretreatment has shown to solubilize about 44% (w/w) lignin from corn stover with high purity and high proportion of intact, native lignin functionality (e.g., β-O-4 linkages) (da Costa Sousa et al. 2016). Although process conditions have to be optimized, EA pretreatment represents a very promising technology since it requires about 60% lower enzyme loadings to reach similar saccharification yields than AFEX, ammonia is an inexpensive commodity chemical with easy recycling, detoxification of pretreated biomass is not necessary, and it offers the possibility of lignin revalorization.

3.2.3.5 Other Physicochemical Pretreatments

With the aim of enhancing biomass digestibility, other physicochemical pretreatment processes, including wet oxidation, the use of microwave energy in the presence of chemical reagents, ultrasound pretreatment, ARP, supercritical fluids, and sulfite pretreatment to overcome recalcitrance of lignocellulose (SPORL) have been also described:

Wet Oxidation Wet oxidation is an oxidative pretreatment method that takes place at high temperatures (170–200 °C) and pressures (10–15 bar), and uses oxygen or air as catalyst (Olsson et al. 2005). It promotes lignin oxidation and disruption of LCCs bonds, leading to 60–70% lignin removal (Olsson et al. 2005). The presence of oxygen at high temperatures makes the process exothermic, thus reducing the energy input. This technology has been applied to agricultural and wood feedstocks and it can be also combined with Na_2CO_3 for enhancement of cellulose and hemicellulose recovery yields (up to 96% and 70%, respectively) (Klinke et al. 2002).

Microwave Pretreatment This method uses electromagnetic waves, with frequencies ranging from 0.3 to 300 GHz, to irradiate lignocellulosic materials. To avoid interference with telecommunications, 915 MHz (or 896 MHz in the United Kingdom) and 2450 MHz are the most common microwave frequencies use for industrial purposes (Xu 2015). By interacting with lignocellulose, microwaves encourage thermal and also non-thermal effects for enhancing the accessibility of cellulose to hydrolytic enzymes. Microwave pretreatment has been applied to a wide range of lignocellulosic biomass feedstocks, including both agricultural and woody biomass. During microwave pretreatment, biomass is submerged in water or in solution with alkalis, acids, ILs, or salts, to increase the effectiveness of pretreatment process. Microwave pretreatement has shown to increase saccharification yields from 1.5 to 4 times in comparison with non-pretreated biomass, obtaining 50–98% of the theoretical glucose (Xu 2015).

Ultrasound Pretreatment Ultrasonication is a pretreatment technology that consists of rapid compression and decompression cycles of sonic waves to generate cavitation (formation, growth, and subsequent collapse of microbubbles, resulting in localized temperatures and pressures of about 5000 °C and 1000 bar). Ultrasound can be applied at low frequencies (<50 kHz) and it affects lignocellulosic biomass through mechanoacoustic (physical) and sonochemical (chemical) effects, allowing lignin extraction (Bussemaker and Zhang 2013). Although working at higher temperatures would improve pretreatment efficiency, the cavitation effect is maximized at temperatures between 30 °C and 70 °C. These temperatures allow integration of ultrasonication and enzymatic hydrolysis, thus decreasing overall process time and increasing saccharification yields. As for microwave pretreatment, ultrasound pretreatment is usually combined with other pretreatment technologies, such as hydrogen peroxide or alkali- and acid-based pretreatments, for increasing delignification efficiency up to 90% (Bussemaker and Zhang 2013).

ARP ARP is based on passing aqueous ammonia at a concentration of 5–15% through a reactor packed with biomass at 140–210 °C (Kim et al. 2008). This process mainly promotes lignin removal and hemicellulose solubilization, with little loss in cellulose content. Depending on the process conditions, about 30–80% of both lignin and hemicellulose can be removed (Kim et al. 2008).

Supercritical Fluids Supercritical fluid pretreatment, also known as CO_2 explosion, uses CO_2 compressed at temperatures above its critical point to liquid-like density (Schacht et al. 2008). It is employed as a delignification method for enhancement of biomass accessibility. In order to increase the delignification efficiency, the use of organic solvents such as ethanol and methanol has been combined with this technology. One of the main advantages of this pretreatment process is the possibility of cost reduction by reutilization of the CO_2 produced during other microbial processes, such as bioethanol production. In addition, CO_2 is non-toxic, non-flammable, and easy to recover.

SPORL This process uses an aqueous sulfite solution at 160–180 °C and low pH (between 2 and 4) for about 30 min, combined with a mechanical step using a disk mill (Bensah and Mensah 2013). It mainly causes hemicellulose removal and lignin sulfonation. Even working at high temperatures, the concentration of degradation products is lower than other pretreatment technologies such as dilute acid or steam explosion. Sulfite pretreatment has been successfully tested at pilot scales with several types of biomass including corn stover, switchgrass, agave stalk or lodgepole pine. In the particular case of corn stover, SPORL pretreatment has shown to remove up to 92% lignin, which enables to obtain 78.2% hydrolysis yield (Bensah and Mensah 2013).

3.2.4 Biological Pretreatments

Several microorganisms, including white rot Basidiomycetes (*Ceriporiopsis subvermispora*, *Trametes versicolor*, *Pycnoporus cinnabarinus*, or *Phanerochaete chrysosporium*), Ascomycetes (*Trichoderma reesei* or *Aspergillus terreus*) and bacteria (*Bacillus macerans*, *Cellulomonas cartae,* or *Zymomonas mobilis*), and/or their ligninolytic enzyme systems (mainly laccases or laccase-mediator systems), have been used as single biological pretreatment methods or combined with other pretreatment technologies to improve biomass hydrolysability (Moreno et al. 2015). This pretreatment process mainly focusses on delignification of lignocellulose and requires mild reaction conditions (15–40 °C and pH 4–5), which promotes few side reactions, lowers the energy demand, and there are no strict reactor requirements to resist pressure and/or corrosion. In addition to lignin removal, partial degradation of hemicellulose and cellulose can be observed. Biological pretreatment does not generate degradation products, but requires long reaction times in case of using microorganisms (up to several weeks). In contrast, when using ligninolytic enzymes the overall process time is reduced (4–24 h), but the addition of external enzymes represents an extra cost. Lignocellulosic feedstocks such as eucalyptus, wheat straw, pine, or corn stover have been subjected to biological pretreatment, showing delignification efficiency up to 97% (Moreno et al. 2015). Although further research is needed to overcome current bottlenecks and optimize these processes, biological pretreatments—especially enzymatic delignification—are a very interesting alternative for future biorefineries.

3.3   Challenges in Converting Pretreated Biomass

Biomass pretreatment usually requires harsh conditions (high temperatures and pressures, the use of solvents or the addition of chemical catalysts), leading to biomass degradation and/or leaving residual chemicals that limit the subsequent saccharification and fermentation steps. Biomass degradation generates several byproducts during pretreatment processes that are inhibitory for hydrolytic enzymes and fermenting microorganisms. These inhibitory compounds can be classified according to their chemical nature into three major groups: (1) weak organic acids, (2) furan derivatives, and (3) phenolic compounds (Taherzadeh and Karimi 2011; Moreno et al. 2015). Furan derivatives include 2-furaldehyde (furfural) and 5-hydroxymethylfurfural (HMF), which come from degradation of pentose and hexose sugars, respectively. Weak acids (acetic acid, formic acid, and levulinic acid) are generated from the hydrolysis of acetyl groups and further degradation of furfural and HMF. Finally, a wide variety of phenolic compounds such as vanillin, syringaldehyde, and hydroxycinnamic acids (the composition varies depending on feedstocks) are released from lignin. In the particular case of wet oxidation pretreatment, phenols are not end-products and are further degraded to carboxylic acids. In addition to the formation of inhibitory compounds, biomass degradation also affects biomass recovery yields. This is another important parameter since any loss in cellulose and/or hemicellulose (the sugar source) has a direct impact in the concentration of the final desired products, resulting in lower revenues. Also, the presence of residual solvents and/or chemical catalysts affects both enzymes and fermenting microorganisms and should be therefore taken into account (Bensah and Mensah 2013; Brandt et al. 2013). In this context, an optimal pretreatment process must compromise biomass hydrolysability, minimizing biomass degradation and maximizing the recovery yields of chemical catalysts. Several physical, chemical, and biological detoxification methods have been also proposed and studied for reducing the inhibitory effect of degradation compounds, and improving the conversion of pretreated biomass (Taherzadeh and Karimi 2011; Moreno et al. 2015). However, these detoxification methods should be avoided whenever possible since they increase the overall process costs. Another strategy that is being implemented is the development of genetic and evolutionary engineering strategies to obtain robust microorganisms with the ability to convert/tolerate higher concentrations of inhibitory compounds (Koppram et al. 2014). In a similar way, the use of extremophillic microorganisms and/or enzymes that can perform the conversion processes on such challenging media (e.g. those microorganisms/enzymes that are able to catalyze reactions in acid or alkali environments) are also very promising alternatives to be considered (Miller and Blum 2010).

3.4 Economic and Environmental Evaluation of Pretreatment Processes

In addition of balancing biomass accessibility with sugar recoveries and biomass degradation, the pretreatment process must be evaluated from the economic and environmental point of view to meet sustainability criteria. Economic evaluations of energy consumption and process costs, and the environmental impact are usually estimated by techno-economic analyses and life-cycle assessments (LCAs), respectively. Although both parameters are independent of each other, there is a tendency to couple them and quantify the impact of research progress from an economic and environmental point of view.

Pretreatment can represent about 30–40% of total costs (Tomás-Pejó et al. 2011). During an economic analysis of pretreatment technologies, the energy demand and the cost of chemicals (including solvents, acids, alkalis, ligninolytic enzymes and/or the nutrients required for microorganism growth) are the major variables to be considered. Thus, pretreatment technologies with low energy and chemical requirements would represent a better choice. Pretreatment methods such as mechanical pretreatment and ozonolysis are considered economically unviable because of the high energy demand and the large amount of ozone required during the process, respectively (Hendriks and Zeeman 2009; Travaini et al. 2015; Kim et al. 2016). Similar drawbacks are found in wet oxidation and other chemical pretreatment technologies such as organosolv, or alkali-based pretreatments. In the case of AFEX or EA, although ammonia recycling is feasible even despite its high volatility, these pretreatments can be hindered by ammonia recovery yields and the environmental concerns derived of the use of this chemical (Wang et al. 2015). A reduction in ammonia concentration together with a decrease in enzyme loading would aid in reducing overall costs of AFEX/EA processes, making these technologies effective alternative for biomass fractionation and revalorization (da Costa Sousa et al. 2016). ILs, COSLIF and CELF are also quite attractive technologies from the biorefinery point of view due to the possibility of obtaining certain high value-added products such as lignin-derived compounds. Nevertheless, these processes are limited by solvent prices, biomass loading, recovery yields, especial reactor requirements to resist corrosion, and the need of neutralizing pretreated materials.

Among all different processes, dilute-acid and steam-explosion have been reported to be cost-effective technologies, and have been commercially used to pretreat several lignocellulosic feedstocks. Steam explosion has been a competitive pretreatment technology since the 1980s (Wang et al. 2015). The effectiveness of steam explosion depends directly on the SF values applied to biomass. SF values of 3–4.5 are considered optimal, but lower SF values can be beneficial for reducing biomass degradation. However, lower SF values results in less hemicellulose solubilization, and higher enzyme loadings are required for reaching similar saccharification yields, which increases the overall costs. Dilute-acid pretreatment is considered to be a simple, low-cost and effective pretreatment technology. In contrast, additional steps after pretreatment, such as neutralization, inhibitor removal, salt disposal, and acid recovery increase final production costs. The use of extremophiles microorganisms and/or enzymes that can tolerate acid and/or salty environments might help in the cost-effectiveness of the process (Miller and Blum 2010). Extrusion is another versatile and energy-efficient technology for lignocellulosic pretreatment (Zheng and Rehmann 2014; Duque et al. 2014). This physical technology produce very low amounts of degradation products, can be adapted to different process configurations, and allows the possibility to add chemical and/or biological catalysts to boost biomass accessibility. The combination of different pretreatment technologies has been also considered to be a suitable choice for reaching high sugar yields, using milder process conditions, lower concentrations of costly solvents, and lower enzyme loadings (Bensah and Mensah 2013). In this context, extrusion has been successfully combined—even at industrial scale—with a continuous steam explosion process (Fang et al. 2011).

Regarding environmental evaluation, greenhouse gas (GHG) emissions (including CO_2, NO_2, SO_2, and CH_4), water requirements, wastewater produced, the use of chemicals, and the energy demand associated with fossil fuels are the main variables to be considered. So far, chemical pretreatments have a higher impact on the environment in comparison with other pretreatment technologies, such as steam explosion and hydrothermal pretreatment. On the contrary, biological delignification is a promising technology with very low environmental impact, high product yields, mild reaction conditions, few side reactions, less energy demand, and reduced reactor requirements (Moreno et al. 2015). However, before scaling up biological pretreatment technologies, shorter reaction times and/or lower prices for ligninolytic enzymes are required to meet the economic needs.

3.5 Concluding Remarks

Lignocellulosic biomass is an appropriate feedstock for developing a bio-based economy, relying on sugar-related products. Different physical, chemical, physicochemical, and biological pretreatments technologies have been developed and

evaluated to alter the highly recalcitrant three-dimensional structure of lignocellulose, and increase its digestibility for optimal biomass conversion. After consider
economic and environmental impact, only few pretreatment methods fulfill the
sustainability criteria and are suitable for their use at commercial scale. Focusing on
the needs of local biorefineries, a common pretreatment technology would be of
benefit to make use of all lignocellulosic feedstocks available in the nearby areas,
thus avoiding transportation costs. Although there is no best pretreatment technology, dilute-acid pretreatment, steam explosion, extrusion, COSLIF, CELF, ILs,
Cu-AHP and certain ammonia-based technologies such as EA, are considered to be
effective methods that can be applied to a wide range of lignocellulosic feedstocks.
In addition, they also offer the possibility of providing other high value-added
product (such as lignin-derived compounds), contributing to the economy of the
process. In contrast, although some of these technologies are now in commercial
scale (e.g. steam explosion and dilute-acid pretreatment), certain parameters such
as biomass degradation, enzyme loadings required in the subsequent saccharification step, or efficient recycling processes must be optimized to make these pretreatments viable from the economic and environmental point of view. The
combination of different pretreatment processes such as extrusion and steam
explosion, alkali-based pretreatment and enzymatic delignification, solvents and
acid-based pretreatment (COSLIF, CELF), ILs and microwave, etc., also offers
possibilities for enhancing the effectiveness of pretreatment processes by promoting synergistic effects (e.g. higher lignin and hemicellulose solubilization in comparison of using single processes, and/or the need of lower enzyme loadings), and
reducing the environmental impact due to milder process conditions. With this
respect, further research at pilot and demonstration scale should be performed to
evaluate the feasibility and full potential for already established pretreatment
technologies, and at laboratory scale to further develop the non-efficient but
promising technologies in order to meet the economic and environmental requirements, giving them the possibility of representing an actual choice.

Acknowledgment Authors are grateful to the Swedish Energy Agency (Energimyndigheten) for
the financial support.

References

Alvira P, Tomás-Pejó E, Ballesteros M, Negro MJ (2010) Pretreatment technologies for an
    efficient bioethanol production process based on enzymatic hydrolysis: a review. Bioresour
    Technol 101:4851–4861
Akhtar N, Gupta K, Goyal D, Goyal A (2015) Recent advances in pretreatment technologies for
    efficient hydrolysis of lignocellulosic biomass. Environ Prog Sustain Energy 35(2):489–511
Akhtar N, Gupta K, Goyal D, Goyal A (2016) Recent advances in pretreatment technologies for
    efficient hydrolysis of lignocellulosic biomass. Environ Prog Sustain Energy 35(2):489–511
Arora A, Carrier DJ (2015) Understanding the pine dilute acid pretreatment system for enhanced
    enzymatic hydrolysis. ACS Sustain Chem Eng 3(10):2423–2428

Bensah EC, Mensah M (2013) Chemical pretreatment methods for the production of cellulosic ethanol: technologies and innovations. Int J Chem Eng 2013:719607

Bhalla A, Bansal N, Stoklosa RJ, Fountain M, Ralph J, Hodge DB, Hegg EL (2016) Effective alkaline metal-catalyzed oxidative delignification of hybrid poplar. Biotechnol Biofuels 9:34

Brandt A, Gräsvik J, Halletta JP, Welton T (2013) Deconstruction of lignocellulosic biomass with ionic liquids. Green Chem 15:550–583

Bussemaker MJ, Zhang D (2013) Effect of ultrasound on lignocellulosic biomass as a pretreatment for biorefinery and biofuel applications. Ind Eng Chem Res 52(10):3563–3580

Cara C, Ruiz E, Oliva JM, Sáez F, Castro E (2008) Conversion of olive tree biomass into fermentable sugars by dilute acid pretreatment and enzymatic saccharification. Bioresour Technol 99:1869–1876

da Costa SL, Jin M, Chundawat SPS, Bokade V, Tang X, Azarpira A, Lu F, Avci U, Humpula J, Uppugundla N, Gunawan C, Pattathil S, Cheh AM, Kothari N, Kumar R, Ralph J, Hahn MG, Wyman CE, Singh S, Simmons BA, Dale BE, Balan V (2016) Next-generation ammonia pretreatment enhances cellulosic biofuel production. Energy Environ Sci 9:1215–1223

Duque A, Manzanares P, Ballesteros I, Negro MJ, Oliva JM, González A, Ballesteros M (2014) Sugar production from barley straw biomass pretreated by combined alkali and enzymatic extrusion. Bioresour Technol 158:262–268

Fang H, Deng J, Zhang X (2011) Continuous steam explosion of wheat straw by high pressure mechanical refining system to produce sugars for bioconversion. Bioresources 6(4):4468–4480

Gharehkhani S, Sadeghinezhad E, Kazi SN, Yarmand H, Badarudin A, Safaei MR, Zubir MNM (2015) Basic effects of pulp refining on fiber properties—a review. Carbohydr Polym 115:785–803

Hendriks ATWM, Zeeman G (2009) Pretreatments to enhance the digestibility of lignocellulosic biomass. Bioresour Technol 100:10–18

Jørgensen H, Kristensen JB, Felby C (2007) Enzymatic conversion of lignocellulose into fermentable sugars: challenges and opportunities. Biofuels Bioprod Biorefin 1:119–134

Karimi K, Taherzadeh MJ (2016) A critical review on analysis in pretreatment of lignocelluloses: degree of polymerization, adsorption/desorption, and accessibility. Bioresour Technol 203:348–356

Kim JS, Kim H, Lee JS, Lee JP, Park SC (2008) Pretreatment characteristics of waste oak wood by ammonia percolation. Appl Biochem Biotechnol 148:15–22

Kim SM, Dien BS, Singh V (2016) Promise of combined hydrothermal/chemical and mechanical refining for pretreatment of woody and herbaceous biomass. Biotechnol Biofuels 9:97

Kim TH, Lee YY (2005) Pretreatment of corn stover by soaking in aqueous ammonia. Appl Biochem Biotechnol 124(1–3):1119–1131

Klinke HB, Ahring BK, Schmidt AS, Thomsen AB (2002) Characterization of degradation products from alkaline wet oxidation of wheat straw. Bioresour Technol 82:15–26

Kootstra AMJ, Beeftink HH, Scott EL, Sanders JPM (2009) Comparison of dilute mineral and organic acid pretreatment for enzymatic hydrolysis of wheat straw. Biochem Eng J 46:126–131

Koppram R, Tomás-Pejó E, Xiros C, Olsson L (2014) Lignocellulosic ethanol production at high-gravity: challenges and perspectives. Trends Biotechnol 32(1):46–53

Mesa L, González E, Cara C, González M, Castro E, Mussatto SI (2011) The effect of organosolv pretreatment variables on enzymatic hydrolysis of sugarcane bagasse. Chem Eng J 168:1157–1162

Miller PS, Blum PH (2010) Extremophile-inspired strategies for enzymatic biomass saccharification. Environ Technol 31(8-9):1005–1015

Moreno AD, Ibarra D, Alvira P, Tomás-Pejó E, Ballesteros M (2015) A review of biological delignification and detoxification methods for lignocellulosic bioethanol production. Crit Rev Biotechnol 35(3):342–354

Nguyen TY, Cai CM, Osman O, Kumar R, Wyman CE (2016) CELF pretreatment of corn stover boosts ethanol titers and yields from high solids SSF with low enzyme loadings. Green Chem 18:1581–1589

Olsson L, Jørgensen H, Krogh KBR, Roca C (2005) Bioethanol production from lignocellulosic material. In: Dimitriu S (ed) Polysaccharides structural diversity and functional versatility. Marcel Dekker, New York, pp 957–993

Pan X, Arato C, Gilkes N, Gregg D, Mabee W, Pye K, Xiao Z, Zhang X, Saddler J (2005) Biorefining of softwoods using ethanol organosolv pulping: preliminary evaluation of process streams for manufacture of fuel-grade ethanol and co-products. Biotechnol Bioeng 90 (4):473–481

Sánchez OJ, Cardona CA (2008) Trends in biotechnological production of fuel ethanol from different feedstocks. Bioresour Technol 99:5270–5295

Satimanont S, Luengnaruemitchai A, Wongkasemjit S (2012) Effect of temperature and time on dilute acid pretreatment of corn cobs. Int J Chem Mol Nucl Mater Metall Eng 6(4):316–320

Schacht C, Zetzl C, Brunner G (2008) From plant materials to ethanol by means of supercritical fluid technology. J Supercrit Fluids 46:299–321

Socha AM, Parthasarathi R, Shi J, Pattathil S, Whyte D, Bergeron M, George A, Tran K, Stavila V, Venkatachalam S, Hahn MG, Simmons BA, Singh S (2014) Efficient biomass pretreatment using ionic liquids derived from lignin and hemicellulose. Proc Natl Acad Sci U S A 111(35): E3587–E3595

Sun Y, Cheng J (2002) Hydrolysis of lignocellulosic materials for ethanol production: a review. Bioresour Technol 83(1):1–11

Taherzadeh MJ, Karimi K (2011) Fermentation inhibitors in ethanol processes and different strategies to reduce their effects. In: Pandey A, Larroche C, Ricke SC et al (eds) Biofuels. Alternative feedstocks and conversion processes. Elsevier, Amsterdam, pp 287–311

Tomás-Pejó E, Alvira P, Ballesteros M, Negro MJ (2011) Pretreatment technologies for lignocellulose-to-bioethanol conversion. In: Pandey A, Larroche C, Ricke SC et al (eds) Biofuels. Alternative feedstocks and conversion processes. Elsevier, Amsterdam, pp 149–176

Travaini R, Marangon-Jardim C, Colodette JL, Morales-Otero M, Bolado-Rodríguez S (2015) Ozonolysis. In: Pandey A, Negi S, Binod P, Larroche C (eds) Pretreatment of biomass. Processes and technologies. Elsevier, Amsterdam, pp 105–135

Wang K, Chen J, Sun SN, Sun RC (2015) Steam Explosion. In: Pandey A, Negi S, Binod P, Larroche C (eds) Pretreatment of biomass. Processes and technologies. Elsevier, Amsterdam, pp 75–104

Wildschut J, Smit AT, Reith JH, Huijgen WJ (2013) Ethanol-based organosolv fractionation of wheat straw for the production of lignin and enzymatically digestible cellulose. Bioresour Technol 135:58–66

Xu J (2015) Microwave pretreatment. In: Pandey A, Negi S, Binod P, Larroche C (eds) Pretreatment of biomass. Processes and technologies. Elsevier, Amsterdam, pp 157–172

Zhang YH, Ding SY, Mielenz JR, Cui JB, Elander RT, Laser M, Himmel ME, McMillan JR, Lynd LR (2007) Fractionating recalcitrant lignocellulose at modest reaction conditions. Biotechnol Bioeng 97(2):214–223

Zhao Y, Wang Y, Zhu JY, Ragauskas A, Deng Y (2008) Enhanced enzymatic hydrolysis of spruce by alkaline pretreatment at low temperature. Biotechnol Bioeng 99(6):1320–1328

Zheng J, Rehmann L (2014) Extrusion pretreatment of lignocellulosic biomass: a review. Int J Mol Sci 15(10):18967–18984

Chapter 4
Approaches for Bioprospecting Cellulases

Baljit Kaur and Bhupinder Singh Chadha

What Will You Learn from This Chapter?

Cellulases are industrially important enzymes with a market share of 500 million dollars that is expected to rise to 1.5 billion dollars by 2018. Cellulases play a crucial role in generating sugar feedstock for lignocellulosic based biorefinery platform. In addition, their demand in textile, paper, feed and food industries is rising steadily. However, these industrial applications require thermostable, catalytically highly efficient cellulases for making the processes commercially viable. Therefore, search and discovery for novel sources of cellulases is continued area of research. This chapter highlights some of the approaches for bioprospecting for novel cellulases.

4.1 Introduction

Lignocellulosics constitutes key component of plant biomass (forestry and agricultural wastes) and municipal solid wastes (MSW) that primarily comprises of polymeric cellulose (40–50%), hemicellulose (20–30%), and lignin (10–25%) that varies according to their sources or genetic makeup (Fig. 4.1). Owing to their relative abundance and being a rich inherent source of sugar moieties, crop residues (corn stover, corn cobs, rice straw, wheat straw, barley husks, sugarcane bagasse, etc.) form an important feedstock for biotechnological intervention to produce

B. Kaur • B.S. Chadha (✉)
Department of Microbiology, Guru Nanak Dev University, Amritsar 143005, India
e-mail: chadhabs@yahoo.com; chadhabs@gmail.com

© Springer International Publishing AG 2017
R.K. Sani, R.N. Krishnaraj (eds.), *Extremophilic Enzymatic Processing of Lignocellulosic Feedstocks to Bioenergy*, DOI 10.1007/978-3-319-54684-1_4

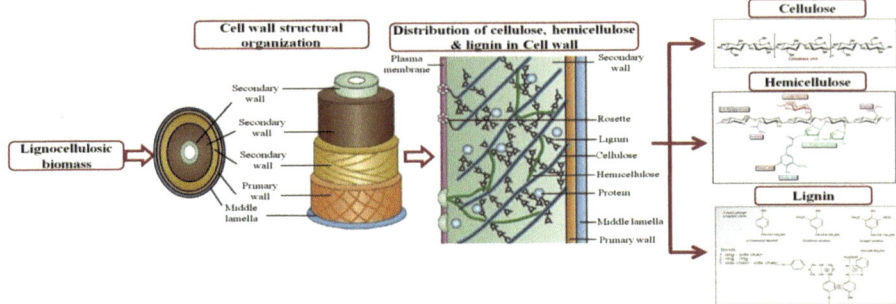

Fig. 4.1 Diagrammatic illustration of the framework of lignocelluloses; cellulose, hemicellulose and lignin (Menon and Rao 2012)

value added products. Being a rich inherent source of sugar moieties, lignocellulosics constitute important feedstock for the production of second generation cellulosic ethanol. However, the release of sugars from complex lignocellulosic substrate requires hydrolysis of pretreated substrates. The pretreatment processes employed bring about structural changes and loosens up the complex have been reported in literature as well as at commercial scale. This chapter would only discuss about the cellulases and enzymatic hydrolysis to achieve sugars for subsequent fermentation to ethanol (Xiao et al. 2012)

4.2 Glycosyl Hydrolases Involved in Hydrolysis of Lignocellulosics

The hydrolysis of lignocellulosics into simpler form of sugars (hexoses/and pentoses) is accomplished by the wide array of enzymes collectively termed as glycosyl hydrolases (GH). These enzymes are responsible for synthesis, modification and degradation of carbohydrates and are clustered as carbohydrate-active enzymes (CAZymes). CAZymes are classified on the basis their similarities in amino acid sequences, enzymatic mechanism and protein folding and include Glycosyl hydrolases (GH), Polysaccharidelyases (PL), Carbohydrate esterases (CE) and Glycosyl transferases (GT). In addition, glycosyl hydrolases do include chitinolytic, pectinolytic and amylolytic enzymes which is not discussed in this chapter. These enzymes are grouped into 133 GH families in the continually updated CAZymes. Interestingly, in spite of a vast diversity of GHs their active site topologies essentially remain the same and can be divided into three groups: pocket, cleft and tunnel (Davies and Henrissat 1995).

4.3   Cellulose Binding Modules

Most cellulolytic enzymes comprises of a catalytic domain connected to a cellulose-binding domain (CBD) through a linker segment rich in Pro/Ser/Thr-amino acids. Bacterial system is rich in proline content whereas, high serine residues are found in eukaryotic linkers. CBM are amino acid sequences with a unique folding patterns involved in recognizing and adhering to polysaccharides. CBMs are appended to CAZymes that degrade insoluble polysaccharides and are classified into 71 families on the basis of amino acid sequences. CBMs actively participate in targeting the enzyme towards specific substrates and enhance the penetration of enzyme to cellulose for efficient hydrolysis. On contrary to this, one of the recent reports suggest that CBM lacking cellulases are advantageous in enzyme recycling as higher share of unbound enzyme can be recovered. Kraulis and co-workers (1989) elucidated the first structure of family 1 CBM from *Trichoderma reesei* (TrCel7A) belonging to family GH7 (Fig. 4.2). The structural analysis of CBM revealed the presence of β-sheet with two disulphide bridges and a flat surface linked with polar and aromatic amino acids.

The catalytic domains can exhibit both *N*- and *O*-linked glycans, where the amino acids in linker contain *O*-linked glycosylation, which has long been implicated in exhibiting protease protection and more recently implicated in substrate binding (Payne et al. 2013). Linkers connecting cellulase domains from different families exhibit different average lengths. It has been reported that linker length has significant influence on enzymatic activity and binding affinity such as in *T. reesei* Cel7A where shorter linker length promotes processivity, whereas the longer linker lengths in Cel6A enable it to search for hydrolytic site.

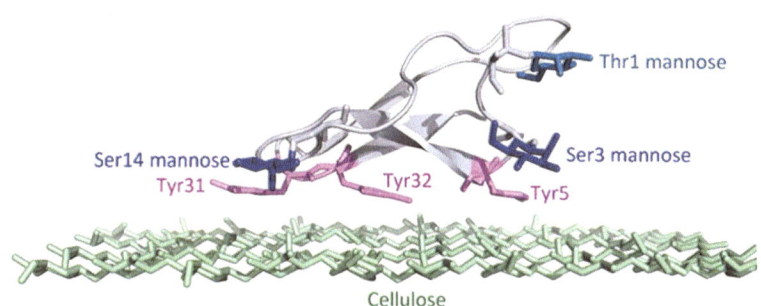

Fig. 4.2 The NMR structure of the Family 1 CBM and the top layer of cellulose (Kraulis et al. 1989). The tyrosine residues are shown in *purple*. The O-linked mannoses are shown in *cyan* and *blue*

4.4 Classification of Cellulases and Their Structural Features

Cellulose degradation is primarily attributed through multi-component cellulolytic enzyme system that works in a synergistic manner to degrade the complex polymeric structure of cellulose. These enzymes include endoglucanases (EC 3.2.1.4), exoglucanases (EC 3.2.1.74 and 3.2.1.91), and β-glucosidases (EC 3.2.1.21). Endoglucanase (EG) mainly attack amorphous cellulose by cleaving the internal glycosidic bonds in a random fashion and releases oligomers cellobiose, cellotriose, cellotetraose as the products (Zhang et al. 2006). The active sites of most EGs are open cleft shaped and may possess carbohydrate binding modules (modulator) or other domains (Sweeney and Xu 2012). Cellobiohydrolases (CBH) act processively on the reducing (CBHI) and non-reducing (CBHII) termini of cellulose fibers to primarily release cellobiose and are abundant in the secretome of cellulolytic fungi. EGs are widely classified on the basis of their structural and functional characterization into glycosyl hydrolases families 5, 6, 7, 8, 9, 12, 45 and 74. Cellobiohydrolases are represented by families GH 6, 7, 8, 9 and 48.

Crystal structures of the cellulase belonging to GH families 5–9, 12, 45 and 48 have been elucidated and revealed variation in folding topology (Table 4.1). Enzymes of all GH families except GH 9 and 12 are modular multi-domain proteins. GH family 5 and 7 enzymes catalyze the hydrolysis of glycosidic bonds with retention mechanism, while GH families 6, 9, and 48 mediated hydrolysis results in inverting configuration (Dodd and Cann 2009; Miotto et al. 2014). The processivity of enzyme action is not only limited to CBH but are also found in GH 5 and GH 9. The processive bifunctional EG (GH5), exhibiting both "endo" and "exo" type activities, have been reported from diverse organisms such as in brown rot basidiomycete *Gloeophyllum trabeum* and the marine bacterium *Saccharophagus degradans* 2–40. The structure of GH5 endoglucanase comprises of catalytic domain solely with β/α topology and seven subsites (−4 to +3) which are present at the C-terminal end of the barrel.

The structure of cellulolytic enzyme of family GH6 is composed of a single domain with a distorted α/β barrel topology; the active site that resides in a long cleft at the C-terminal end of the parallel β-strands that make up the barrel (Harhangi et al. 2003). The solved structures of catalytic domains of cellulases belonging to family GH7, suggested the presence of two distinct groups i.e. EG and CBHI. The solved structure showed the presence of common β-jelly roll fold with two face-to-face packed antiparallel β-sheets which form a curved β-sandwich. Nakamura and co-workers (2014) solved the crystal structure of cellulase of *T. reesei* (TrCel7A, ascomycete) and *Phanerochaete chrysosporium* (PcCel7C and PcCel7D, basidiomycete) belonging to family GH7. The structural and functional relationship between CBH and EG was resolved and found that in ascomycete CBHs possess four loops that cover the active subsite resulting in stronger binding affinity and increased processivity. These enzymes possess the ability to degrade crystalline cellulose "processively" as they have tunnel like active sites.

Table 4.1 Structure and folding topologies among different GH families of cellulases.

GH families	Organism	Folding topology	Structure	Mechanism	Enzyme activities
1	*Caldicellulosiruptor bescii*	$(\alpha/\beta)_8$-TIM barrel fold		Retaining	β-glucosidase, β-mannosidase, β-xylosidase, exo β-1-4glucanase
2	*Thermoascus auranticus*	$(\beta/\alpha)_8$		Retaining	EG, CBH, Xylanase, β-mannosidase, β-mannanase
3	*Humicola insolens*	α/β-barrel		Inverting	EG, CBH
4	*Trichoderma reesei*	β-jelly roll		Retaining	EG,CBH
5	*Clostridium thermocellum*	$(\alpha/\alpha)_6$		Inverting	CBH, EG, xylanase
6	*Clostridium cellulolyticum*	$(\alpha/\alpha)_6$		Inverting	EG, CBH
7	*Humicola grisea*	β-jelly roll		Retaining	EG
8	*Melanocarpus albomyces*	Six stranded β-barrel		Inverting	EG
9	*Clostridium cellulolyticum*	$(\alpha/\alpha)_6$-helix barrel		Inverting	EG, CBH

Such processive reactions and insolubility of the cellulose substrate makes CBH kinetics deviant from Michaelis Menton model and show significant fractal and "local jamming effect" (Igarashi et al. 2011). Whereas, EGs lack two of these loops and exhibited open cleft shaped active sites. In comparison CBHs in basidiomycete lack one of the loops that cover the active subsite.

The members of GH8 are evolutionary related to members of GH48 as they share a similar tertiary structure and displays $(\alpha/\alpha)6$ folding topology. CBM has proven to be a critical for catalyzing hydrolysis by processive EGs of GH9 except Cel9A of *Cytophaga hutchensonii*, which is devoid of CBM. The solved structure of cellulase from *Thermomonospora fusca* (CelE4, GH9) showed the presence of catalytic domain with $(\alpha/\alpha)_6$-barrel topology CBD with antiparallel β-sandwich fold. It was observed that deletion of CBM3c led to loss in the processivity and cellulose degradation ability (Irwin et al. 1998). Whereas, processivity of EGs in *S. degradans* 2–40 was found to be independent of CBM. Enzymes from family GH12 hydrolyze β-1-4 and β-1,3-1,4 linkages with retaining mechanism possessing catalytic domains and lacks CBMs.

β-glucosidases (βG) exhibit an exo-type action that hydrolyze cellobiose to glucose from the non-reducing ends. βG catalytic core features pocket shaped active sites, topology that equips βG to act on non-reducing ends. β-glucosidases belong to the GH1, 3 and 9 families and play a key role in the efficient hydrolysis as its action on cellobiose mitigates product inhibition on CBH and EG (Opassiri et al. 2007). The family1 beta-glucosidases are also classified as members of the 4/7 super family with a common $(\alpha/\beta)_8$ fold barrel motif. Family 3 of glycosyl hydrolases consists of nearly 44 beta-glucosidases and hexosaminidases of bacterial, mold, and yeast origin. Most of the fungal beta-glucosidases studied belong to the family 3 of glycosyl hydrolases. GH family 3 is one of the most abundant families of carbohydrate active enzymes and includes members that possess distinct enzymatic   activities,   including   β-D-glucosidase,   β-D-xylosidase,   α-L-arabinofuranosidase, and *N*-acetyl-β-D-glucosaminidase activities. βGs are well characterized, biologically important enzymes that catalyze the transfer of glycosyl group between oxygen nucleophiles either reverse hydrolysis or transglycosylation (Bhatia et al. 2002).

Recent discovery suggests the role of novel class of oxidative enzymes responsible for enhancing cellulose degradation which are identified and classified as auxiliary activity (AA) enzymes in CAZy database (Fig. 4.3). These oxidative enzymes are referred as lytic polysaccharide monooxygenases (LPMOs), a term coined by Horn and co-workers (2012) and classified into three families in CAZy database. The first one is AA9, formerly known as GH 61 that belongs to fungal enzymes specifically. The second family is AA10 which was previously known as carbohydrate binding module 33 (CBM 33) and it is dominated in bacteria and viruses, the third family is AA11 which acts on chitin and shares some structural and spectroscopic characteristics with AA9 and AA10. The fourth family is starch degrading LPMO referred as AA13.

4.5 Metagenomics for Isolating Novel Cellulases

In addition to culturable microbes there is growing understanding of the fact that the source of novel cellulases lies with enormous unculturable microbial diversity that resides within unique ecological niches and, therefore, screening of metagenomic/ environmental DNA libraries that potentially harbour many of the open reading frames (ORF) including those coding for the desired novel enzymes including cellulases is being advocated (Voget et al. 2003). In last one-decade commercialization of number of metagenomic technologies by companies such as Diversa, Cubist, BRAIN, BASF, Genecor, Prokaria for enzyme production has come

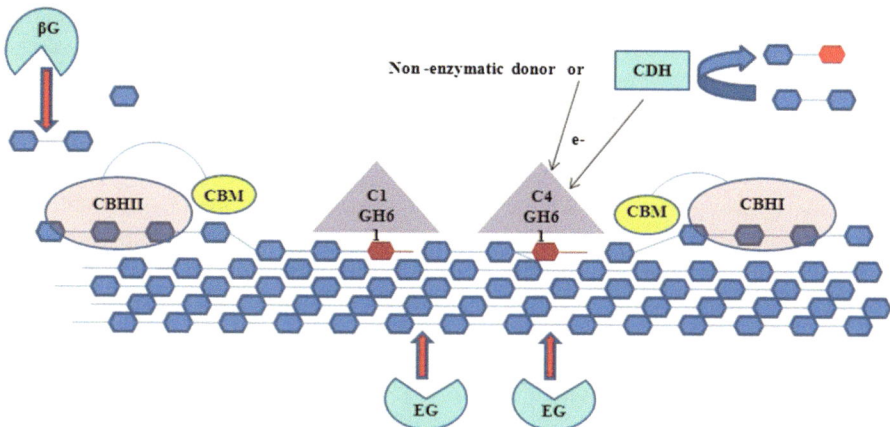

Fig. 4.3 Fungal enzymatic degradation of cellulose. EG, endoglucanase; CBH cellobiohydrolase; βG, betaglucosidase; CDH, cellobiose dehydrogenase; CBM, carbohydrate binding module. The picture shows a C1 and a C4 oxidizing GH61 which would generate optimal (i.e. non-oxidized) ends for the CBH2 and the CBH1, respectively (oxidized sugars are colored *red*)

up. Some of the recent reviews on metagenomic approaches for mining industrial enzymes have been published. Keeping in view the current and near foreseeable future, the market and demand for enzymes like cellulases is expected to rise dramatically. The metagenomic libraries can be screened to select highly efficient cellulose degrading genes that can be compared to the known sequences in the databases to determine their novel nature. To date, only few metagenome-derived cellulases genes have been identified, with biochemical characterization of the protein products (Kim et al. 2008). The potential use in industry for the cellulases cloned from metagenome has not been fully explored.

For making genomic library it is important that the method for isolation of good quality intact DNA is standardized. Most often soil DNA preparations are contaminated with humic acid which interferes in carrying out subsequent molecular protocols effectively. Use of CTAB in the extraction buffer to remove polysaccharides and humic acid impurities is recommended. The partially digested DNA is cloned in either suitable plasmid like pUC18/19 that can harbour up to 12 Kbp or in some cases cosmid/fosmid libraries and bacterial artificial chromosome (BAC) libraries containing insert size of up to 40 and 150 Kbp, respectively, are prepared. Generally, BAC libraries are preferred for cloning genes coding for entire pathway such as for secondary metabolites (antibiotics). However, for cloning genes of hydrolytic enzymes like cellulases an insert size of 5–8 Kbp may be sufficient to clone entire ORF. The clones selected on the basis of blue/white colony selection on lauria bertini (LB) ampicillin medium are further replica plated on to CMC containing LB ampicillin medium plates for function-driven analysis of uncultured microorganisms. The clones positive for cellulases are picked on the basis of pale/clear zones around the colonies in a plate flooded with 0.2% Congo Red. Using these approaches halo tolerant cellulases from soil, sediments and surrounding soda lakes of Wadi el Natrun in Libyan Desert and multifunctional hybrid glycosyl hydrolases from metagenomic library of the ruminant gut and uncultured microorganisms in rabbit cecum has been discovered (Palackal et al. 2007). Recent report on the metagenomic and proteogenomic analyses of a compost-derived bacterial consortium adapted to switchgrass at elevated temperature with high levels of glycoside hydrolase activities have been isolated. Near-complete genomes were reconstructed for the most abundant populations, which included composite genomes for populations closely related to sequenced strains of *Thermus thermophilus* and *Rhodothermus marinus*, and for novel populations that are related to thermophilic *Paenibacilli* and an uncultivated subdivision of the little studied *Gemmatimonadetes* phylum. Partial genomes were also reconstructed for a number of lower abundance thermophilic *Chloroflexi* populations. Identification of genes for lignocellulose processing and metabolic reconstructions suggested *Rhodothermus, Paenibacillus* and *Gemmatimonadetes* as key groups for deconstructing biomass. Mass spectrometry-based proteomic analysis of the consortium was used to identify 3000 proteins in fractionated samples from the cultures, and confirmed the importance of proteins from *Paenibacillus* and *Gemmatimonadetes* in biomass deconstruction.

4.6 Cellulase Producing Microorganisms

Complete cellulose degradation involves a dense interconnection between different cellulolytic microbial populations. Broad range of microbial diversity, that includes fungi, bacteria, and protozoans, is known that secretes these hydrolytic enzymes either freely or in complexed form that efficiently degrades such complex polysaccharides. Bioprospecting of diverse range of ecological niches for isolating novel cellulolytic microorganisms including the human, herbivore, arthropods and termite gut, terrestrial and aquatic environments have been explored. Such as wide array of bacterial strains which include *Cellulomonas* sp., *Bacillus cereus*, *Bacillus licheniformis*, *Bacillus pumilus* were isolated from mangrove areas of Philippines. A novel bacterial isolate, identified as *Cellulomonas composti* sp., showing high sequence similarity (98.5%) to *Cellulomonas uda* DSM 20107(T), was isolated from compost at a cattle farm near Daejeon, Republic of Korea and possesses endoglucanase andβ-glucosidase activities. Potent novel species were isolated from forest soil, which were found to be closely related to *Betaproteo bacteria* and *Pseudo gulbenkiania* on the basis of 16SrRNA sequencing. *Marinobacter* sp. (MSI032), isolated from the marine sponge *Dendrillanigra*, produces an extracellular alkaline cellulase at pH 9.0 at early stage of growth which facilitates it for industrial process. Another potent cellulase producing strain named as *Clostridium phytofermentans* was isolated from forest soil and was found to be a processive endoglucanase, active on both crystalline cellulose and soluble CMC (Warnick et al. 2002) that was later developed for consolidated bioprocessing of cellulose into ethanol and has been demonstrated for commercialization by Qteros at Masuchesset institute of technology. In order to obtain high-purity cellulase and facilitate its production, the *cel9* gene from *C. phytofermentans* was expressed in *Escherichia coli*, and the recombinant protein was purified and characterized. *Caulobacteria* sp. FMC-1, a facultative mesophilic strain, was isolated from shallow freshwater and was observed to produce cellulase under aerobic and anaerobic conditions.

The marine isolate *S. degradans* is presently being researched for spectrum of cell wall degrading enzymes. The genome of this bacterium has been sequenced to completion, and more than 180 open reading frames have been identified that encode carbohydrases. There is a distinct difference in cellulolytic strategy between aerobic and anaerobic bacteria. The anaerobic bacteria such as *Clostridium, Butyrivibrio fibrisolvens, Acetovibrio cellulolyticus, Bacteroides cellulosolvens, Ruminococcus albus* and *Ruminococcus flavefaciens* utilizes multiprotein complexes for achieving total degradation of cellulose.

Cellulolytic fungi are the most suited option for cellulose degradation as they have capability to produce copious amounts of extracellular enzymes and this significant characteristic of cellulolytic fungi attracts researcher's interest over bacteria. The best known cellulase producing fungi include *Trichoderma* sp., *Aspergillus* sp., *Fusarium* sp., *Penicillium* sp., *Rhizopus* sp., and *Alternaria* sp. Most of the commercial cellulases available are produced from *T. reesei* and *Aspergillus niger* but *T. reesei* lack sufficient amount of β- glucosidase to perform a proper and complete hydrolysis. In the search of indigenous cellulose degrading fungi, Barro and coworkers (2010) revealed the production of cellulases and xylanases by thermotolerant fungi *Acrophialophora nainiana* and *Ceratocystis paradoxa* using shake flask culture. The novel isolate *Aspergillus glaucus* XC9 produced extracellular cellulase and shares common characteristics with those from industrial cellulase-producing fungi, such as *A. niger* and *T. reesei* suggesting its possible use in industry. The enzyme was found to be stable over a wide pH range (3.5–7.5) and at temperatures below 55 °C. Two extracellular endoglucanases, named RCE1 and RCE2, produced by *Rhizopus oryzae* FERM BP-6889 isolated from soil, were identified and purified. The molecular masses of the two enzymes were 41.0 and 61.0 kDa, respectively. Like bacterial cellulosomes, anaerobic fungi anaerobic produce large multienzyme complexes which can depolymerise both amorphous and crystalline cellulose. Cellulosome-type complexes with endoglucanase, xylanase, mannanase, and β-glucosidase activities containing at least 10 proteins have been found in *Neocallimastix frontalis, Piromyces,* and *Orpinomyces.*

The ecological niches having high temperatures such as hot springs and composting heaps are also an attractive source for thermophilic cellulase producing microorganisms. A thermophilic bacterium *Aneurinibacillus thermoaerophilus* WBS2 which produces extracellular thermophilic cellulases was isolated from hot spring in India and the strain was subjected to optimization to enhance cellulase production (Acharya and Chaudhary 2012). Cellulases isolated from various thermophilic fungi include, *Chaetomium thermophilum, Humicola insolens, Humicola grisea, Myceliopthora thermophila, Talaromyces emersonii, Thermoascus auranticus, Sporotrichum thermophile, Melanocarpus albomyces.*

In addition to the this, certain ruminants, termites and herbivorous animals e.g. cattle, goats, sheep, buffalo, deer and camels etc., contain fungal and bacterial group of microorganisms including *Trichonympha, Clostridium, Actinomycetes, Bacteroides succinogenes, B. fibrisolvens, R. albus, and Methanobrevibacter ruminantium* have been documented for cellulose degradation (Ni and Tokuda 2013).

4.7 Methods for Isolation of Cellulase Producers and Cellulase Activity Assays

The physical heterogeneity of the cellulosic substrate and complexity of enzyme system makes cellulase assay and the interpretation of results a formidable problem (Wood and Bhat 1988). The measurement of cellulases has been subject of intense lab experiments. Various workers have published details of standardized methods (Wood and Bhat 1988) to name a few. The work of Mary Mandel has been revisited in a recent review republished in Biotechnology for Biofuels issue of September 2009 (Eveleigh et al. 2009). An excellent review on methodology by Zhang and Lynd (2004) critically examines the methods being employed for assay of enzymes.

For isolating and selecting diverse cellulase producing microorganisms, screening of metagenomic libraries or clones, the most commonly employed method is based on plate assay. This method relies on flooding the plates with Congo Red to detect visible solubilization of substrate particles and the formation of halos on petri dishes. Clearing zones around ball milled filter paper containing plates has also been used for isolating cellulolytic microorganisms. A method involving use of mixture of dyed polysaccharides has been used for simultaneous detection of carbohydrase activities (Ten et al. 2004). Dyed cellulose is prepared by mixing cellulose with a variety of dyes, such as Remazol Brilliant Blue, Reactive Orange, Reactive Blue 19, and fluorescent dye 5-4,6- dichlorotriazinyl) aminoflurescein. Because of large variations in the surface areas of cellulose and the binding conditions, the quantitative relationship between released dye and reducing sugars must be established for each batch of dyed cellulose. Insoluble cellulose derivatives, such as slightly substituted CMC, can be mixed with a variety of dyes, including Cibacron Blue 3GA and Reactive Orange 14 to produce insoluble dyed-CMC (Ten et al. 2004). Insoluble cellulose derivatives can also be chemically substituted with trinitrophenyl groups to produce chromogenic trinitrophenyl-carboxymethyl cellulose (TNP-CMC) and fluorophoric fluram cellulose. These methods are semi-quantitative, and are well suited to monitoring large numbers of samples. The plate assay is convenient and easy to perform, rapid, and more adaptable for screening of a large number of samples but does not show quantitative analyses of protein. Whereas, *quantitative approaches* make it possible to accurately estimate the saccharifying *activities* of crude *cellulase*.

Quantitative cellulase activity assays can be divided into three types: (1) the formation of products after hydrolysis, (2) the reduction in substrate quantity, and (3) the change in the physical properties of substrates. The two basic approaches to measuring cellulase activity are (1) measuring the individual cellulase (endoglucanases, exoglucanases, and β-glucosidases) activities, and (2) measuring the total cellulase activity. In general, hydrolase enzyme activities are expressed in form of initial hydrolysis rate for individual enzyme component within a short time, or the end-point hydrolysis for the total enzyme mixture to achieve a fixed hydrolysis degree within a given time. For cellulase activity assays, there is always a gap between initial cellulase activity assays and final hydrolysis measurement. Endoglucanases cleave intramolecular β-1,4-glucosidic linkages randomly, and their activities are often measured on a soluble high degree of polymerization (DP) cellulose derivative, such as CMC with the lowest ratio of FRE/Fa (the fraction of β-glucosidic bond accessible to cellulase (Fa), the fraction of reducing ends (FRE), and relative ratio of FRE/Fa). The modes of actions of endoglucanases and exoglucanases differ in that endoglucanases decreases the specific viscosity of CMC significantly with little hydrolysis due to intramolecular cleavages, whereas exoglucanases hydrolyze long chains from the ends in a processive process (Zhang and Lynd 2004). Endoglucanase activities can be measured based on a reduction in substrate viscosity and/or an increase in reducing ends determined by a reducing sugar assay. Because exoglucanases also increase the number of reducing ends, it is strongly recommended that endoglucanase activities be measured by both methods (viscosity and reducing ends). Soluble oligosaccharides and their chromophore substituted substrates, such as p-nitrophenyl glucosides and methylumbelliferyl-β-D-glucosides, are also used to measure endoglucanase activities based on the release of chromophores or the formation of shorter oligosaccharide fragments, which are measured by HPLC or TLC. The methodologies that are based on polarography for detection of cellulase activity by coupling the liberation of glucose with oxygen consumption have been described by various researchers. The polarographic assay couples cellulase with an excess mixture of β-1,4-glucosidase (EC 3.2.1.21), glucose oxidase (EC 1.1.3.4), mutarotase (EC 5.1.3.3), and catalase (EC 1.11.1.6). Glucose oxidase couples β-D-glucose formation, the end product of cellulose hydrolysis, with the consumption of oxygen. Thus, the enzyme-coupled system provides a means of continuously monitoring cellulase activity, or the susceptibility of cellulosic substrates to enzymatic degradation, by measuring oxygen consumption in the reaction medium.

Exoglucanases cleave the accessible ends of cellulose molecules to liberate glucose and cellobiose. *T. reesei* cellobiohydrolase (CBH) I and II act on the reducing and non-reducing cellulose chain ends, respectively (Zhang and Lynd 2004). Avicel has been used for measuring exoglucanase activity because it has the highest ratio of FRE/Fa among insoluble cellulosic substrates. During chromatographic fractionation of cellulase mixtures, enzymes with little activity on soluble CMC, but showing relatively high activity on avicel, are usually identified as exoglucanases. Unfortunately, amorphous cellulose and soluble cellodextrins are substrates for both purified exoglucanases and endoglucanases. Therefore, unlike endoglucanases and β- glucosidases, there are no substrates specific for exoglucanases within the cellulase mixtures (Wood and Bhat 1988). In some reports it has been described that 4-methylumbelliferyl-β-D-lactoside was an effective substrate for *T. reesei* CBH I, yielding lactose and phenol as reaction products, but it was not a substrate for *T. reesei* CBH II. Some endoglucanases of *T. reesei* EG I, structurally homologous to CBH I, also cleaves 4-methylumbelliferyl- β-D-lactoside, yet these enzymes can be differentiated by adding cellobiose, an inhibitor that strongly suppresses cellobiohydrolase activity. *T. reesei* CBH II does not hydrolyze 4-methylumbelliferyl-β-D-aglycones of either glucose or cellobiose units, but does cleave 4-methylumbelliferyl-β-D-glycosides with longer glucose chains.

β-D-glucosidases predominantly act on cellobiose to release glucose as the major product which can be measured employing glucose oxidase and peroxidase (GOD-POD) kit. Whereas other simple sensitive assay methods are based on colored or fluorescent products released from para nitrophenyl β-D-1,4-glucopyranoside, β-naphthyl-β-D-glucopyranoside, 6-bromo-2-naphthyl-β-D-glucopyranoside, and 4-methylumbelliferyl-β-D-glucopyranoside. Also, β-D-glucosidase activities can be measured using cellobiose, which is not hydrolyzed by endoglucanases and exoglucanases, and using longer cellodextrins, which are hydrolyzed by endoglucanases and exoglucanases (Zhang and Lynd 2004).

The total cellulase system consists of endoglucanases, exoglucanases, and β-D-glucosidases, all of which hydrolyze crystalline cellulose synergically. Total cellulase activity assays are always measured using insoluble substrates, including pure cellulosic substrates such as Whatman No. 1 filter paper, cotton linter, microcrystalline cellulose, bacterial cellulose, algal cellulose; and cellulose-containing substrates such as dyed cellulose, α-cellulose, and pretreated lignocellulose. The heterogeneity of insoluble cellulose and the complexity of the cellulase system cause formidable problems in measuring total cellulase activity. Experimental results show that the heterogeneous structure of cellulose (filter paper and bacterial cellulose) gives rise to a rapid decrease in the hydrolysis rate within a short time (less than an hour), even when the effects of cellulase deactivation and product inhibition are taken into account. The most common total cellulase activity assay is the filter paper assay (FPA) using Whatman No. 1 filter paper as the substrate, which was established and published by the International Union of Pure and Applied Chemistry (IUPAC). This assay requires a fixed amount (2 mg) of glucose released from a 50-mg sample of filter paper (i.e., 3.6% hydrolysis of the substrate), which ensures that both amorphous and crystalline fractions of the substrate are hydrolyzed. A series of enzyme dilution solutions is required to achieve the fixed degree of hydrolysis. The strong points of this assay are (1) it is based on a widely available substrate, (2) it uses a substrate that is moderately susceptible to cellulases, and (3) it is based on a simple procedure (the removal of residual substrate is not necessary prior to the addition of the DNS reagent).

Recent trends show that automated methods for screening large microbial/ libraries are being employed. Decker et al. (2003) reported automated filter paper assay method based on a Cyberlabs C400 robotics deck which is equipped with customized incubation, reagent storage and plate-reading capabilities that allow rapid evaluation of cellulases acting on cellulose. A functional proteomic assay in a multiplexed setting on an integrated plasmid-based robotic work cell for high-throughput screening of clones expressing mutants with improved endoglucanase F from the anaerobic fungus *Orpinomyces* PC-2 was reported by Hughes et al. (2006).The multiplex method using an integrated automated platform for high-throughput screening in a functional proteomic assay allows rapid identification of plasmids containing optimized clones ready for use in subsequent applications including transformations to produce improved strains or cell lines. A high-throughput assay employing microplate method to study the digestibility of ligno-cellulosic biomass as a function of biomass composition, pre-treatment severity, and enzyme composition was optimized for crystalline cellulose (Avicel) by Chundawat et al. (2008). The method is most suitable for delivering milled biomass to the microplate through multi-pipetting slurry suspensions. Similarly, miniaturization of DNS based detection method employing microtitre plate for screening cellulase producing clones of entire gut of *Reticulitermes flavipes* have been employed. Xia et al. (2005) reported microplate-based carboxymethylcellulose assay for endoglucanase activity. A rapid bio-enzymatic, spectrophotometric assay can be used to determine fermentable sugars. The entire procedure was automated using a robotic pipetting workstation. A new method to determine the activity of cellulase has been developed using a quartz crystal microbalance (QCM) technique. The QCM technique provides results closer to those obtained by measuring the actual reducing sugars.

4.8 Secretome Analysis of Glycosyl Hydrolases

Recent progress in various -omics tools has recently established a fundament for a system biological approach towards an understanding of cellulase and hemicellulose production. Proteomics is an excellent tool in profiling, discovering, and identifying proteins produced in response to a particular cellular environment as well as to study the distinct glycosyl hydrolases produced by industrially important cellulolytic strains. Secretome analysis in the past has focused on few of the representative fungal genera belonging to Ascomycetes (*Aspergillus*, *Penicillum*, *Trichoderma*) and basidiomycetes (*P. chrysosporium*). The list of basidiomycetes has expanded with the reports of the secretomes from *Pleurotus sapidus*, *Trametes versicolor* and *Coprinopsis cinerea* and *Phanerocheate carnosa*. Perhaps most well studied fungi is *Aspergillus* where the whole-genome sequencing projects has been completed for *A. aculeatus Aspergillus carbonarius A. clavatus A. fumigatus, A. terreus, A. parasiticus, A. flavus*. In one of the initial researches on secretome 73 secreted proteins of *A. flavus* were identified. Many of these proteins were proteases, metabolic proteins or proteins involved in electron/proton transport. Comparative secretome analyses is usually focused on fungal strains grown under submerged and solid state fermentation, or grown in the presence of different carbon sources or for comparisons among wood-rotters or phytopathogens. This comprehensive approaches been used to reveal the differential protein expression profiling in the parent and heterokaryons, developed through inter-specific proto-plast fusion between *Aspergillus nidulans* and *Aspergillus tubingensis* (Kaur et al. 2013). It has also been reported in one of the study that expression of β-glucosidase activity in the 2DE gel can be detected using methylumbelliferyl glucoside. Two novel β-glucosidases of *A. fumigatus* were identified by this *in situ* activity staining method, and the gene coding for a novel β-glucosidase (EAL88289) was cloned and heterologously expressed. The expressed β-glucosidase showed far superior heat stability to the previously characterized β-glucosidases of *A. niger* and *Aspergillus oryzae*. In a recent report Sharma et al. (2010) has reported a methodology for detection of (CBHI/EGI) in secretome of *A. fumigatus* using methyl umbelliferyl β-D lactopyranoside as substrate.

Chundawat et al. (2011) explored the protein composition of several commercial cellulase and xylanase preparations from *T. reesei* using a proteomics approach with high throughput quantification using liquid chromatography–tandem mass spectrometry (LC–MS/MS). As expected, Cel7A (former CBH1) was the predom-inant cellulase in all major commercial enzymes, followed by Cel6A (former CBH 2). Interestingly, proteomic approach has been employed to detect several intracel-lular enzymes in the culture filtrate of *T. reesei* which indicates that enzyme secretion is accompanied by considerable autolysis or mycelia fragmentation, whose role for high enzyme production has not yet been investigated. The secretome of white rot fungus *Bjerkandera adusta* produced in the presence of water-soluble olive mill extractives and the influence of the latter on the oxidore-ductase expression pattern was investigated. Distinct changes in the protein com-position of oxidoreductases, namely diverse class-II peroxidases and aryl alcohol oxidases were observed. The secretome analysis of anaerobic bacteria *Clostridium cellulolyticum* and *Cellulomonas flavigena* has also been reported. In addition to its capability of complementing transcriptome level changes, proteomics can also detect translational and post-translational regulations, thereby providing new insights into complex biological phenomena such as abiotic stress responses in plant roots. Recently, a novel technique termed as iTRAQ (isobaric tags for relative and absolute quantitation) has been adapted for proteomic quantitation which overcomes some of the limitations of 2D gel electrophoresis and also improves the throughput of proteomic studies (Liu et al. 2014).

The secretome analysis is not limited to fungi; few of bacterial systems have also been studied. The secretome including that of *Bacillus stereothermophilus, C. cellulolyticum, C. flavigena, Thermobifida fusca* have also been reported. The secretome analysis of *C. cellulolyticum* showed presence ofat least 30 dockerin-containing proteins (designated cellulosomal proteins) and 30 noncellulosomal components. Most of the known cellulosomal proteins, including CipC, Cel48F, Cel8C, Cel9G, Cel9E, Man5K, Cel9M, and Cel5A, were identified by using two-dimensional Western blot analysis with specific antibodies, whereas Cel5N, Cel9J, and Cel44O were identified by using N-terminal sequencing. Unknown enzymes having carboxymethyl cellulase or xylanase activities were detected by zymogram analysis of two-dimensional gels. Using Trap-Dock PCR and DNA walking, seven genes encoding new dockerin-containing proteins were cloned and sequenced. Some of these genes belonging to glycoside hydrolase families GH2, GH9, GH10, GH26, GH27, and GH59 are clustered.

4.9   Strain Improvement Programme

Some of the wild type fungal strains are known to produce copious amount of cellulolytic/hemicellulolytic enzymes. However, they are fall of being considered ideal for commercialization as the enzyme titres are low from industrial point of view. These cellulolytic strains *T. reesei*, *Penicillium decumbens*, *Acremonium cellulolyticus* have been subjected to continuous strain improvement programme, primarily involving repeated mutagenesis, selection and rational screening approaches (Liu et al. 2013). These procedures involve physical/chemical mutagens followed by rational screening strategies for enhancing the levels of cellulase and other cell wall degrading enzymes. Ethane methane sulphate (EMS) N-methyl-N′nitro-N-nitrosoguanidine (NTG), acriflavin are widely used chemical mutagens and UV radiation is primarily employed for physical treatment. The genealogy of these strain development program shows that either these mutagens have been used alone or in combination. In some cases, combined mutagenesis involving both chemical and physical treatments have resulted in the overproduction of cellulolytic enzymes. The developed mutants are picked randomly or on the basis of phenotypic traits such as resistance to antimetabolite, morphological differences, developmental (lacking cell differentiation to spores). The synthesis of cellulases is known to be primarily regulated through carbon catabolite repression (CCR) where expression of various transcriptional factors and expression of kinases and phosphatases that senses different carbon sources have important role to control cellular energy states. 2-Deoxy-D-glucose (2DG), a toxic analogue of glucose is widely used as selection marker for selecting deregulated mutants. Catabolite derepression is associated with the mutation in *creI* gene as described in *T. reesei* and *P. decumbens*. Expression of cellulolytic and hemicellulolytic genes in the presence of glucose in the developed mutants carries truncated creI gene encoded only one zinc finger protein. Whereas, *A. nidulans* in addition to CreA also contain CreB, CreC and CreD genes which actively participate in CCR. It has also been reported that high protein secretion is generally associated with alterations in certain organelles that are involved in the secretory pathway such as six to seven fold higher endoplasmic reticulum (ER) was observed in one the developed mutant (RUT-C30) of *T. reesei*. Several reports are available on the characterization of the selected mutants at genomic and proteomic level. To uncover the genetic changes in the mutants sequencing approach is widely employed for comparative analysis of parent and mutant strains to indicate the deletions or insertions in the nucleotide sequence. The expression profiling through SDS-PAGE of wild and developed strains also indicated the up/downregulated proteins which were identified through peptide mass fingerprinting (PMF) (Kaur et al. 2014).

Protoplast fusion is a powerful approach through which potential strains with desirable properties could be obtained with minimal disturbance in their physiology (Savitha et al. 2010). Most of the laboratories engaging in fungal genetics are using gene manipulation procedures based on protoplasts. Isolation of protoplasts are carried out by the digestion of cell wall with the aid of different hydrolytic enzymes, such as Novozyme234, chitinase, lysozyme etc., in the presence of osmotic stabilizer. Osmotic stabilizers play crucial role in maintaining the integrity of the protoplasts and protect the lysis of protoplasts. Inorganic osmotic stabilizers have been reported to be effective in *Thermomyces lanuginosus*, *Graphium putredinis*, *Trichoderma harzianum*, *Aspergillus* sp. Genetic manipulations can successfully be achieved through fusion of protoplasts in filamentous fungi that lack the capacity for sexual reproduction. The protoplast fusion is possible at intra-specific, inter-specific and inter-generic level involving two or more complex parental genomes. Therefore, a number of desirable genes from divergent strains can potentially be recombined into a fusant strain by this method. This technique was successfully employed as useful tool for genetic manipulation of desirable traits, for enhancing the cellulase production in *T. ressei* by inter-specific protoplast fusion. It has been studied that morphological markers (colony morphology and spore size and shape) and genetical markers like, mycelial protein pattern, restriction digestion pattern and random amplified polymorphic DNA (RAPD) analysis indicates the genetic recombination. RAPD method has proven to be efficient in describing DNA polymorphism in fusants obtained through the inter-specific protoplast fusion between *A. carbonarius* and *A. niger* for overproduction of pectinase. Protein expression analysis through SDS-PAGE is also effectively used as one of the markers which indicates the genetic recombination between parents and fusants. The presence or absence of protein bands between the parents and the hybrids (fusants) confirm the hybrid formation (Savitha et al. 2010). Similar genomic and proteomic approaches were carried to study the genetic relatedness in the heterokaryons developed through inter-specific protoplast fusion between two cellulolytic strains of *A. nidulans* (AN) and *A. tubingensis* (Dal8) (Kaur et al. 2013). One of the extensions of this breeding approach is genome shuffling that uses alternative cycles of genome recombination and selection to combine the useful alleles of many parental strains into single cells showing the desired phenotype. This technique offers the advantage of simultaneous genetic changes at different positions throughout the entire genome without the necessity for genome sequence information.

Cloning of genes by complementation cloning techniques has been used for the isolation of some enzyme genes in yeast and bacteria. However, complementation cloning is dependent on host strains having appropriate mutations and is therefore not of general use in the cloning of non-essential enzyme genes. In contrast expression cloning results in cloning full length genes using good quality mRNA and cDNA derived thereof is fast and efficient method of cloning of enzyme genes from fungi that are known express a variety of glycosyl hydrolases under inducible conditions. The advancement in genome sequencing has further made this technique more powerful and relevant where specific primers can amplify the entire ORF coding for the gene. It is possible to over-express individual enzyme components in different host systems using strong promoters such as AOX1 in *Pichia pastoris*, GLA in *A. oryzae*, CBH in *T. reesei* to name a few (Dalboge 1997).

4.10 Future Perspectives

The search for catalytically efficient thermostable cellulases from diverse extremophilic niches (culturable/unculturable) using wet lab and bioinformatics tools for mining data for prospecting unique and novel cellulases is foreseen as an area of future research. The use of system biology approaches for understanding and discovery of new set of proteins that can usher new concepts as deviation from the existing paradigm of cellulose degradation is also anticipated. Developing new and more efficient expression platforms for commercial exploitation of the identified genes is also an area that would draw the attention of the researchers and industrial houses alike.

Take Home Message

- Lignocellulosic wastes are primarily composed of polymeric cellulose (40–50%), hemicellulose (20–30%), and lignin (10–25%) that varies depending on their sources. Glycosyl hydrolases (GH) are those enzymes that can be used for the hydrolysis of lignocellulosics into simpler form of sugars (hexoses/and pentoses).
- These enzymes are responsible for synthesis, modification and degradation of carbohydrates and are clustered as carbohydrate-active enzymes (CAZymes). It includes Glycosyl hydrolases (GH), Polysaccahridelyases (PL), Carbohydrate esterases (CE), Glycosyl transferases (GT). It also includes chitinolytic, pectinolytic and amylolytic enzymes. The cellulases includes endoglucanases (EC 3.2.1.4), exoglucanases (EC 3.2.1.74 and 3.2.1.91), and β-glucosidases (EC 3.2.1.21). Endoglucanase acts on amorphous cellulose by cleaving the internal glycosidic bonds in a random fashion and releases oligomers such as cellobiose, cellotriose, and cellotetraose. β-glucosidases hydrolyze cellobiose to glucose from the non-reducing ends. *Cellobiohydrolase* hydrolyze the 1,4-β-D-glycosidic bonds of cellulose. Cellulolytic fungi are the most suited option for cellulose degradation as they have capability to produce greater amounts of extracellular enzymes when compared with the other group of microorganisms.

References

Acharya S, Chaudhary A (2012) Alkaline cellulase produced by a newly isolated thermophilic *Aneurinibacillus thermoaerophilus* WBS2 from hot spring, India. Afr J Microbiol Res 6 (26):5453–5458

Barros RR, Oliveira RA, Gottschalk LM, Bon EP (2010) Production of cellulolytic enzymes by fungi *Acrophialophora nainiana* and *Ceratocystis paradoxa* using different carbon sources. Appl Biochem Biotechnol 161:448–454

Bhatia Y, Mishra S, Bisaria VS (2002) Microbial β-glucosidases: cloning, properties, and applications. Crit Rev Biotechnol 22(4):375–407

Chundawat SPS, Balan V, Dale BE (2008) High-throughput microplate technique for enzymatic hydrolysis of lignocellulosic biomass. Biotechnol Bioengineer 99:1281–1294

Chundawat SP, Lipton MS, Purvine SO, Uppugundla N, Gao D, Balan V, Dale BE (2011) Proteomics-based compositional analysis of complex cellulase/hemicellulase mixture. J Proteome Res 10:4365–4372

Dalboge H (1997) Expression cloning of fungal enzyme genes; a novel approach for efficient isolation of enzyme genes of industrial relevance. FEMS Micribiol Rev 21:29–42

Davies G, Henrissat B (1995) Structures and mechanisms of glycosyl hydrolases. Structure 3:853–859.

Decker SR, Adney WS, Jennings E, Vinzant TB, Himmel ME (2003) Automated filter paper assay for determination of cellulase activity. Appl Biochem Biotechnol 105–108:689–703

Dodd D, Cann IO (2009) Enzymatic deconstruction of xylan for biofuel production. GCB Bioenerg 1:2–17

Eveleigh DE, Mandels M, Andreotti R, Roche C (2009) Measurement of saccharifying cellulase. Biotechnol Biofuels 2:21. doi:10.1186/1754-6834-2-21

Harhangi HR, Freelove ACJ, Ubhayasekera W, Dinther MV, Steenbakkers PJM, Akhmanova A, van der Drift C, Jetten MSM, Mowbray SL, Gilbert HJ, Op den Camp HJM (2003) Cel6A, a major exoglucanase from cellulosome of the anearobic fungi *Piromyces* sp. E2 and *Piromyces equi*. Biochem Biophys Acta 1628:30–39

Horn SJ, Vaaje-Kolstad G, Westereng B, Eijsink VGH (2012) Novel enzymes for the degradation of cellulose. Biotechnol Biofuels 5:45

Hughes SR, Riedmuller SB, Mertens A, Li XL, Bischoff KM, Qureshi N, Cotta MA, Philip J, Farrelly PJ (2006) High-throughput screening of cellulase F mutants from multiplexed plasmid sets using an automated plate assay on a functional proteomic robotic work cell. Proteome Sci 4:1–14

Igarashi K, Uchihashi T, Koivula A, Wada M, Kimura S et al (2011) Traffic jams reduce hydrolytic efficiency of cellulase on cellulose surface. Science 333:1279–1282

Irwin D, Shin DH, Zhang S, Barr BK, Sakon J, Karplus PA, Wilson DB (1998) Roles of catalytic domain and two cellulose binding domains of *Thermomonospora fusca* E4 in cellulose hydrolysis. J Bacterio 180(7):1709–1714

Kaur B, Sharma M, Soni R, Oberoi HS, Chadha BS (2013) Proteome-based profiling of hypercellulase – producing strains developed through interspecific protoplast fusion between *Aspergillus nidulans* and *Aspergillus tubingensis*. Appl Biochem Biotechnol 169:393–407

Kaur B, Oberoi HS, Chadha BS (2014) Enhanced cellulase producing mutants developed from heterokaryotic *Aspergillus* strain. Bioresour Technol 156:100–107

Kim Y, Nandakumar MP, Marten MR (2008) The state of the proteome profiling in the fungal genus *Aspergillus*. Brief Funct Genomic Proteomic 7:87–94

Kraulis J et al (1989) Determination of the three-dimensional solution structure of the C-terminal domain of cellobiohydrolase I from *Trichoderma reesei*. A study using nuclear magnetic resonance and hybrid distance geometry-dynamical simulated annealing. Biochem 28 (18):7241–7257

Liu GD et al (2013) Long-term strain improvements accumulate mutations in regulatory elements responsible for hyper-production of cellulolytic enzymes. Sci Rep 3:1569. doi:10.1038/srep01569

Liu GT, Ma LDW, Wang BC, Li JH, Xu HG, Yan XQ, Yan BF, Li SH, Wang LJ (2014) Differential proteomic analysis of grapevine leaves by iTRAQ reveals responses to heat stress and subsequent recovery. BMC Plant Biol 14:110

Menon V, Rao M (2012) Trends in bioconversion of lignocellulose: biofuels, platform chemicals and biorefinery concept. Prog Energy Combust Sci 38(4):522–550

Miotto LS, de Rezende CA, Bernardes A, Serpa VI, Tsang A et al (2014) The characterization of the endoglucanase Cel12A from *Gloeophyllum trabeum* reveals an enzyme highly active on β-Glucan. PLos One 9(9):e108393

Nakamura A, Watanabe H, Ishida T, Uchihashi T, Wada M, Ando T, Igarashi K, Samejima M (2014) Trade-off between processivity and hydrolytic velocity of cellobiohydrolases at the surface of crystalline cellulose. J Am Chem Soc 136:4584–5492

Ni J, Tokuda G (2013) Lignocellulose-degrading enzymes from termites and their symbiotic microbiota. Biotechnol Adv 31:838–850

Opassiri R, Pomthong B, Akiyama T, Nakphaichit M, Onkoksoong T, Ketudat-Cairns M, Ketudat CJR (2007) A stress-induced rice beta-glucosidase represents a new subfamily of glycosyl hydrolase family 5 containing a fascin-like domain. Biochem J 408:241–249

Palackal N, Lyon CS, Zaidi S, Luginbuhl P, Dupree P, Goubet F, Macomber JL, Short JM, Hazlewood GP, Robertson DE, Steer BA (2007) A multifunctional hybrid glycosyl hydrolase discovered in an uncultured microbial consortium from ruminant gut. App Microbiol Biotechnol 74:113–124

Payne CM et al (2013) Glycosylated linkers in multimodular lignocellulose-degrading enzymes dynamically bind to cellulose. Proc Natl Acad Sci USA 110(36):14646–14651

Savitha S, Sadhasivam S, Swaminathan K (2010) Regeneration and molecular characterization of an intergeneric hybrid between *Graphium putredinis* and *Trichoderma harzianum* by protoplasmic fusion. Biotehnol Adv 28:285–292

Sharma M, Soni R, Nazir A, Oberai HS, Chadha BS (2010) Evalution of glycosyl hydrolases in secretome of *Aspergillus fumigatus* and saccharification of Alkali- Treated Rice Straw. Appl Biochem Biotechnol 163(5):577–591

Sweeney MD, Xu F (2012) Biomass converting enzymes as industrial biocatalysts for fuels and chemicals: recent developments. Catalysts 2:244–263

Ten LN, Im WT, Kim MK, Kang MS, Lee ST (2004) Development of a plate technique for screening of polysaccharide-degrading microorganisms by using a mixture of insoluble chromogenic substrates. J Microbiol Methods 56:375–382

Voget S, Leggewie C, Uesbeck A, Raasch C, Jaeger KE, Streit WR (2003) Prospecting for novel biocatalysts in a soil metagenome. Appl Environ Microbiol 69:6235–6242

Warnick TA, Methe BA, Leschine SB (2002) *Clostridium phytofermentans* sp. nov., a cellulolytic mesophile from forest soil. Int J Sys Evol Micr 52:1155–1160

Wood TM, Bhat KM (1988) Methods for measuring cellulase activities. In: Wills AM, Scott TK (eds) Methods enzymol 160. Academic, San Diego, CA, pp 87–112

Xia Z, Storms R, Tsang A (2005) Microplate-based carboxymethylcellulose assay for endoglucanase activity. Anal Biochem 342:176–178

Xiao W, Wang Y, Xia S, Ma P (2012) The study of factors affecting the enzymatic hydrolysis of cellulose after ionic liquid pretreatment. Carbohydr Polym 87:2019–2023

Zhang YHP, Lynd LR (2004) Toward an aggregated understanding of enzymatic hydrolysis of cellulose: noncomplexed cellulase systems. Biotechnol Bioeng 88:797–824

Zhang YHP, Himmel ME, Mielenz JR (2006) Outlook for cellulase improvement: Screening and selection strategies. Biotechnol Adv 24:452–481

Chapter 5
Extremophilic Xylanases

Hemant Soni, Hemant Kumar Rawat, and Naveen Kango

What Will You Learn from This Chapter?

Xylanolytic enzymes find wide range of applications in pulp and paper, food and feed and textile industries. Xylanases replace chlorine based bleaching in pulp industry thus avoiding release of harmful organo-chloro chemicals. Realization of industrial applications necessitates that xylanase should be optimally active under process conditions. Pertaining to high operational temperature and pH of the industrial processes, xylanases showing high thermal and alkali stability are desirable. A few such xylanases are reported from different groups of microorganisms including thermophilic fungi, extremophilic bacteria and archaea (e.g. *Thermomyces lanuginosus, Geobacillus thermodenitrificans, Thermotoga maritima* etc.). These domains are further being explored for identifying hyperthermophilic microorganisms harboring robust xylanases. Culture-independent or metagenomic approaches have also yielded some thermostable xylanases. Another approach of obtaining industrially useful extremophilic xylanases involves designing of tailored xylanases using protein engineering and computational tools. This chapter focuses on sources, properties and development of extremophilic xylanases.

5.1 Introduction

Efficient enzymatic lignocellulose hydrolysis for generation of value-added products such as fermentable sugars and various other industrial objectives has been a matter of intense research (Subramaniyan and Prema 2002). Lignocellulose is

H. Soni • H.K. Rawat • N. Kango (✉)
Department of Microbiology, Dr. Harisingh Gour Vishwavidyalaya (A Central University),
Sagar, MP 470003, India
e-mail: nkango@gmail.com

© Springer International Publishing AG 2017
R.K. Sani, R.N. Krishnaraj (eds.), *Extremophilic Enzymatic Processing of Lignocellulosic Feedstocks to Bioenergy*, DOI 10.1007/978-3-319-54684-1_5

composed of three major structural polymers *viz.* lignin, cellulose and hemicellulose and hence demands action of ligninases, cellulases and hemicellulases (Sluiter et al. 2010). Among these, hydrolysis of heteropolymeric hemicelluloses such as xylan and mannan, on account of a variety of potential industrial applications, has attracted attention of several workers (Kulkarni et al. 1999; Kango and Jain 2005; Juturu and Wu 2012).

Xylanases have potential applications in a wide range of industrial processes, covering food industry (fruit and vegetable processing, brewing, wine production, baking), animal feeds (monogastric-swine and poultry and ruminant feeds), paper and pulp industry (biobleaching of kraft pulps, bio-mechanical pulping), coffee extraction, preparation of soluble coffee, detergents, production of pharmacologically active polysaccharides for use as antimicrobial agents etc. (Collins et al. 2005). Biobleaching applications in pulp and paper industry require cellulase-free xylanase, while in many other applications combination with other hydrolases such as proteases, oxidases, isomerases is desirable (Verma and Satyanarayana 2012; Thomas et al. 2014). Xylanases from extremophilic sources have tremendous utility in many biotechnological processes. In particular, thermostable and alkalistable xylanases could be used in industrial applications where high temperature and alkaline pH are an integral part of process (Viikari et al. 1994). For instance, pulping in paper industry and pelleting process in animal feed industry, where high temperature (70–90 °C) is required, necessitates use of thermostable xylanases. Acidophilic and alkaliphilic enzymes would obviously be beneficial in processes where extreme pH conditions are required or where adjustment of pH to neutral conditions is uneconomical. Alkaliphilic xylanases are also required for detergent applications where high pH is typically used (Viikari et al. 1994; Verma and Satyanarayana 2012). Archaea, bacteria and fungi isolated from extreme environments such as Antarctic region, hot water springs and oceans depths are understood to be a possible source of psychrophilic, thermophilic and halophilic xylanases, respectively (Collins et al. 2005). Some microbial sources of extremophilic xylanases and their properties are listed in Table 5.1.

As newer enzymes and their applications are being explored, their scope and market continues to grow. Hydrolases cover approximately 75% of the market share of industrial enzymes and among them xylanases make a significant part. Industrial relevance of xylanases can be appreciated on account of various patents being filed in the developed and developing countries (Soni and Kango 2013).

Although conventional screening methods have led to the identification of several potential xylanase producers, newer approaches, such as amino acid modification, functional PCR screening of environmental DNA libraries, DNA shuffling, error prone PCR and site directed mutagenesis are now being explored routinely for improvement of xylanases enabling their applicability at very high temperature and pH (Verma and Satyanarayana 2012).

Table 5.1 Some microbial sources of extremophilic xylanases

S.N.	Microorganism	Xylanase characteristics	Reference
Thermophilic			
1	*Myceliophthora thermophile*	Thermostable 1 h at 65 °C	Maijala et al. (2012)
2	*Chaetomium thermophilum*	Thermostable	Hakulinen et al. (2003)
3	*Talaromyces thermophiles*	Thermostable (4 h at 80 °C, 2 h at 90 °C, and 1 h at 100 °C)	Romdhane et al. (2010)
4	*Talaromyces cellulolyticus*	Thermostable	Inoue et al. (2015)
5	*Thermoauscus aurantiacus*	Stable 6 h at 70 °C	Jain et al. (2015)
6	*Humicola insolens*	Stable below 50 °C	Du et al. (2013)
7	*Thermoanaerobacterium thermosaccharolyticum*	Thermostable for 1 h at 71 °C with broad pH range (4–8)	Li et al. (2014)
8	*Caldicoprobacter algeriensis*	Stable at 50, 60, 70, and 80 °C with half-life of 10, 9, 8, and 4 h	Bouacem et al. (2014)
9	*Geobacillus thermodenitrificans*	Thermostable and alkalistable	Verma et al. (2013)
10	*Clostridium beijerinckii* G117	Thermostable at 40–50 °C	Ng et al. (2015)
11	*Streptomyces thermovulgaris*	Xylanase cultivation at 50 °C	Chaiyaso et al. (2011)
12	*Geobacillus thermoleovorans*	Thermostability at 40–100 °C and stable at broad pH range 6–12	Verma and Satyanarayana (2012)
13	*Thermotoga maritima*	Stable at 95 °C for 22 min and had a half-life of 57% and stable at broad pH range 5.4–8.5	Xue and Shao (2004)
Halophilic			
14	*Halorhabdus utahensis*	Highly active xylanase and β-xylosidase over a broad NaCl range (0–30% NaCl)	Wainø and Ingvorsen (2003)
15	*Bacillus halodurans* S7	Alkalistable (pH 9) and thermo-stable at 65 °C	Mamo et al. (2006)
16	*Bacillus halodurans* TSEV1	Stable at pH 4.0 and 11.0 and thermostable at 70 and 80 °C had half-life of 40 and 15 min respectively	Kumar and Satyanarayana (2014)
Psychrophilic			
17	*Anthrobacter* sp.	Active at low temperature 0–30 °C	Zhou et al. (2015)
Alkalophilic			
18	*Penicillium* sp.	Xylanase production at alkaline pH (8–10). Maximum xylanase activity at 60 °C (65.4%)	Bajaj et al. (2011)
19	*Bacillus pumilus*		Asha Poorna and Prema (2007)
20	*Cellulosimicrobium cellulans*		Walia et al. (2015)

5.2   Xylan: Structure and Occurrence

Next to cellulose, hemicellulose is the most abundant structural polysaccharide. In plant hemicelluloses, xylan content in hardwood is about 15–30% (e.g. *Betula verrucosa*—27.5%) with a degree of polymerization of 150–200 while it is upto 7–15% in softwood (e.g. *Picea abies* 8.6%) with a degree of polymerization of 70–130 (Casey 1980; Kulkarni et al. 1999; Ek et al. 2007). In this polysaccharide, xylopyranose residues are linked by β-1,4 linkages and also have various other side chain substituents such as 4-*O*-methyl glucuronic acid (glucuronoxylan). It is an important component of dicots (15%–30%) in which 4-*O*-methyl glucuronic acid is linked with α-1,2 linkage at C-2 position (Collins et al. 2005). When substituted with L-arabinofuranose, xylan is called arabinoxylan in which L-arabinofuranose residues are attached by α 1 → 2 or α 1 → 3 linkages to the xylose units of main chain at C-3 position.

5.3   Xylanases and Their Applications

Xylan is acted upon by endoxylanases, β-xylosidases, arabinofuranosidases and acetyl xylan esterases (Fig. 5.1). Endoxylanase (1,4-β-D-Xylan xylanohydrolase, EC 3.2.1.8) is understood to be the chief enzyme hydrolyzing the backbone of xylan macromolecule while other enzymes act upon side groups of the heteropolysaccharide (Collins et al. 2005). Endoxylanases belong to family GH 10 or 11 (a few also belong to GH 5, 8, 43) and hydrolyze bonds of xylan backbone in a random way and yielding xylooligosaccharides. These are the best studied enzymes among the group of xylanolytic enzymes. Xylosidases (1,4-β-D-Xylan

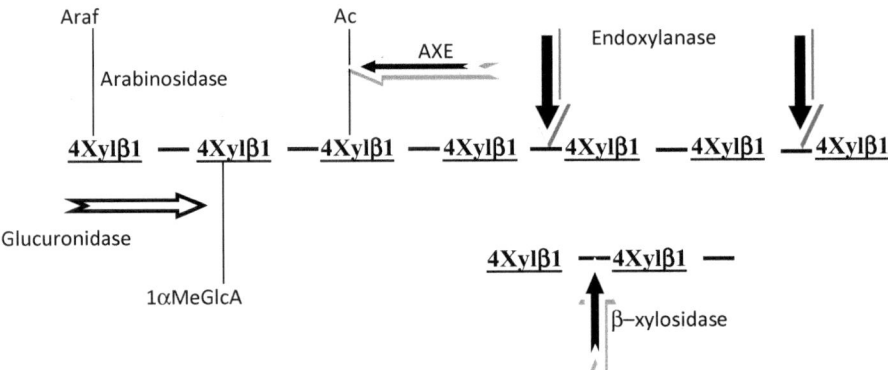

Fig. 5.1 Xylan being acted upon by xylanolytic enzymes (*Xyl* xylose; *Araf* arabinofuranose; *Ac* acetyl; *MeglcA* methylglucuronic acid; *AXE* acetyl xylan esterase)

xylohydrolases, EC 3.2.1.37, GH 3, 39, 43, 52 and 54) hydrolyze xylobiose or xylooligosaccharides and yield D-xylose. Arabinofuranosidases (α-L-arabinofuranosidase, EC 3.2.1.55) hydrolyze α-L-arabinofuranosidic linkages in arabinoxylan. Glucuronidase (α-4-o-methyl-D-glucuronidase, EC 3.2.1.1) catalyze cleavage of α-4-o-methyl-D-glucuronic side chains of glucuronoxylan. Another accessory enzyme, acetyl xylan esterase (EC 3.1.1.6) releases acetic acid from acetylxylan (Shallon and Shoham 2003).

Several patents disclosing enhancement of bleachability of pulp by microbial xylanases indicate their potential in biobleaching (Soni and Kango 2013). Several benefits of xylanase pretreatment of pulp include higher brightness ceilings, significant reductions in the amounts of chlorine based bleaching thus avoiding consequential release of organo-chlorine compounds in bleach plant effluents (Collins et al. 2005). Xylanase pretreatment of food and feed with significant xylan content brings improvement in their nutritional and textural properties. A potential application of xylanases in combination with pectinolytic enzymes lies in debarking of bast fibres such as flax, jute and hemp (Kango et al. 2003).

Xylanases also find use in manufacturing of functional xylo-oligosaccharides, saccharification and bioconversion of lignocellulosic waste for production of value-added products (e.g. ethanol, xylitol, single cell protein), clarification of fruit juices, improvement in consistency of beer, enhancement in digestibility of animal feed and quality upgradation of fibres, production of surfactant, coffee extract and preparation of instant coffee, detergent industry, production of antioxidants (Juturu and Wu 2012). Bleaching process involves many steps which are carried out under high pH and temperature regimes and therefore alkalistable and thermostable xylanases are required to achieve better results. With non-thermostable enzymes, the pulp must be cooled to reaction temperature which consumes time and energy (Kulkarni et al. 1999; Juturu and Wu 2012). The stability of xylanase at higher temperature and alkaline condition is very important for realizing its application. Xylanases with these two key features can increase efficacy of enzyme during biobleaching of pulp (Viikari et al. 2007). Xylanases produced from *Thermotoga maritima* had temperature optima more than 80 °C suiting to bleaching of pulp but the alkali stability of these enzymes was less than needed for bleaching process (Yoon et al. 2004). Xylanase from *Bacillus halodurans* (Mamo et al. 2006) had sufficient activity under alkaline conditions but thermostability associated with this xylanase was less biobleaching of pulp. Apart from being thermo- and alkali-stable, low-molecular weight xylanases are preferable because of their easy diffusion into pulp fibers to perform catalytic activity (Mamo et al. 2006). For application in feed and food industry, xylanase should be thermostable to facilitate the catalytic reaction during pellet making (Viikari et al. 1994). Advent of metagenomics has enabled workers to explore unculturable microbial diversity (Handelsmen 2004) and complete set of genomes harvested from environmental samples collected from heated and alkaline sites are being looked for poly-extremophilic xylanases (Anand et al. 2013; Sadaf and Khare 2014).

5.4 Production and Properties of Extremophilic Xylanases

Realization of enzyme application at large scale necessitates its cost-effective production using potential isolates elaborating copious amounts of thermostable xylanase using cheap industrial media. Thus the strategy involves screening, optimization and characterization of useful enzymes from myriad microbial forms. Archaea and bacteria sourced from extreme niches have helped in identifying poly-extremophilic xylanases. Thermophilic and thermotolerant fungi being natural colonizers of lignocelluloses, occur in self-heated environments such as composts and many of them (such as *Thermomyces lanuginosus, Melanocarpus albomyces*) are known for producing thermostable xylanolytic enzymes (Maijala et al. 2012).

Yoon et al. (2004) have used *Thermotoga maritima* producing 170 U/ml xylanase in medium containing oat spelt xylan as carbon source. Xylanase was optimally active at 90 °C with half life of 1 h at 90 °C. Xylanase from *Thermococcus zilligii* had half-life of 8 min at 100 °C (Uhl and Daniel 1999). Xylanase production by hyperthermophilic bacteria *Geobacillus thermodenitrificans* TSAA1 was found to be 2.750 U/ml on medium containing wheat bran under submerged conditions at 60 °C (Anand et al. 2013). The thermostable xylanase exhibited half-life of 30 min at 80 °C. Bouacem et al. (2014) have used an alkalophilic bacteria, *Caldicoprobacter algeriensis* for xylanase production yielding 140 U/ml alkali-thermostable xylanase optimally active at pH 11.0 and 70 °C. Recently, Sun et al. (2015) screened out xylanase from environmental DNA and produced 106 U/ml in heterologous host. Xylanase was optimally active pH 7.0 and 75 °C with 24 h half-life activation at 50 °C.

Bacillus halodurans produced 625 U/ml xylanase under submerged condition at 42 °C with xylanase showing pH and temperature optima at 7.0 and 70 °C, respectively. *Bacillus pumilus* produced highly alkali-stable xylanase active at pH (12.0) and 60 °C for 1 h (Thomas et al. 2014). These reports suggest occurrence of thermo and alkalistable xylanases in a variety of archaea and eubacteria. The temperature range for optimum activity ranged between 65 and 90 °C while the optimum pH range was 6.0–11.0. Various xylanases from different alkaliphilic and thermophilic microorganisms with their yield and properties are summarized in Table 5.2. Properties of various β-xylosidases from extremophilic organisms are given in Table 5.3.

Use of statistical optimization (response surface methodology) for xylanase production from thermophilic *T. lanuginosus* SDYKY-1 resulted in 144% increase of xylanase activity (3078 U/ml) suggesting the role of process parameters in enzyme production (Yishan et al. 2011). Many instances suggested use of cheaper hemicellulosic substrates in media for xylanase production from thermophilic and thermotolerant fungi (Kango and Jain 2005). Corncob supported production of 131 U/ml xylanase in case of thermophilic mold *Chaetomium* sp. CQ31 while *Paecilomyces thermophila* J18 produced 18580 U/g on wheat straw (Yang et al. 2006). Utilization of deoiled *Jatropha curcas* seed-cake as solid substrate for

Table 5.2 Production and properties of thermo-alkali-stable xylanases from microorganisms

Microorganism	Yield, substrate, incubation Temp, RPM	Optimum pH	Optimum Temp (°C)	Mr (kDa)	Half-inactivation period (τ1/2)	Reference
Archaea and Bacteria						
Thermotoga sp. FjSS3-B.1					90 min at 95 °C	Simpson et al. (1991)
Thermococcus zilligii		6.0	75	95	8 min at 100 °C	Uhl and Daniel (1999)
Sulfolobus solfataricus		7.0	90	57	47 min at 100 °C	Cannio et al. (2004)
Geobacillus thermodenitrificans TSAA1	2.75 U/ml, WB, 60 °C, 200	7.5	70	43	30 min at 80/75 °C	Anand et al. (2013)
Geobacillus sp. strain WSUCF1		6.5	70	37	60 h at 60 °C	Bhalla et al. (2014)
Caldicoprobacter algeriensis	140.0 U/ml	11	70		8/4 h at 70/80 °C	Bouacem et al. (2014)
Geobacillus sp. WBI		6.0–9.0	50–90	47	100% active at 65° for 1 h (pH-10.0)	Mitra et al. (2015)
Thermotoga maritima	170 U/ml, oatspeltsarabinoxylan	6.0	90	119	1 h at 90 °C	Yoon et al. (2004)
Caldicellulosiruptor owensensis (Coxyn A)		7.0	75	40	1 h at 70 °C	Mi et al. (2014)
Bacillus halodurans TSEV1	40 U/ml	9.0	80	36	40/15 min at 70 /80 °C	Kumar and Satyanarayana (2011)
Bacillus pumilus		9.0–10	50–60		100% for 1 h at 60 °C and 12 pH	Thomas et al. (2014)
Bacillus halodurans S7		9.0/10	75/70	43		Mamo et al. (2006)
Actinomadura sp. strain Cpt20		10	80	20	2/1 h at 90/100 °C	Taibi et al. (2012)
Fungi						
Malbranchea cinnamomea(McXyn10)		6.5	80	43.5	76 min at 70 °C	Fan et al. (2014)
Myceliophthora fergusii(MTCC 9293)	27 nkat/ml, WB, 45 °C,140				1 h at 65 °C	Maijala et al. (2012)
Uncultured microorganism	106 IU/ml	7.0	75		24 h at 50 °C	Sun et al. (2015)
Uncultured microorganism		6.0	100		2.5/10 min at 100/90 °C	Sunna and Bergquist (2003)

WB wheat bran; *U* Unit of xylanase activity stands for μmols of xylose liberated per minute under standard conditions

Table 5.3 Properties of thermo-alkali-stable β-xylosidase from various microorganisms

| Microorganism | Optimum | | kDa | PI | Half-life inactivation period (τ1/2) | Reference |
	pH	Temp (°C)				
Geobacillus sp. SUCF1	6.5	70	230		9 days at 70 °C	Bhalla et al. (2014)
Caldicellulosiruptor owensensis (Coxyn A)	5.0	75	55		>60% at 70, 75, 80 °C	Mi et al. (2014)
Geobacillus thermodenitrificans TSAA1	7.0	60			3 h at 70 °C	Anand et al. (2013)
Bacillus stearothermophilus	6.0	70	150	4.2	No effect till 1 h at 60 °C	Nanmori et al. (1990)
Geobacillus stearothermophilus T-6	6.5	65				Shallom et al. (2005)
Thermoanaerobacterethanolicus	5.9	93	165	4.6	3 h at 82 °C	Shao and Wiegel (1992)
Thermotoga maritime	6.0	90			22 min/1 h at 95/80 °C	Xue and Shao (2004)
Thermotoga sp. FjSS3-B.1			92		Many h at 98 °C	Ruttersmith and Daniel (1993)

production of xylanase from thermophilic *Scytalidium thermophilum* and *Sporotrichum thermophile* under optimized conditions resulted in production of 1025 U/gds and 1455 U/gds of xylanase (Sadaf and Khare 2014).

5.5   Approaches for Improving Microbial Xylanases

Pertaining to extreme conditions of industrial operations, xylanases active at high pH and temperature are desirable. Owing to the few number of microorganisms producing such robust xylanases, various approaches such as protein engineering, metagenomics and molecular re-shuffling have been used as tools for both search and evolution of thermo-alkali stable xylanases.

Some approaches like directed evolution, amino acid modification, molecular dynamics, N and C terminal mutagenesis, rational design strategy, computational designing, site directed mutagenesis, DNA shuffling etc. (Joo et al. 2011; Stephens et al. 2014; Zouari et al. 2015) have been used for producing robust microbial xylanases.

Very recently, Song et al. (2015) have evolved a single domain GH10 xylanase, from *Aspergillus niger* (Xyn10A_ASPNG) to improve its thermostability. This effort involved computational analysis and random mutagenesis through rounds of iterative saturation mutagenesis (ISM). ISM generated quintuple mutant 4S1

(R25W/V29A/I31L/L43F/T58I) exhibiting thermal inactivation half-life ($t_{1/2}$) at 60 °C that was prolonged by 30-fold in comparison with wild-type enzyme.

In another case, rational design strategy involved replacement of the N-terminus of mesophilic xylanase with that of thermophilic xylanase. The first 31 residues of the thermophilic xylanase TfxA from *Thermomonospora fusca* were successfully implemented into a mesophilic xylanase to produce several thermostable hybrid xylanases, such as StxAB and ATx (Zhang et al. 2010). These hybrid xylanases exhibited higher thermostabilities than their corresponding mesophilic parents. The optimum temperature of StxAB (80 °C) was 15 °C higher than SoxB (65 °C) and half-life of StxAB at 70 and 80 °C were 8 h and 21 min, respectively.

Palackal et al. (2004) and Ruller et al. (2008) have also demonstrated that mutation in N-terminus region of xylanase plays an important role in conferring hyperthermostability. Recently, Zhang et al. (2014) experimentally proved that seven N-terminal residues of a thermophilic xylanase are sufficient to confer hyperthermostability on its mesophilic counterpart. They studied mesophilic SoxB (from *Streptomyces olivaceoviridis*) and thermophilic TfxA (from *Thermomonospora fusca*) and concluded that at region 4 two mutations (N32G and S33P) were responsible in improving the thermostability of mesophilic SoxB. The mutant (M2-N32G-S33P) had a melting temperature (T_m) much higher than that of SoxB. More importantly, it showed more thermal stability than thermophilic TfxA and had 9 °C higher T_m. This property was not exhibited by hybrid xylanases (StxAB and ATx) created by rational design strategy of Zhang et al. (2010).

Directed evolution has also been successfully applied to confer thermostability and hyperthermostability in mesophilic xylanases. Ruller et al. (2008) improved the melting temperature (T_m) of xylanase A from *Bacillus subtilis* from 59 to 76.5 °C with four mutations using directed evolution. Palackal et al. (2004) successfully obtained a thermostable variant (9X) with a high T_m (95.6 °C). More recently, through error prone PCR and DNA shuffling, Ruller et al. (2014) created xylanase (XynA) containing eight mutations (Q7H/G13R/S22P/S31Y/T44A/I51V/I107L/S179C) that exhibited temperature and pH optimum of 80 °C and 8.0 as compared to wild type which had temperature and pH optima of 50 °C and 6.0, respectively. The enzyme also had improved half-life inactivation about of 60 min at 80 °C (wild type <2 min). Another example of DNA shuffling is S340 and S325 mutant xylanase (from *T. lanuginosus*) created by DNA shuffling using the StEP recombination method (Stephens et al. 2014). Protein S340 retained 54% stability at 80 °C and 60% stability at pH 10 and another recombinant, S325, displayed 85% stability at 80 °C and 60% stability at pH 10.

Protein engineering has become a regular feature for designing thermo-alkaline stable xylanases. Various approaches employed for imparting extremophilic properties in xylanases are summarized in Table 5.4. US patent 5759840 (Sung et al. 1998) described modification of xylanase protein to improve thermophilicity, alkalophilicity and thermostability. Modification of xylanase involved three types of modifications. Firstly, amino acids 10, 27, and 29 of xylanase from *Trichoderma reesei* were replaced with histidine, methionine and leucine, respectively. Secondly, substitution at N-terminal with another xylanase enzyme (xylanase from

Table 5.4 Various approaches employed towards development of extremophilic xylanases

Approach	Xylanase source	Reference
Amino acid modification	*Trichoderma reesei* xylanase-II	Zouari et al. (2015)
	B. circulans	Joo et al. (2011)
	B. circulans (Bcx)	Li and Wang (2011)
Directed evolution	*B. subtilis* (Xyn A)	Ruller et al. (2014)
Molecular dynamics	*B. circulans (Bcx)*	Joo et al. (2010)
Modification in C or N terminal region	*A. niger*	Sun et al. (2015)
	Streptomyces olinaceovirdis	Zhang et al. (2010)
Computational designing	*B. circulans (Bcx)*	Joo et al. (2011)
X-ray crystallography	*C. thermophilium* (CTX) *Nonomuraeaflexuosa*	Hakulinen et al. (2003)
Disulfide bonds	*B. stearthermophilus*236	Jeong et al. (2007)
DNA shuffling Error prone PCR	*T. lanuginosus* DSM 5826 *xylanase* (xynA) *T. lanuginosus*	Stephens et al. (2014)
Metagenomic	*G. stearothermophilus* Xylanase gene (Xyn-b39)	Zhao et al. (2013)
Error prone PCR	*G. stearothermophilus*	Wang et al. (2013)

Thermomonospora fusca) was made to yield a chimeric xylanase which showed higher thermophilicity and alkalophilicity. Third modification involved an extension of the N-terminus of the xylanase with glycine-arginine-arginine amino acids (tripeptide) to improve its performance.

Recently, Zouari et al. (2015) created thermostable xylanase-II to replace serine on the surface of *T. reesei* xylanase with threonine residues and mutant exhibited half-life inactivation in about 37 min at 55 °C while wild-type xylanase had half-life time of 20 min at 55 °C. Joo et al. (2011) demonstrated fivefold increase in half-life with slight increase in T_m with amino acid modification in xylanase of *Bacillus circulans*. Li and Wang (2011) observed improvement in catalytic activity when they substituted asparagine at position 35 in *B. circulans* xylanase with aspartic acid. Similarly, substitution of arginine in place of Ser/Thr in *A. niger* BCCI 144505 enhanced xylanase themostability. Table 5.4 lists some of the approaches used for development of thermostable xylanases. Half-life of xylanase of *B. subtilis* was enhanced by point mutagenesis and DNA shuffling (Miyazaki et al. 2006). Wang and Xia (2008) and Stephens et al. (2014) made significant changes in xylanase of *Thermobifida fusca* and *Thermomyces lanuginosus*, respectively, using error prone PCR. Similarly, Wang et al. (2013) improved catalytic efficiency of xylanase cloned from *G. stearothermophilus* by error prone PCR method. Introduction of disulfide bond between Cys100 and Cys154 improved thermostability of xylanase of *T. lanuginosus* DSM5726 and thermostability of xylanase of *B. stearothermophilus* 236 was enhanced by 5 °C by introducing disulfide bond between Ser and Cys100 and Asn to Cys150 (Stephens et al. 2014; Jeong et al.

2007). Zhao et al. (2013) have cloned xylanase gene Xyn-b39 from alkaline waste water sludge using metagenomic approach. They claimed it can be useful in paper industry to reduce consumption of chlorine dioxide because xylanase was highly active (80%) at pH 9.0 and was thermostable at 55 °C. Denaturation temperature midpoint of xylanase (XYL7747) was raised from 61 to 96 °C by gene site saturation mutagenesis (GSSM) (Palackal et al. 2004), GSSM also enhanced optimum temperature by 25 °C of mutant EvXyn11^{TS} from GH11 xylanase as compared to parent enzyme (Dumon et al. 2008).

5.6 Conclusion

Extremophilic xylanases from extremely thermophilic bacteria are more likely to be useful for industrial applications. Extreme thermophiles such as *Caldicellulosiruptor* have recently drawn attention (Peng et al. 2015). Some extremophilic xylanases are fully characterized for catalytic domain, hydrolytic capability, thermostability and their efficacy in lignocellulose hydrolysis. Use of enzymatic repertoire consisting of cellulases and hemicellulases from extremophiles can be directly employed for lignocellulose hydrolysis thus avoiding any pre-treatments. Exclusive production of cellulase-free thermostable xylanase has been achieved by over-exploring the xylanase gene of extremophilic bacterium (e.g. *T. thermosaccharolyticum*) in *E. coli*. Engineering of N-terminus of mesophilic xylanases, directed evolution and metagenomic approach has lead to development of some promising extremophilic xylanases and also holds the key for future endeavors.

Take Home Message

- The xylan content in plant hemicelluloses is about 15–30% (with a degree of polymerization of 150–200) and 7–15% (with a degree of polymerization of 70–130) in hardwood and softwood respectively.
- Xylan is acted upon by endoxylanases, β-xylosidases, arabinofuranosidases and acetyl xylan esterases. Endoxylanase hydrolyze the backbone of xylan macro-molecule whereas the other xylanolytic enzymes act upon side groups of the heteropolysaccharide. Xylosidases hydrolyze xylobiose or xylooligosaccharides and yield D-xylose. Arabinofuranosidases hydrolyze α-L-arabinofuranosidic linkages in arabinoxylan. Glucuronidase catalyze cleavage of α-4-*o*-methyl-D-glucuronic side chains of glucuronoxylan.
- Xylanolytic enzymes find wide range of applications in pulp and paper, food and feed and textile industries. Alkaliphilic xylanases are also required for detergent applications where high pH is typically used. Thermostable xylaneses are produced by different groups of microorganisms including thermophilic fungi, extremophilic bacteria and archaea (e.g. *Thermomyces lanuginosus*, *Geobacillus thermodenitrificans* and *Thermotoga maritima*.

- Approaches such as directed evolution, amino acid modification, molecular dynamics, N and C terminal mutagenesis, rational design strategy, computational designing, site directed mutagenesis, DNA shuffling etc. have been used for producing robust microbial xylanases.

Acknowledgment Authors (HS and NK) thank University grants commission, New Delhi for financial support (Grant no.42-474/2013SR).

References

Anand A, Kumar V, Satyanarayana T (2013) Characteristics of thermostable endoxylanase and β-xylosidase of the extremely thermophilic bacterium *Geobacillus thermodenitrificans* TSAA1 and its applicability in generating xylooligosaccharides and xylose from agro-residues. Extremophiles 17:357–366

Asha Poorna C, Prema P (2007) Production of cellulase-free endoxylanase from novel alkalophilic thermotolerant *Bacillus pumilus* by solid-state fermentation and its application in waste paper recycling. Bioresour Technol 98:485–490

Bajaj BK, Sharma M, Sharma S (2011) Alkalistable endo-β-1,4-xylanase production from a newly isolated alkali tolerant *Penicillium* sp. SS1 using agro-residues. 3 Biotech 1:83–90

Bhalla A, Bischoff KM, Sani RK (2014) Highly thermostable GH39 β-xylosidase from a *Geobacillus* sp. strain WSUCF1. BMC Biotechnol 14:963

Bouacem K, Bouanane-Darenfed A, Boucherba N, Joseph M, Gagaoua M, Ben Hania W, Kecha M, Benallaoua S, Hacène H, Ollivier B, Fardeau ML (2014) Partial characterization of xylanase produced by *Caldicoprobacter algeriensis*, a new thermophilic anaerobic bacterium isolated from an Algerian hot spring. Appl Biochem Biotechnol 174:1969–1981

Cannio R, Di Prizito N, Rossi M, Morana A (2004) A xylan-degrading strain of *Sulfolobus solfataricus*: isolation and characterization of the xylanase activity. Extremophiles 8:117–124

Casey JP (1980) Pulp and paper chemistry and chemical technology, vol 1, 3rd edn. Wiley Interscience Publication, New York

Chaiyaso T, Kuntiya A, Techapun C, Leksawasdi N, Seesuriyachan P, Hanmoungjai P (2011) Optimization of cellulase-free xylanase production by thermophilic *Streptomyces thermovulgaris* TISTR1948 through Plackett-Burman and response surface methodological approaches. Biosci Biotechnol Biochem 75:531–537

Collins T, Gerday C, Feller G (2005) Xylanases, xylanase families and extremophilic xylanases. FEMS Microbiol Rev 29:3–23

Du Y, Shi P, Huang H, Zhang X, Luo H, Wang Y, Yao B (2013) Characterization of three novel thermophilic xylanases from *Humicola insolens* Y1 with application potentials in the brewing industry. Bioresour Technol 130:161–167

Dumon C, Alexander VA, Wall MA, Flint JE, Lewis RJ, Lakey JH, Morland C, Luginbuhl P, Healey S, Todaro T, DeSantis G, Sun M, Gessert LP, Tan X, Weiner DP, Gilbert HJ (2008) Engineering hyperthermostability into a GH11 xylanase is mediated by subtle changes to protein structure. J Biol Chem 283:22557–22564

Ek M, Gellerstadt G, Henriksson G (2007) Ljungberg textbook. Pulp and paper chemistry and technology. Book 1, Wood chemistry and wood biotechnology. KTH, Stockholm. isbn:572 777 3143

Fan G, Yang S, Yan Q, Guo Y, Li Y, Jiang Z (2014) Characterization of a highly thermostable glycoside hydrolase family 10 xylanase from *Malbranchea cinnamomea*. Int J Biol Macromol 70:482–489

Hakulinen N, Turunen O, Janis J, Leisola M, Rouvinen J (2003) Three dimensional structures of thermophilic b-1,4-xylanases from *Chaetomium thermophilum* and *Nonomuraea flexuosa*. Europ J Biochem 270:1399–1412

Handelsman J (2004) Metagenomics: application of genomics to uncultured microorganisms. Microbiol Mol Biol Rev 68:669–685

Inoue H, Kishishita S, Kumagai A, Kataoka M, Fujii T, Ishikawa K (2015) Contribution of a family 1 carbohydrate-binding module in thermostable glycoside hydrolase 10 xylanase from *Talaromyces cellulolyticus* toward synergistic enzymatic hydrolysis of lignocellulose. Biotechnol Biofuels 8:77

Jain KK, Bhanja Dey T, Kumar S, Kuhad RC (2015) Production of thermostable hydrolases (cellulases and xylanase) from *Thermoascus aurantiacus* RCKK: a potential fungus. Bioprocess Biosyst Eng 38:787–796

Jeong MY, Kim S, Yun CW, Choi YJ, Chob SG (2007) Engineering a *de novo* internal disulfide bridge to improve the thermal stability of xylanase from *Bacillus stearothermophilus* No. 236. J Biotechnol 127:300–309

Joo JC, Pohkrel S, Packb SP, Yoo YJ (2010) Thermostabilization of *Bacillus circulans* xylanase via computational design of a flexible surface cavity. J Biotechnol 146:31–39

Joo JC, Pack SP, Kim YH, Yoo YJ (2011) Thermo stailization of *Bacillus circulans* xylanase: computational optimization of unstable residues based on thermal fluctuation analysis. J Biotechnol 151:56–65

Juturu V, Wu JC (2012) Microbial xylanases: engineering, production and industrial applications. Biotechnol Adv 30:1219–1227

Kango N, Jain PC (2005) Production and application of fungal xylanases. In: Rai MK, Deshmukh SK (eds) Fungi: diversity and biotechnology. Scientific Publishers, India, pp 251–281

Kango N, Agrawal SC, Jain PC (2003) Production of xylanase by *Emericella nidulans* NK-62 on low-value lignocellulosic substrates. World J Microbiol Biotecnol 19:691–694

Kulkarni N, Shendye A, Rao M (1999) Molecular and biotechnological aspects of xylanases. FEMS Microbiol Rev 23:411–456

Kumar V, Satyanarayana T (2011) Applicability of thermo-alkalistable and cellulase-free xylanase from a novel thermo-haloalkaliphilic *Bacillus halodurans* in producing xylooligosaccharides. Biotechnol Lett 33:2279–2285

Kumar V, Satyanarayana T (2014) Production of thermo-alkali-stable xylanase by a novel polyextremophilic *Bacillus halodurans* TSEV1 in cane molasses medium and its applicability in making whole wheat bread. Bioprocess Biosyst Eng 37:1043–1053

Li J, Wang L (2011) Why substituting the asparagine at position 35 in *Bacillus circulans* xylanase with an aspartic acid remarkably improves the enzymatic catalytic activity? A quantum chemistry-based calculation study. Poly Degrad Stab 96:1009–1024

Li X, Shi H, Ding H, Zhang Y, Wang F (2014) Production, purification, and characterization of a cellulase-free thermostable endo-xylanase from *Thermoanaerobacterium thermosaccharolyticum* DSM 571. Appl Biochem Biotechnol 174:2392–2402

Maijala P, Kango N, Szijarto N, Viikari L (2012) Characterization of hemicellulases from thermophilic fungi. Antonie Van Leeuwenhoek 101(4):905–917

Mamo G, Delgado O, Martinez A, Mattiasson B, Kaul R (2006) Cloning, sequencing analysis and expression of a gene encoding an endoxylanase from *Bacillus halodurans* S7. Mol Biotechnol 33:149–159

Mi S, Jia X, Wang J, Qiao W, Peng X, Han Y (2014) Biochemical characterization of two thermostable xylanolytic enzymes encoded by a gene cluster of *Caldicellulosiruptor owensensis*. PLoS One 9(8):e105264

Mitra S, Mukhopadhyay BC, Mandal AR, Arukha AP, Chakrabarty K, Das GK, Chakrabartty PK, Biswas SR (2015) Cloning, overexpression, and characterization of a novel alkali-thermostable xylanase from *Geobacillus* sp. WBI. J Basic Microbiol 55(4):527–537

Miyazaki K, Takenouchi M, Kondo H, Noro N, Suzuki M, Suda ST (2006) Thermal stabilization of *Bacillus subtilis* family-11 xylanase by directed evolution. J Biol Chem 281:10236–10242

Nanmori T, Watanabe T, Shinke R, Kohno A, Kawamura Y (1990) Purification and properties of thermostable xylanase and b-xylosidase produced by a newly isolated *Bacillus stearothermophilus* strain. J Bacteriol 172:6669–6672

Ng CH, He J, Yang KL (2015) Purification and characterization of a GH11 xylanase from biobutanol-producing *Clostridium beijerinckii* G117. Appl Biochem Biotechnol 175:2832–2844

Palackal N, Brennan Y, Callen WN, Dupree P, Frey G et al (2004) An evolutionary route to xylanase process fitness. Protein Sci 13:494–503

Peng X, Qiao W, Mi S, Jia X, Su H, Han Y (2015) Characterization of hemicellulase and cellulase from the extremely thermophilic bacterium *Caldicellulosiruptor owensensis* and their potential application for bioconversion of lignocellulosic biomass without pretreatment. Biotechnol Biofuel 28:131–144

Romdhane IB, Achouri IM, Belghith H (2010) Improvement of highly thermostable xylanases production by *Talaromyces thermophilus* for the agro-industrials residue hydrolysis. Appl Biochem Biotechnol 162:1635–1646

Ruller R, Deliberto L, Ferreira TL, Ward RJ (2008) Thermostable variants of the recombinant xylanase A from *Bacillus subtilis* produced by directed evolution show reduced heat capacity changes. Proteins 70:1280–1293

Ruller R, Alponti J, Deliberto LA, Zanphorlin LM, Machado CB, Ward RJ (2014) Concommitant adaptation of a GH11 xylanase by directed evolution to create an alkali-tolerant/thermophilic enzyme. Protein Eng Des Sel 27(8):255–262

Ruttersmith LD, Daniel RM (1993) Thermostable a-glucosidase and β-xylosidase from *Thermotoga* sp. strain FjSS3-B.1. Biochim Biophys Acta 1156:167–172

Sadaf A, Khare SK (2014) Production of *Sporotrichum thermophile* xylanase by solid state fermentation utilizing deoiled *Jatropha curcas* seed cake and its application in xylooligosachharide synthesis. Bioresour Technol 153:126–130

Shallom D, Leon M, Bravman T, Ben-David A, Zaide G, Belakhov V, Shoham G, Schomburg D, Baasov T, Shoham Y (2005) Biochemical characterization and identification of the catalytic residues of a family 43 b-D-xylosidase from *Geobacillus stearothermophilus* T-6. Biochemistry 44:387–397

Shallon D, Shoham Y (2003) Microbial hemicellulases. Current Opinion Microbiol 6:219–228

Shao W, Wiegel J (1992) Purification and characterization of a thermostable β-xylosidase from *Thermoanaerobacter ethanolicus*. J Bacteriol 174:5848–5853

Simpson HD, Haufler UR, Daniel RM (1991) An extremely thermostable xylanase from the thermophilic eubacterium *Thermotoga*. Biochem J 277(Pt 2):413–417

Sluiter JB, Ruiz RO, Scarlata CJ, Sluiter AD, Templeton DW (2010) Compositional analysis of lignocellulosic feedstocks. 1. review and description of methods. Agric Food Chem 58:9043–9053

Song L, Tsang A, Sylvestre M (2015) Engineering a thermostable fungal GH10 xylanase, importance of N-terminal amino acids. Biotechnol Bioeng 112(6):1081–1091

Soni H, Kango N (2013) Hemicellulases in lignocellulose biotechnology: recent patents. Recent Pat Biotechnol 7:207–218

Stephens DE, Khan FI, Singh P, Bisetty K, Singh S, Permaul K (2014) Creation of thermostable and alkaline stable xylanase variants by DNA shuffling. J Biotechnol 187:139–146

Subramaniyan S, Prema P (2002) Biotechnology of microbial xylanases: enzymology, molecular biology, and application. Crit Rev Biotechnol 22:33–64

Sun MZ, Zheng HC, Meng LC, Sun JS, Song H, Bao YJ, Pei HS, Yan Z, Zhang XQ, Zhang JS, Liu YH, Lu FP (2015) Direct cloning, expression of a thermostable xylanase gene from the metagenomic DNA of cow dung compost and enzymatic production of xylooligosaccharides from corncob. Biotechnol Lett 37:1877–1886

Sung WL, Yaguchi M, Ishikawa K (1998) Modification of xylanase to improve thermophilicity, alkalophilicity and thermostability. US Patent 5,759,840

Sunna A, Bergquist PL (2003) A gene encoding a novel extremely thermostable 1,4-β-xylanase isolated directly from an environmental DNA sample. Extremophiles 7:63–70

Taibi Z, Saoudi B, Boudelaa M, Trigui H, Belghith H, Gargouri A, Ladjama A (2012) Purification and biochemical characterization of a highly thermostable xylanase from *Actinomadura* sp. strain Cpt20 isolated from poultry compost. Appl Biochem Biotechnol 166:663–679

Thomas L, Ushasree MV, Pandey A (2014) An alkali-thermostable xylanase from *Bacillus pumilus* functionally expressed in *Kluyveromyces lactis* and evaluation of its deinking efficiency. Bioresour Technol 165:309–313

Uhl AM, Daniel RM (1999) The first description of an archaeal hemicellulase: the xylanase from *Thermococcus zilligii* strain AN1. Extremophiles 3:263–267

Verma D, Satyanarayana T (2012) Cloning, expression and applicability of thermo-alkali-stable xylanase of *Geobacillus thermoleovorans* in generating xylooligosaccharides from agro-residues. Bioresour Technol 107:333–338

Verma D, Anand A, Satyanarayana T (2013) Thermostable and alkalistable endoxylanase of the extremely thermophilic bacterium *Geobacillus thermodenitrificans* TSAA1: cloning, expression, characteristics and its applicability in generating xylooligosaccharides and fermentable sugars. Appl Biochem Biotechnol 170:119–130

Viikari L, Kantelinen A, Sundquist J, Linko M (1994) Xylanases in bleaching: from an idea to the industry. FEMS Microbial Rev 13:335–350

Viikari L, Alapuranen M, Puranen T, Vehmaanperä J, Siika-aho M (2007) Thermostable enzymes in lignocellulose hydrolysis. Advan Biochem Eng Biotechnol 108:121–145

Wainø M, Ingvorsen K (2003) Production of beta-xylanase and beta-xylosidase by the extremely halophilic archaeon *Halorhabdus utahensis*. Extremophiles 7:87–93

Walia A, Mehta P, Guleria S, Shirkot CK (2015) Modification in the properties of paper by using cellulase-free xylanase produced from alkalophilic *Cellulosimicrobium cellulans* CKMX1 in biobleaching of wheat straw pulp. Can J Microbiol 61:671–681

Wang Q, Xia T (2008) Enhancement of the activity and alkaline pH stability of *Thermobifida fusca* xylanase A by directed evolution. Biotechnol Lett 30:937–944

Wang Y, Feng S, Zhan T, Huang Z, Wu Z, Liu Z (2013) Improving catalytic efficiency of endo-β-1, 4-xylanase from *Geobacillus stearothermophilus* by directed evolution and H179 saturation mutagenesis. J Biotechnol 168:341–347

Xue YM, Shao WL (2004) Expression and characterization of a thermostable β-xylosidase from hyperthermophile *Thermotoga maritima*. Biotechnol Lett 26:1511–1515

Yang SQ, Yan QJ, Jiang ZQ, Li LT, Tian HM, Wang YZ (2006) High-level of xylanase production by the thermophilic *Paecilomyces themophila* J18 on wheat straw in solid-state fermentation. BioresourTechnol 97:1794–1800

Yishan S, Zhang X, Hou Z, Zhu X, Guo X, Ling P (2011) Improvement of xylanase production by thermophilic fungus *Thermomyces lanuginosus* SDYKY-1 using response surface methodology. New Biotechnol 28(1):40–46

Yoon HS, Han NS, Kim CH (2004) Expression of *Thermotoga maritima* endo-β-1,4-xylanase gene in *E. coli* and characterization of the recombinant enzyme. Agric Chem Biotechnol 47:157–160

Zhang S, Zhang K, Chen X, Chu X, Sun F, Dong Z (2010) Five mutations in N-terminus confer thermostability on mesophilic xylanase. Biochem Biophys Res 395:200–206

Zhang S, He Y, Yu H, Dong Z (2014) Seven N-terminal residues of a thermophilic xylanase are sufficient to confer hyperthermostability on its mesophilic counterpart. PLoS One 9(1):e87632

Zhao Y, Meng K, Luo H, Huang H, Yuan T, Yang P, Yao B (2013) Molecular and biochemical characterization of a new alkaline active multidomain xylanase from alkaline wastewater sludge. World J Microbiol Biotechnol 29:327–334

Zhou J, Liu Y, Shen J, Zhang R, Tang X, Li J, Wang Y, Huang Z (2015) Kinetic and thermody-
    namic characterization of a novel low-temperature-active xylanase from *Arthrobacter*
    sp. GN16 isolated from the feces of *Grus nigricollis*. Bio engineered 6:111–114
Zouari AD, Hmida SA, Ben HH, Ben MS, Mezghani M, Bejar S (2015) Improvement of
    Trichoderma reesei xylanase II thermal stability by serine to threonine surface mutations. Int
    J Biol Macromol 72:163–170

Chapter 6
Lytic Polysaccharide Monooxygenases

Madhu Nair Muraleedharan, Ulrika Rova, and Paul Christakopoulos

What Will You Learn from This Chapter?
Lytic Polysaccharide Monooxygenases have now been evolved as one of the most promising enzymes, attracting huge research attention due to their potential use in saccharification of lignocellulosic biomass for the production of fuels and value added chemicals. In the presence of molecular oxygen, these copper depended enzymes break the recalcitrant cellulose chain by a combined oxidative and hydrolytic action, and increase the substrate accessibility for other cellulases to work. This 'boosting effect' and ability to act in synergy makes them important subject to research, towards the future goal of sustainable bioeconomy. Diversity of this enzyme group ranges from early discovered chitin and cellulose active ones, to the recently identified hemicellulose and starch active ones. In this chapter we present a brief summary about LPMOs and the findings related to them from their discovery to the recent developments.

6.1 Introduction

The inspiration from nature in many revolutionary discoveries of mankind is not trivial. The strategy of 'learn from or imitating nature' ended up in many classical discoveries from gravity to penicillin. It is however not a different story when comes to the discovery of enzymes that degrade cellulose which is the most abundant and recalcitrant biopolymer in the earth. Search for those astonishing

---

The original version of this chapter was revised. An erratum to this chapter can be found at https://doi.org/10.1007/978-3-319-54684-1_15.

M.N. Muraleedharan • U. Rova • P. Christakopoulos (✉)
Biochemical and Chemical Process Engineering, Division of Chemical Engineering,
Department of Civil, Environmental and Natural Resources Engineering, Luleå University
of Technology, 97187 Luleå, Sweden
e-mail: paul.christakopoulos@ltu.se

biocatalysts through the life cycle of cellulose, ended up in the identification and study of nature's own bio-decomposers comprise of many fungi groups and bacterial species. Significance of these enzymes has increased tremendously in this last decade, as lignocellulose has been considered as the most potential resource as an alternative to fossil fuels, for the production of energy and high value chemicals, on the context of depletion of fossil fuels and increase of global warming.

Upon an efficient refinement, the abundance of this glucose polymer has the capability to diminish the dependency on fossil fuels to a great extent (Ragauskas et al. 2006). Currently there are many ways of refining lignocellulose, such as thermochemical and biochemical methods. The former mainly involves severe treatments that disrupt the cellulose chains and its carbohydrate structure whereas the biochemical treatment with enzymes is rather mild, that it breaks only the cellulose chain, preserving the monomeric sugars intact. This method is more advantageous, as this 'sugar platform' is flexible enough to channel towards the synthesis of different chemicals or fuels, using microbes (Ragauskas et al. 2006).

However, this wide potential of lignocellulose is still limited from full level exploitation due to its incredible stability against chemical and mechanical degradation, the result of the remarkable arrangement for glucose monomers with β-1,4 linkage to form long chains up to 10,000 units. When nature engineered this amazing structure, making it resistant to most kind of bio-, chemical-, and mechanical stress, certain class of organisms got the ability and skills on breaking this structure as a part of their survival process; the decomposers in fungi and bacteria, with their remarkable consortium of cellulolytic enzymes.

Traditionally, cellulose hydrolysis was attributed to the synergetic activities of three complimentary enzymes such as (i) endoglucanases or 1,4-β-d-glucan-4-glucanohydrolases (EC 3.2.1.4), the endo acting enzymes that randomly cut at internal amorphous sites of cellulose chain to produce oligosaccharides of various lengths, (ii) Exocellulases or Cellobiohydrolases (EC3.2.1.91) which are progressive exo-acting enzymes that act from reducing or non-reducing ends of cellulose chain, generating cellobioses and (iii) β-glucosidases or β-glucoside glucohydrolases (EC 3.2.1.21), which cleaves cellobiose to glucose monomers (Lynd et al. 2002).

However, when compared to the remarkable cellulolytic properties of natural saprophytes, there had been assumptions that their inventory is much more advanced from this three enzyme categories known to the scientists, which facilitated further exploration of the genome of these organisms to identify the unknown actors. Earliest assumption dated to 1950 where Reese and his coworkers suggested the possibility of a non-hydrolytic factor that could disrupt the tightly packed crystalline region of the chain, making the substrate more accessible to the hydrolases (Reese et al. 1950). This was followed by assumptions about small oxidative molecules and their Fenton type reactions in the presence of CDH (Cellobiose dehydrogenase) or another oxidase (Baldrian and Valášková 2008).

Families CBM33 (Carbohydrate Binding Module 33) and GH61 (Glycoside Hydrolases 61) of CAZy (Henrissat 1991), have always been a puzzle to scientists due to their weak or no endoglucanase activities. CBM33s are primarily from bacterial and viruses while GH61 are produced by fungal species. Their actual mechanism remained unknown until 2005, when Vaaje-Kolstad reported a small non-catalytic protein named CBP 21 from *Serratia marcescens*, which belonged to CBM family 33 made structural changes in crystalline chitin via non-hydrolytic disruption. This enzyme increased the accessibility of substrate for other chitinases and thereby enhanced the total degradation of chitin (Vaaje-Kolstad et al. 2005a, b). Later in 2007, it was showed that certain proteins from the fungi *Thielavia terrestris* with homology to family GH61 of CAZy, enhanced the activity of cellulases on lignocellulose (Merino and Cherry 2007). In early 2010, Harris et al. reconfirmed the boosting effects, studying the influence of various divalent metal ions. They also found out that GH61 proteins lack the conserved structural properties of other canonical hydrolases and suspected that they are unlikely to be glycoside hydrolases (Harris et al. 2010).

Later in 2010, in a landmark discovery, Vaaje-Kolstad et al. first reported that CBP 21 was an oxidative enzyme which, when applied on chitin, released even number of oxidized oligosaccharides and the activity was boosted in the presence of a reducing agent. It was also proposed that, due to the structural homology and other similarities, GH61 proteins could have the same activity as CBP 21 (Vaaje-Kolstad et al. 2010). This finding opened a new phase in the research for the novel cellulases, and the following years witnessed many important discoveries in this matter. In 2011, Forsberg et al. showed that CelS2, a CBM33 protein from *Streptomyces coelicolor* could also cleave crystalline cellulose, proving the versatility and mystery of these novel oxidases (Forsberg et al. 2011). Both studies showed the dependence of these enzymes to divalent metal ions. Later the same year, Quinlan and co-workers solved the crystal structure of a GH61 from *Thermoascus auranticus* and for the first time showed that these enzymes are copper dependent (Quinlan et al. 2011), which were later confirmed by other research groups in GH61s from *Neurospora crassa* (Phillips et al. 2011) and *Phanerochaete chrysosporium* (Westereng et al. 2011). In 2012, Vaaje-Kolstad and group solved the structure of a chitin-active CBM33 and showed its copper dependency too (Vaaje-Kolstad et al. 2012). Thus these new enzyme candidates started being collectively called as Polysaccharide Monooxygenases (PMOs) or Lytic Polysaccharide Monooxygenases (LPMOs) (Phillips et al. 2011).

6.2 Lytic Polysaccharide Monooxygenases

Lytic Polysaccharide Monooxygenases are class of copper dependent enzymes, found in the organisms that survive on dead biomass, which facilitates oxidative cleavage of glycosidic bonds in cellulose (Phillips et al. 2011; Quinlan et al. 2011; Westereng et al. 2011), hemicellulose (Agger et al. 2014), chitin (Vaaje-Kolstad et al. 2010) and starch (Vu et al. 2014) in the presence of a reducing agent. They are unique from other hydrolases secreted by these organisms due to their distinctive mode of substrate disruption. They have been historically known as Glycoside

Hydrolase 61 (for fungal entries) and Carbohydrate Binding Module 33 (for bacterial and viral candidates) in the constantly updating CAZy database and, currently at the time of writing, have been classified as auxiliary activities (AA) 9, 10, 11 and 13.

6.2.1 Occurrence in Nature and CAZy Classification

LPMOs are present in abundance in the genome of most of the biomass degrading and pathogenic microorganisms such as different types of fungi and bacteria with few exceptions. AA9 family members are almost entirely from *ascomycetous* and *basidiomycetous* fungi with exceptions from *Zea mays* and few 'unidentified eukaryotes'. AA9 member enzymes are active on cellulose and hemicellulose, which explains why the majority of them are from plant cell wall degrading fungi. AA10 represents LPMOs active on chitin and cellulose, isolated from a variety of organisms, primarily bacteria but also virus, few eukaryotes and an archaebacterium. AA11 mostly comprises of fungal species with one bacterial exception and are all chitin active. AA11 members differs from previous two families in their difference in active site details and was first discovered in *Aspergillus oryzae* LPMO (Hemsworth et al. 2014) which is so far the only LPMO characterized in this family. The fourth and the latest is the AA13 family, which are fungal LPMOs active on starch (Vu et al. 2014). Among all these microbial candidates listed in CAZy, the presence of LPMO genes vary from as low as 1 (*Talaromyces stipitatus*) to a plentiful of 44 (*Chaetomium globosum*) (Žifčáková and Baldrian 2012). This makes the assumptions strong that the multiple LPMOs in one organism is not meant for degrading only one kind of substrate, but designed for the versatility of the native organism to survive on various substrate sources.

6.2.2 Structure

LPMOs are small metallo enzymes with a type two copper atom in their active site, with molecular masses ranging from 18 to 80 kDa (Dimarogona et al. 2012; Vaaje-Kolstad et al. 2005a, b). The first high resolution 3D structure of an LPMO was determined in 2005 in a CBM33 member (now AA10), CBP21 from *S. marcescens* at 1.55 Å resolution (Vaaje-Kolstad et al. 2005a, b), which showed two β sheets arranged as a sandwich fold with an alpha helical loop. It was found that the conserved aromatic residues that are supposed to play the role in chitin binding were present inside the protein, and the structure lacked typical carbohydrate binding architecture of groove, cleft or tunnel like arrangement with aromatic residues that are required for binding, as in cellulases.

Later in 2008, the first structure of an AA9 LPMO was studied in Cel61B from *Hypocrea jecorina*, and it showed the absence of conserved carboxylic residues of hydrolytic cellulases (Karkehabadi et al. 2008). In 2010, crystal structure of GH61E from *T. terrestris* was also published (Harris et al. 2010). Both enzymes had a β-sandwich structure with a metal ion coordinated by two highly conserved histidine residues located near their N-terminus. One striking feature observed was the structural homology of these fungal proteins to the previously solved CBP 21. Like the CBP21 structure, these enzymes also lacked the traditional structure of cellulases, confirming the homology between them. Some LPMOs have a cellulose binding motif (CBM) near the C-terminal with conserved sequences, but the role of this domain to the efficiency could not be identified (Harris et al. 2010).

In general LPMOs share a tertiary structure in which the core consists of immunoglobulin like β-sandwich fold connecting with alpha helical loop. The active site is unique from other hydrolases as it has a flat face with aromatic residues, which enables the protein to interact with crystalline or ordered substrate surfaces (Vaaje-Kolstad et al. 2010; Westereng et al. 2011). The N-terminal active site has the Cu (II) ion which is coordinated by two highly conserved histidine residues, called histidine braces. One significant difference from AA9 LPMO with AA10 is, in case of former, the N-terminal histidine is methylated but its biological role is yet unknown (Quinlan et al. 2011).

6.2.3 Types and Mechanism

According to the amino acid sequence and their polysaccharide bond preferences, LPMOs are classified into three types. Type 1 LPMO oxidizes C1 carbon of the glucose moiety to generate a reducing end oxidized product, lactone, which will be hydrolyzed to gluconic acid. Type 2 acts on C4 carbon and cause non-reducing end oxidization to produce 4-ketoaldoses (Fig. 6.1). The third type is less specific, as it cleaves polysaccharide chains, oxidizing both C1 and C4 carbons but not on the same glucose molecule. Existence of additional C6 oxidative LPMOs that produces 6-hexodialdoses have been claimed previously based on mass spectrometry analysis, but the same molecular weights of 4-ketoaldoses and 6-hexodialdoses questioned the credibility of mass spectrometry data (Quinlan et al. 2011).

The importance of copper in the activity of LPMO was shown by Westereng and group where they added EDTA to chelate the copper which made the enzyme inactive and activity was restored only with the addition of copper (Westereng et al. 2011). A reducing agent is required for the action of LPMO, so a supporting enzyme 'cellobiose dehydrogenase' that generates electrons is also produced by the same organism (Phillips et al. 2011). However, in the case of chitin degrading organisms, an in situ electron donor is unknown as an equivalent of cellobiose dehydrogenase is not yet discovered that is active on chitobiose. Nevertheless, various chemical reducing agents like ascorbic acid or Gallic acid can also be

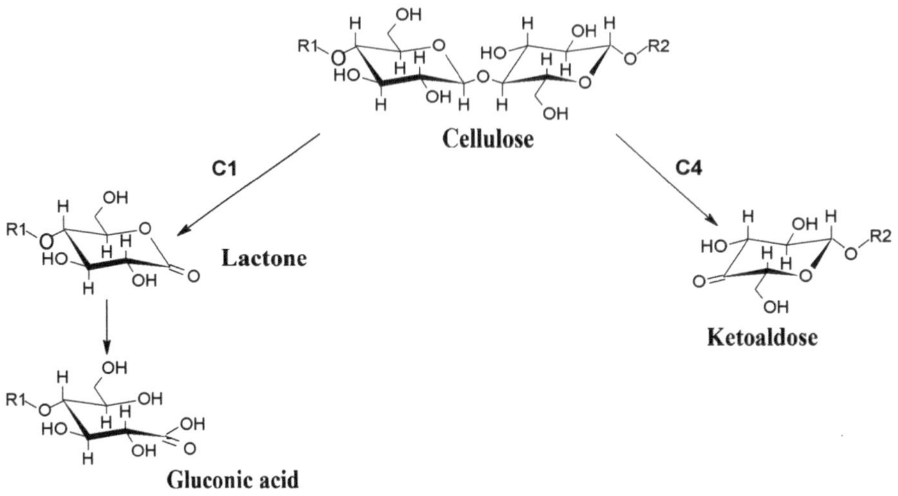

Fig. 6.1 Schematic representation of cellulose oxidation by LPMOs. C1 oxidation produces lactones when will be hydrolysed to gluconic acid. C4 oxidation produces 4-ketoaldose

used, but when it comes to natural substrates compounds like lignin serves the role of electron donor (Dimarogona et al. 2012).

The reaction of LPMO on substrate involves an oxidative step and hydrolysis step. Presence of molecular oxygen is crucial too for this activity. Isotope labeling experiments shows that one oxygen atom that is inserted at the oxidized chain end comes from water whereas the other oxygen atom is from molecular oxygen. Inhibition of LPMO activity on removal of dissolved oxygen from the reaction mixture proved this (Vaaje-Kolstad et al. 2010). Phillips and coworkers, in 2011, explained the mechanism of direct oxidation of cellulose by LPMO, disproving the popular belief till then that LPMO produces reactive oxygen species which attacks substrate randomly (Phillips et al. 2011).

6.2.4 Synergism and Boosting Effect

Long sought questions about how the saprophytes perform the initial attack on crystalline cellulose or chitin are now answered with the discovery of LPMOs, the powerful oxidative enzymes. These enzymes are highly upregulated along with other cellulases and necessary accessory enzymes like cellobiose dehydrogenases, when these organisms are grown in its natural substrates (Forsberg et al. 2014). It is now understood that, in the presence of molecular oxygen and reducing agent, these 'monooxygenases' breaks the crystalline part of the long chained substrate, making oxidative ends either at reducing or non-reducing side depends on the LPMO type (Fig. 6.2). At the same time Endoglucansaes (EGs) hydrolytically cleave the amorphous part of the substrate. Cellobiohydrolases (CBHs) enters at the nick

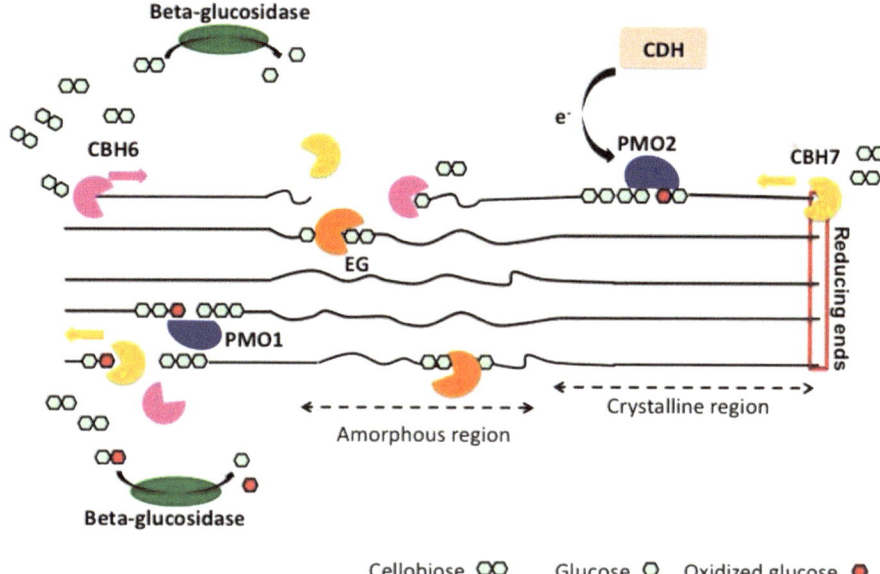

Fig. 6.2 Synergistic action of LPMO (*purple*) with other cellulases on cellulose. LPMO 1 and LPMO 2 both act on crystalline region where LPMO1 produce a reducing end oxidized chain where LPMO2 generates a non-reducing end oxidized chain. Simultaneously endoglucanases (*yellow*), cleaves cellulose at the amorphous region. Cellobiohydrolases (*red* and *green*) enters the reaction from the openings made by endoglucanases and progressively act to produce cellobioses (or oxidized cellobioses), which are sliced by betaglucosidases to individual glucoses (or oxidized glucoses). Picture taken from Dimarogona et al. (2013)

made by EGs and LPMOs and progressively generate cellobiose or oxidized cellobiose. Cellobioses and oxidative cellobioses are subsequently hydrolyzed by Betaglucosidases to produce glucose monomers (Dimarogona et al. 2012).

LPMOs themselves have very weak endoglucanase activity and should not be considered as sole cellulose degrading factor. However, what makes it attractive is its ability to work in synergy with other cellulose and to enhance total hydrolysis by increasing substrate accessibility. By effectively acting at crystalline region of substrates, they considerably reduce the load on other hydrolases (Harris et al. 2010; Quinlan et al. 2011; Vaaje-Kolstad et al. 2010). It is for this reason; LPMO is one of the inevitable ingredients in modern commercial cellulose mixtures.

6.2.5 Applications

Abundant possibilities to use glucose or sugars via various microorganisms for biosynthesis of several value added products and biofuels makes the concept of 'sugar platform' highly advantageous (Ragauskas et al. 2006). This is the main driving factor, when it comes to the use of lignocellulose as sustainable resource. Traditional hydrolytic enzymes have been used and researched for the goal of saccharification of lignocellulose and LPMO has been added as the new player in this, which increased the saccharification efficiency by considerably reducing the enzyme loading of other cellulases (Dimarogona et al. 2012). Interestingly, the newly discovered starch active LPMO (AA13), is shown to have direct application in food industries due to its action on amylose and amylopectin containing substrates (Isaksen et al. 2014). The synthesis of specific products such as glucono-δ-lactone or 4-keto-aldose from different LPMO types is yet to be considered as the main use of the enzymes.

6.3   Conclusions and Future Perspective

It will not be too fictional to state that, with the discovery of LPMO the attempts towards the dream of complete hydrolysis and exploitation of lignocellulose is one step more close. This remarkable class of enzymes has been proved to increase the saccharification efficiency of traditional hydrolases on recalcitrant substrates like lignocellulose and chitin, which opens wide opportunities towards better use of renewable biomass. Ever since its discovery in 2010, these enzymes have been constantly studied for better answers to the questions on its structure and functions. Originally thought to be active alone on crystalline substrates, these enzymes now have been shown to have activities on wide range of substrates such as soluble cello-oligosaccharides, hemicelluloses and most recently starch (Agger et al. 2014; Isaksen et al. 2014; Vu et al. 2014), which shows the depth of research needed towards this enigmatic class and the possibility of new CAZY additions.

A fair amount of focus has been given so far towards identification, characterization and structural studies of LPMOs but very little work has been done for developing a most efficient way to incorporate this enzyme in industrial applications. Though different LPMOs produce industrially important chemicals such as glucono-δ-lactone or 4-keto-aldose, the lack of an optimal downstream processing makes it difficult to exploit these features.

It needs additional research to develop an optimized usage of LPMOs with other cellulases and to exploit their full potential. Perhaps the current enzymatic hydrolysis steps need to be re-designed for the action of LPMOs. Specific needs of LPMO such as good aerobic conditions and its probability to get inhibited from the products of other cellulases (Cannella et al. 2012), are factors that need to be rectified for effective use of LPMOs.

Take Home Message

• Lytic Polysaccharide Monooxygenases are class of copper dependent enzymes, found in the organisms that survive on dead biomass, which facilitates oxidative cleavage of glycosidic bonds in cellulose. They are unique from other hydrolases secreted by these organisms due to their distinctive mode of substrate disruption.

- LPMOs are present in abundance in the genome of most of the biomass degrading and pathogenic microorganisms such as different types of fungi and bacteria with few exceptions. LPMOs are small metallo enzymes with a type two copper atom in their active site, with molecular masses ranging from 18 to 80 kDa.
- LPMOs share a tertiary structure in which the core consists of immunoglobulin like β-sandwich fold connecting with alpha helical loop. The active site is unique from other hydrolases as it has a flat face with aromatic residues, which enables the protein to interact with crystalline or ordered substrate surfaces.
- According to the amino acid sequence and their polysaccharide bond preferences, LPMOs are classified into three types. Type 1 LPMO oxidizes C1 carbon of the glucose moiety to generate a reducing end oxidized product, lactone, which will be hydrolyzed to gluconic acid. Type 2 acts on C4 carbon and cause non-reducing end oxidization to produce 4-ketoaldoses. The third type is less specific, as it cleaves polysaccharide chains, oxidizing both C1 and C4 carbons but not on the same glucose molecule. The reaction of LPMO on substrate involves an oxidative step and hydrolysis step. Presence of molecular oxygen is crucial too for this activity. starch active LPMO (AA13), is shown to have direct application in food industries due to its action on amylose and amylopectin containing substrates.

References

Agger JW, Isaksen T, Várnai A et al (2014) Discovery of LPMO activity on hemicelluloses shows the importance of oxidative processes in plant cell wall degradation. Proc Natl Acad Sci USA 111:6287–6292

Baldrian P, Valášková V (2008) Degradation of cellulose by basidiomycetous fungi. FEMS Microbiol Rev 32:501–521

Cannella D, Hsieh C-W, Felby C et al (2012) Production and effect of aldonic acids during enzymatic hydrolysis of lignocellulose at high dry matter content. Biotechnol Biofuels 5:26

Dimarogona M, Topakas E, Olsson L et al (2012) Lignin boosts the cellulase performance of a GH-61 enzyme from Sporotrichum thermophile. Bioresour Technol 110:480–487

Dimarogona M, Topakas E, Christakopoulos P (2013) Recalcitrant polysaccharide degradation by novel oxidative biocatalysts. Appl Microbiol Biotechnol 97:8455–8465

Forsberg Z, Vaaje-Kolstad G, Westereng B et al (2011) Cleavage of cellulose by a CBM33 protein. Protein Sci 20:1479–1483

Forsberg Z, Røhr ÅK, Mekasha S et al (2014) Comparative study of two chitin-active and two cellulose-active AA10-type lytic polysaccharide monooxygenases. Biochemistry 53:1647–1656

Harris PV, Welner D, Mcfarland K et al (2010) Stimulation of lignocellulosic biomass hydrolysis by proteins of glycoside hydrolase family 61: structure and function of a large, enigmatic family. Biochemistry 49:3305–3316

Hemsworth GR, Henrissat B, Davies GJ et al (2014) Discovery and characterization of a new family of lytic polysaccharide monooxygenases. Nat Chem Biol 10:122–126

Henrissat B (1991) A classification of glycosyl hydrolases based on amino acid sequence similarities. Biochem J 280:309–316

Isaksen T, Westereng B, Aachmann FL et al (2014) A C4-oxidizing lytic polysaccharide monooxygenase cleaving both cellulose and cello-oligosaccharides. J Biol Chem 289:2632–2642

Karkehabadi S, Hansson H, Kim S et al (2008) The first structure of a glycoside hydrolase family 61 member, Cel61B from Hypocrea jecorina, at 1.6 Å resolution. J Mol Biol 383:144–154

Lynd LR, Weimer PJ, Van Zyl WH et al (2002) Microbial cellulose utilization: fundamentals and biotechnology. Microbiol Mol Biol Rev 66:506–577

Merino ST, Cherry J (2007) Progress and challenges in enzyme development for biomass utilization. In: Biofuels. Springer, Berlin, pp 95–120

Phillips CM, Beeson Iv WT, Cate JH et al (2011) Cellobiose dehydrogenase and a copper-dependent polysaccharide monooxygenase potentiate cellulose degradation by Neurospora crassa. ACS Chem Biol 6:1399–1406

Quinlan RJ, Sweeney MD, Leggio LL et al (2011) Insights into the oxidative degradation of cellulose by a copper metalloenzyme that exploits biomass components. Proc Natl Acad Sci USA 108:15079–15084

Ragauskas AJ, Williams CK, Davison BH et al (2006) The path forward for biofuels and bio-materials. Science 311:484–489

Reese ET, Siu RG, Levinson HS (1950) The biological degradation of soluble cellulose derivatives and its relationship to the mechanism of cellulose hydrolysis. J Bacteriol 59:485

Vaaje-Kolstad G, Horn SJ, Van Aalten DM et al (2005a) The non-catalytic chitin-binding protein CBP21 from Serratia marcescens is essential for chitin degradation. J Biol Chem 280:28492–28497

Vaaje-Kolstad G, Houston DR, Riemen AH et al (2005b) Crystal structure and binding properties of the Serratia marcescens chitin-binding protein CBP21. J Biol Chem 280:11313–11319

Vaaje-Kolstad G, Westereng B, Horn SJ et al (2010) An oxidative enzyme boosting the enzymatic conversion of recalcitrant polysaccharides. Science 330:219–222

Vaaje-Kolstad G, Bøhle LA, Gåseidnes S et al (2012) Characterization of the chitinolytic machinery of Enterococcus faecalis V583 and high-resolution structure of its oxidative CBM33 enzyme. J Mol Biol 416:239–254

Vu VV, Beeson WT, Span EA et al (2014) A family of starch-active polysaccharide monooxygenases. Proc Natl Acad Sci USA 111:13822–13827

Westereng B, Ishida T, Vaaje-Kolstad G et al (2011) The putative endoglucanase PcGH61D from Phanerochaete chrysosporium is a metal-dependent oxidative enzyme that cleaves cellulose. PLoS One 6:e27807

Žifčáková L, Baldrian P (2012) Fungal polysaccharide monooxygenases: new players in the decomposition of cellulose. Fungal Ecol 5:481–489

Chapter 7
Recent Advances in Extremophilic α-Amylases

Margarita Kambourova

What Will You Learn from This Chapter?

Industrial requirements for α-amylase active at harsh industrial conditions have determined the interest in extremophilic producers suggesting unusual properties of their enzymes. This article discusses last ten years advances in knowledge on its synthesis from extremophiles, including thermophilic/thermoacidophilic, psychrophilic, and halophilic bacterial and archaeal producers. The examples of commercially exploited amylases from extremophiles are limited due to the special conditions for their production and low level of enzyme yield in the case of thermophiles. However, the industrial requirement for enzymes active at harsh industrial conditions as well as developments in cultivation of extremophiles renewed interest towards the biocatalytic applications of amylolytic extremozymes.

Examples of successful gene expression for circumventing the problem of insufficient expression in the natural extremophilic hosts are given. Potential for exploration of newly described enzymes is discussed.

7.1 Introduction

Starch is the most easily available source of carbon and energy on the Earth. An interest to enzymes degrading starch is determined by a broad array of industrial applications such as starch hydrolysates, glucose syrups, fructose, maltodextrin derivatives or cyclodextrins, used in food industry; in the textile, paper, brewing, and distilling industries; ethanol production; clinical, medical, and analytical

M. Kambourova (✉)
Institute of Microbiology, Bulgarian Academy of Sciences, 'Acad. G. Bonchev' Str. 26, 1113 Sofia, Bulgaria
e-mail: margikam@microbio.bas.bg

© Springer International Publishing AG 2017
R.K. Sani, R.N. Krishnaraj (eds.), *Extremophilic Enzymatic Processing of Lignocellulosic Feedstocks to Bioenergy*, DOI 10.1007/978-3-319-54684-1_7

chemistries. Amylases are widely distributed in plants, animals and microorganisms however microbial enzymes are known to fulfill industrial demands in the highest degree. As starch liquefaction and saccharification processes occurred at 80–90 °C, it is desirable that enzymes should be active at high temperature, and therefore, there has been a need for thermophilic and thermostable enzymes.

The current biotechnological interest in extremophilic enzymes is motivated by their ability to work under conditions in which mesophilic enzymes are not active. Extremophiles are organisms that can grow and thrive in extreme environments from an anthropocentric point of view, which were formerly considered too hostile to support life. They live at extreme environments such as geothermal sites (55–120 °C), polar regions and cold oceans (-20–$+20$ °C), acidic (pH < 5) and alkaline (pH > 8) springs, saline lakes (>1.0 M NaCl) and correspondingly, their habitants are thermophiles, psychrophiles, acidophiles, alkaliphiles, and halophiles. Additionally, various extremophiles can tolerate other extreme conditions including high pressure, high levels of radiation or toxic compounds, low water activity, low oxygen tension etc. and still the limits at which life can thrive have not been precisely defined. These peculiar biotopes have been successfully colonized by numerous organisms, mainly extremophilic bacteria and archaea. Extremophiles are structurally adapted at the molecular level to withstand the harsh conditions by a number of mechanisms, one of the most important being by adaptation of their enzymes called extremozymes. Such unique microorganisms and enzymes are promising sources for biotechnological and industrial discovery. Due to their extreme stability, extremozymes offer new opportunities for biocatalysis and biotechnological industry. Whereas conventional enzymes are irreversibly inactivated by heat, the enzymes from these extremophiles show not only great thermostability, but also enhanced activity in the presence of common protein denaturants such as detergents, organic solvents and proteolytic enzymes. Despite of the few number of commercially exploited extremozymes, the new developments in the cultivation of extremophiles and success in the cloning and expression of their genes in mesophilic hosts renewed the interest in the biocatalytic applications of extremozymes. Based on the unique stability of their enzymes at high temperature, extremes of pH and high pressure, combined with their salt, organic solvent and metal tolerance, they are expected to be a powerful tool in industrial biotransformation processes that run at harsh conditions. The benefits of using enzymes from thermophiles are manifold, including reduced risk of contamination, improved transfer rates, lower viscosity and higher solubility of substrates. Extremozymes from other types of extremophilic microorganisms have also received industrial appreciation of their unique properties, although in more limited areas.

Extensive scientific efforts on the screening and investigation of new α-amylases for industrial application have been done in last several decades of the last century in aim to overcome the weakness of the existing amylases, and several excellent reviews on extremophilic amylases became available one decade ago (Bertoldo and Antranikian 2002; Egorova and Antranikian 2005). Despite of the tremendous interest towards this group and description of variety of enzymes still most α-amylases do not meet all industrial criteria for the processes, especially those at high temperature. That's why the interest in identifying amylases with thermosta-bility, pH profile, pH stability, desirably Ca^{2+} independence, which are similar to those with other amylolytic enzymes will permit their synergetic action in one stage process, lowering the cost of sugar syrup production.

Current review summarizes the last decade progress in search of new α-amylases satisfying the specific requirements of various industrial sectors.

7.2   Starch and Starch Degrading Enzymes

Starch is the second most abundant carbohydrate in nature after cellulose synthe-sized as storage product of terrestrial plants. Among them industrial sources of starch are maize, tapioca, potato, and wheat. Starch is composed of two high molecular weight compounds, amylose and amylopectin, and both of these contain glucose as the sole monomer linked to another one through the glycosidic bond. Amylose (15–25%) is a linear, water-insoluble polymer of up to 6000 glucose subunits that are joined by α-1,4-bonds. Amylopectin (75–85%) is a branched, water-soluble polysaccharide composed of short linear chains (10–60 glucose units) linked by α-1,4-bond and α-1,6-linked branch points side chains with 15–45 glucose units. The complete amylopectin molecule contains on average about 2,000,000 glucose units, thereby being one of the largest molecules in nature.

The complex starch structure requires an appropriate combination of enzymes for its depolymerization to oligosaccharides and smaller sugars, such as glucose. Until the nineteenth century, starch saccharification was achieved by acid hydro-lysis; today, starch saccharification is totally enzyme-based. Enzymes that act on starch and related polymers include α-amylase, glucoamylase, α-glucosidase, cyclodextrin glucanotransferase, cyclodextrinase, and pullulanase. During the con-ventional starch processing, starch slurry is firstly gelatinized by heating followed by two enzymatic steps, liquefaction (by α-amylase) and saccharification (mainly by glucoamylase) steps. The commercial α-amylases are active at 90–95° and pH 6.0–6.5, while pH of native starch is 3.2–4.5 and commercial glucoamylases have temperature and pH optima correspondingly 60–65 °C and 4.0–4.5. Changes of these parameters are time and energy consuming and increase the product cost. Furthermore, Ca^{2+} is added to enhance the activity and/or stability of α-amylases and salts removal for the next step is required. Another practical problem in this process is that the glucoamylase is specialized in cleaving α-1,4 glycosidic bonds and slowly hydrolyzes α-1,6 glycosidic bonds present in maltodextrins. This will result in the accumulation of isomaltose. A solution to this problem is to use a pullulanase that efficiently hydrolyzes α-1,6 glycosidic bonds. A prerequisite is that the pullulanase has the same pH and temperature optima as the glucoamylase.

The amylolytic enzymes differ in their amino acid sequences, reaction mecha-nisms, catalytic activities and structural characteristics. Based on their mode of action, the enzymes are divided into two main categories: endoamylases

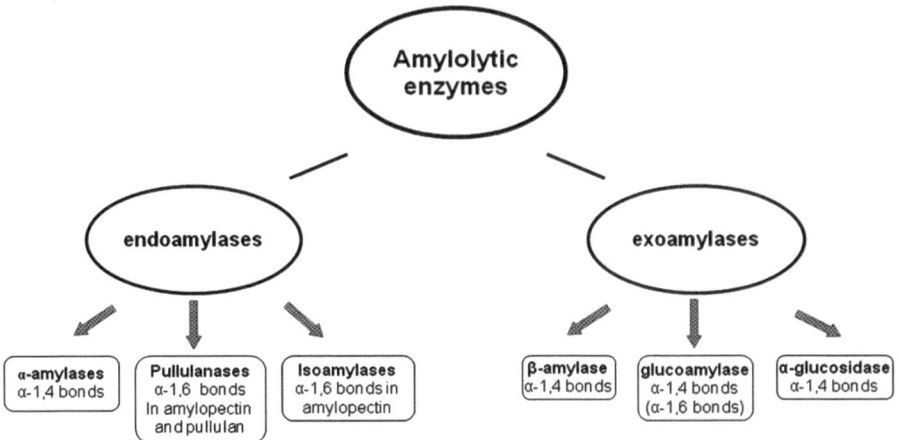

Fig. 7.1 Main types of amylolytic enzymes according to their mechanism of action

(α-amylases, pullulanases, isoamylases) and exoamylases (β-amylase, glucoamylase) (Fig. 7.1). Endo-acting enzymes hydrolyze linkages in the interior of the starch molecule producing linear and branched oligosaccharides of various chain lengths. α-Amylase (EC 3.2.1.1) is a well-known endoamylase whose end products are oligosaccharides with varying length with an α-configuration and α-limit dextrins, which constitute branched oligosaccharides. Endo-acting enzymes are also debranching enzymes that exclusively hydrolyze α-1,6 glycosidic bonds: isoamylase (EC 3.2.1.68), hydrolyzes only the α-1-6 bond in amylopectin and pullulanase (EC 3.2.1.41), hydrolyzes the α -1-6 glycosidic bond in both, pullulan and amylopectin.

Based on their end product α-amylases are categorized as saccharifying which produce free sugars but reduce the viscosity of starch pastes slowly, and liquefying ones which rapidly reduce the viscosity of starch pastes almost without producing free sugars. Alpha-amylases belong to a large clan of families 13, 57, 70 and 77 consisting of about 30 different enzyme activities and substrate specificities acting on α-glycosidic bonds. Most of the microbial α-amylases belong to the family 13 glycosyl hydrolases, which is the largest sequence-based family of glycoside hydrolases. The differences in specificities are based not only on subtle differences within the active site of the enzyme but also on the differences within the overall architecture of the enzymes. The family shares four conserved catalytic regions and adopts a common $(\alpha/\beta)_8$-barrel structure, although the overall sequence similarity is low (Lim et al. 2007). Continuous updates of CAZy show that family GH13 grew exponentially from 40 entries in 1991 to 22,857 bacterial and 282 archaeal entities in June 2015 (http://www.cazy.org/GH13_bacteria.html).

7.3   Recently Described Extremophilic Producers of α-Amylases

Amylases isolated from different sources have a wide range of properties with respect to positional specificity, thermo-stability, pH optimum etc. Polyextremophilic microorganisms are those that can survive in more than one of these extreme conditions. The vast majority of extremophilic organisms belong to the prokaryotes, and are therefore, microorganisms belonging to the Archaea and Bacteria domains.

7.3.1   Thermophiles/Thermoacidophiles

Among the extremophilic microorganisms, those living at extreme temperatures have attracted much attention largely because of the considerable biotechnological potential of their enzymes. Microorganisms that are adapted to grow optimally at high temperatures (60–120 °C) have been isolated from high temperature terrestrial and shallow hot springs, solfataric fields, submarine hydrothermal systems, at which the temperature can reach up to 400 °C. Thermophiles can be generally classified into moderate thermophiles (growth optimum 50–60 °C), extreme thermophiles (growth optimum 60–80 °C) and hyperthermophiles (growth optimum higher than 80 °C). Some of the species belonging to the biotechnologically most explored genus Bacillus and related genera referred to the group of moderate thermophiles, others are extreme thermophiles. Extreme thermophiles are also distributed among the genera *Clostridium*, *Thermoanaerobacter*, *Thermus*, *Fervidobacterium*. The group of hyperthermophiles includes the genera *Aquifex* and *Thermotoga* within the Bacteria and several genera most popular being *Pyrodictium*, *Pyrobaculum*, *Thermoproteus*, *Desulfurococcus*, *Sulfolobus*, *Methanopyrus*, *Pyrococcus*, *Thermococcus*, *Methanococcus*, and *Archaeoglobus* within the Archaea. Many thermophilic archaea inhabit environments (as solfataric fields) with high temperature and high acidity. Such genera like *Sulfolobus*, *Acidianus*, *Thermoplasma* and *Picrophilus* have growth optima between 60 °C and 90 °C and pH 0.7–5.0.

Some of the recently reported thermophilic bacterial α-amylases are listed in Table 7.1.

As can be seen from Table 7.1, the recently described α-amylases display activity at slightly acidic to neutral pH like most of already known thermophilic α-amylases. Some industrial needs in thermoalkaliphilic enzymes resulted in description of alkalitolerant α-amylases from thermophiles, like those from *T. maritima* MSB8 (Ballschmiter et al. 2006) and *Anoxybacillus* (Chai et al. 2012). The low pH optimum of amylase from *B. acidicola* suggests a possibility for simultaneous action of amylase and exo-acting enzymes. An acidic α-amylase purified from thermoacidophilic *Alicyclobacillus* sp. A4 had strong ability to digest

Table 7.1 Properties of some recently described thermostable α-amylases produced by thermophilic bacteria and archaea

Domain	Organism	Enzyme properties					References
		$T_{opt}/$ °C	pH_{opt}	Stability	Mw/ kDa	[a]Ca^{2+}	
Bacteria	*Thermotoga maritima* MSB8	90	8.5	80% after 6 h at 90 °C	241	–	Ballschmiter et al. (2006)
	Thermotoga neapolitana	75	6.5	$t_{1/2}$ 28 h at 100 °C	48	–	Park et al. (2010)
	Geobacillus stearothermophilus L07	70	6.0	$t_{1/2}$ 1 h at 90 °C	55.7	+	Park et al. (2014)
	Geobacillus sp.	60–65	5.5	$t_{1/2}$ 4.25 h at 80 °C	62	–	Jiang et al. (2015)
	Anoxybacillus strains	60	8.0	$t_{1/2}$ 48 h at 65 °C	50	+	Chai et al. (2012)
	Anoxybacillus flavithermus	55	7.0	$t_{1/2}$ 2 h at 55 °C	60	–	Agüloglu et al. (2014)
	Bacillus sp. strain CU-48	60	7.0	n.d.	45	n.d.	Khodayari et al. (2014)
	Alicyclobacillus sp. A4	75	4.2	>95% at 75 °C, for 1 h	64	–	Bai et al. (2012)
	Bacillus licheniformis JAR-26	85	5.5	55% after 30 min at 100 °C	n.d.	+	Jyoti et al. (2009)
	Bacillus acidicola	60	4.0	$t_{1/2}$ 18 min at 80 °C	66	+	Sharma and Satyanarayana (2013)
	Amphibacillus sp. NM-Ra2	54	8.0	70% after 15 min at 55 °C	50	–	Mesbah and Wiegel (2014)
	Bacillus subtilis DR8806	70	5.0	$t_{1/2}$ 125 min at 70 °C	76	+	Emtenani et al. (2015)
Archaea	*Staphylothermus marinus*	100	5.0	[b]T_m 109 °C	82.5	n.d.	Li et al. (2010)
	Thermoplasma volcanium GSS1	80	5.5	$t_{1/2}$ 24 min at 80 °C	72	n.d.	Kim et al. (2007)
	Pyrococcus furiosus	100	5.0	100% after 4 h at 100 °C		–	Wang et al. (2016)

[a]Requirement for both, activity and/or stability
[b]Melting temperature

raw starch (96.71%) with commercial glucoamylase in one step (Bai et al. 2012). Unusually high specific activity from 16.0 kU per mg of protein was reported for the enzyme from *G. stearothermophilus* L07 (Park et al. 2014).

Thermostable amylases were also synthesized by some thermotolerant microorganisms, like alkali-and thermostable amylase from a polyextremophile *Amphibacillus* sp. NM-Ra2 (Mesbah and Wiegel 2014) and thermostable and acidophilic amylase from thermophilic *B. licheniformis* JAR-26 (Jyoti et al. 2009); or even from mesophilic microorganisms like organic-tolerant α-amylase from *B. subtilis* (Emtenani et al. 2015) and a high maltose-forming and acid-stable enzyme from *B. acidicola* (Sharma and Satyanarayana 2013).

7.3.2   Psychrophiles/Psychrohalophiles

Earth is primarily a cold marine planet; 85% of it is occupied by cold ecosystems including the ocean depths, polar and alpine regions. Cold-adapted microorganisms called psychrophiles are a potential source of cold-active amylases possessing high specific activity at low temperatures.

Cold active amylases are mostly extracellular and are highly influenced by nutritional and physicochemical factors such as temperature, agitation, pH, nitrogen source, carbon source, inducers, inorganic sources and dissolved oxygen. Alpha-amylases that can tolerate both cold and salt conditions were recently described (Table 7.2). The enzyme from *Pseudoalteromonas haloplanktis* showed 80% of its initial activity at 4.5 M of NaCl at 10 °C (Srimathi et al. 2007). Aerobic, Gram-negative bacterium *Zunongwangia profunda* was isolated from deep-sea and presented a cold adapted and salt tolerant alpha-amylase (Qin et al. 2014).

7.3.3   Alkaliphiles

The alkaliphiles have been found in carbonate-rich springs and alkaline soils, where pH can be around 10.0 or even higher, although the internal pH is maintained around 8.0. pH values between 11.0 and 12.0 were established in the soda lakes representing a typical habitat where alkaliphilic microorganisms can be isolated. Alkaliphilic α-amylase was isolated from *B. licheniformis* strain AS08E (Table 7.2) growing and producing at extremely alkaline pH (12.5) (Roy and Mukherjee 2013). The purified enzyme showed optimum activity at pH 10.0 and 80 °C, and demonstrated stability toward various surfactants, organic solvents, and commercial laundry detergents. In presence of 5 mm Ca^{2+} ions, the thermostability of enzyme reached up to 80 °C, suggesting that it is metal dependent. The alkaline α-amylase gene from *B. alcalophilus* JN21 (CCTCC NO. M 2011229) was cloned and expressed in *B. subtilis* (Yang et al. 2011). The recombinant alkaline α-amylase was stable at pH from 7.0 to 11.0 and temperature below 40 °C. The optimum pH and temperature of alkaline α-amylase was 9.0 and 50 °C, respectively.

Table 7.2 Properties of some recently described α-amylases produced by psychrophilic, alkaliphilic, and halophilic bacteria

Extremophylic type	Organism	Enzyme properties				References
		$T_{opt}/$ °C	pH_{opt}	Stability at	Mw/ kDa	
Psychrophiles	*Pseudoalteromonas haloplanktis*	30	7.2	$t_{1/2}$ 1 h at 45 °C	n.d.	Srimathi et al. (2007)
	Zunongwangia profunda	35	7.0	<40% after 30 min at 35 °C	66	Qin et al. (2014)
Alkaliphiles	*B. licheniformis* strain AS08E	80	10	100% after 60 min at 60 °C	55	Roy and Mukherjee (2013)
	B. alcalophilus JN21	50	9.0	$t_{1/2}$ 1.5 h at 40 °C	56	Yang et al. (2011)
Halophiles	*Nesterenkonia* sp. strain F	45	7.5	67% after 60 min at 55 °C	100–106	Shafiei et al. (2010)
	Marinobacter sp. EMB8	45	7.0	$t_{1/2}$ 80 min at 80 °C	72	Kumar and Khare (2012)

7.3.4 Halophiles

Halophilic microorganisms require high salt (NaCl) concentrations for growth. They are found in lakes, oceans, salt pans or salt marshes. Recently, an extracellular halophilic α-amylase was isolated from *Nesterenkonia* sp. strain F (Table 7.2) growing well at various NaCl concentrations ranging from 3% to 17% with an optimum growth at 7.5% (w/v) (Shafiei et al. 2010). The enzyme showed maximal activity at pH 7.5 and 45 °C. The amylase was active in a wide range of salt concentrations (0–4 M) with its maximum activity at 0.5–1 M NaCl and was stable at the salt concentration between 1 M and 4 M however it was Ca^{2+} dependent. The amylase hydrolyzed 38% of raw wheat starch and 20% of corn starch in a period of 48 h. Halotolerant bacterium *B. vallismortis* (HQ992818) produced amylase optimally active in the temperature range of 40–70 °C and pH 8 (Suganthi et al. 2015).

7.3.5 Archaea

Enzymes derived from extremophic Archaea surpass in many cases their bacterial homologs in higher stability towards heat, pressure, detergents and solvents. A variety of starch-degrading hydrolases from extremophilic Archaea (genera *Sulfolobus*, *Thermofilum*, *Desulfurococcus*, *Pyrococcus*, *Thermococcus* and *Staphylothermus*) has been reported in the near past and their role in glycogen degradation was assumed as starch is rare in deep-sea hydrothermal vents (Egorova and Antranikian 2005). The optimal temperatures for most of the archaeal amylases range between 80 and 105 °C as well as remarkable thermostability even in the absence of substrate and calcium ions, has been observed.

A novel extremely thermophilic maltogenic amylase (SMMA) which hydrolyzes α-1,4 glycosyl linkages in cyclodextrins and in linear malto-oligosaccharide with an optimal temperature of 100 °C was isolated from *Staphylothermus marinus* (Li et al. 2010). Despite the fact that most thermophilic archaeal α-amylases do not differ from their mesophilic counterparts in their molecular weight and amino acid composition, the SMMA structure analysis provides a molecular basis for the functional properties that are unique to hyperthermophilic maltogenic amylases from archaea and that distinguish SMMA from moderate thermophilic or mesophilic bacterial enzymes.

Biochemical properties of a thermostable archaeal maltogenic amylase from *Thermoplasma volcanium* GSS1 (TpMA) with a preference for cyclodextrins over starch was reported by Kim et al. (2007). TpMA showed high thermostability and optimal activity at 75 and 80 °C for β-CD and soluble starch, respectively. TpMA shared characteristics of both bacterial MAases and alpha-amylases, and located in the middle of the evolutionary process between alpha-amylases and bacterial MAases. A new thermostable amylopullulanase from *S. marinus* with degradation activity towards pullulan and cyclodextrin at 105 °C was recently described (Li et al. 2013). The enzyme from *Thermococcus* sp., which is optimally active at 100 °C, produces a series of cyclic dextrins from starch, amylose and other polysaccharides (Callen et al. 2012).

7.4 Expression and Engineering of Amylase Genes

Difficulty of isolating and growing the vast number of extremophilic microorganisms complicates the characterization and use of their enzymes. Additionally, it has been shown that insufficient number of cells in biotechnological processes with many thermophilic and most hyperthermophilic microorganisms, resulted in too low level of starch-hydrolyzing enzymes preventing their large-scale industrial production. A modern direction in developing radically different and novel biocatalysts is through development of effective gene expression tools, and alternative host-vector systems. Genes encoding several enzymes from extremophiles have been cloned in mesophilic hosts, with the objective of overproducing the enzyme and altering its properties to suit commercial applications. Various strategies were suggested for production of recombinant enzymes at high levels—the use of different host organisms by exploration of novel host bacteria like *Thermus thermophilus, Pseudomonas antarctica, Rhodobacter capsulatus, Gluconobacter oxydans*; the improvement of the heterologous expression; the exploring of fluorimetric assays as more sensitive as compared to chromogenic ones (Liebl et al. 2014).

The new high-throughput–omics methods suggest additional advantages of thermostable protein expression such as resistance to host proteolysis and easy purification by using thermal denaturation of the mesophilic host proteins. The degree of enzyme purity obtained is generally suitable for most industrial applications. In most cases, the thermostable proteins expressed in mesophilic hosts maintain their thermostability. Among them thermophilic α-amylases genes from two *Anoxybacillus* species (Chai et al. 2012); a novel gene (amyZ) encoding a cold-active and salt-tolerant α-amylase (AmyZ) from marine bacterium *Zunongwangia profunda* (Qin et al. 2014) were cloned in *E. coli*. A heterologous expression in a host *B. subtilis* was reported for the α-amylase gene from *G. stearothermophilus* L07 (Park et al. 2014) and the alkaline α-amylase gene from *B. alcalophilus* JN21 (Yang et al. 2011). The yield of the cloned alkaline α-amylase was 79 times that of native alkaline α-amylase of *B. alcalophilus* JN21. Archaeal gene for maltogenic amylases of *T. volcanium* was successfully cloned in *E. coli* (Kim et al. 2007).

Genomic analysis of a hyperthermophilic archaeon, *Thermococcus onnurineus* NA1 (Lim et al. 2007), revealed the presence of an open reading frame consisting of 1377 bp similar to α-amylases from *Thermococcales*. Alpha-amylase was cloned and the recombinant enzyme was characterized. The optimum activity of the enzyme occurred at 80 °C and pH 5.5. Surprisingly, the enzyme was not highly thermostable, with half-life ($t_{1/2}$) values of 10 min at 90 °C, despite the high similarity to α-amylases from *Pyrococcus*.

Wang et al. (2016) reported for pattern-mimicking strategy to express *Pyrococcus furiosus* α-amylase in *B. amyloliquefaciens* using the expression and secretion elements of the host organism, including the codon usage bias and mRNA structure of gene, promoter, and signal peptide. Transcription factor modification and the use of exogenous sigma factors in expression host strains, which has also been called "transcriptional engineering", can be useful for the improvement of the productivity of valuable compounds in recombinant bacteria.

Protein engineering was successfully applied for changes the properties of amylases in terms of increasing their activity and/or thermostability, altering their pH activity profile, Ca^{2+} requirements, product specificity. Higher enzyme activity and more alkaliphilic pH optima were observed for a mutant α-amylase from an alkaline, halophilic and thermophilic *Bacillus* sp. strain CU-48 (Khodayari et al. 2014).

Extremophilic metagenomics expose biotechnological potential to unlock the vast amount of genetic information from the uncultured microbial majority and such a way to accelerate the exploration of microbial diversity. An approach to find new and potentially interesting enzymes is to use the nucleotide or amino acid sequence of the conserved domains in designing degenerated PCR primers. These primers can then be used to screen microbial genomes or metagenomes for the presence of genes putatively encoding the enzyme of interest. Sharma et al. (2010) discovered a novel amylase from a soil metagenome that retained 90% of activity even at low temperature. Sequence-independent functional screening also bears the potential to discover novel enzymes with low level of homology to already known enzyme sequences (Liebl et al. 2014). A highly thermoactive and salt-tolerant α-amylase was isolated from a pilot-plant biogas reactor by screening of metagenome library for starch-degrading enzymes (Jabbour et al. 2013).

7.5 Industrial Application

Today a large number of microbial amylases are available commercially and they have replaced chemical hydrolysis of starch in starch processing industry. Production of high fructose corn syrup from starch is a $1 billion business sharing about 30% of the industrial enzyme market of the world (Adrio and Demain 2014). Commercial α-amylases are mainly derived from the genus *Bacillus* (*B. licheniformis*, *B. stearothermophilus*, and in the past from *B. amyloliquefaciens*) due to their remarkable thermostability and high efficient expression systems available. Still starch industry needs in α-amylases whose properties are entirely compatible with properties of other amylolytic enzymes that could permits development of one-step industrial process of starch hydrolysis.

7.5.1 Food Industry

The most widespread applications of α-amylases are in the starch liquefaction process that converts starch into fructose and glucose syrups. Among the demand enzyme properties the most important are thermoactivity/thermostability, activity at low pH and Ca^{2+} independence. Enzymes from thermoacidophiles suggest such promising properties, and therefore, they can be used in starch as well as in textile and fruit juice industries. These enzymes can be added to the dough of bread to degrade the starch in the flour into smaller dextrins, which results in improvements in the volume and texture of the product. Currently, a thermostable maltogenic amylase of *G. stearothermophilus* is used commercially in the bakery industry (Souza and Magalhães 2010). Pressure resistant enzymes could be of use in food industry where high pressure is applied for processing and sterilization of food products.

Maltose received after starch hydrolysis is commonly used as sweetener and also as intravenous sugar supplement. It has a great value in food industry since it is non-hygroscopic and does not easily crystallize. Maltooligomer mix is a novel commercial product that prevents crystallization of sucrose in foods and keeps a certain level of hardness of the texture during storage. The sweetness of maltotetraose syrup (G4 syrup) is as low as 20% of sucrose. Therefore in foods, G4 syrup can be successfully used in place of sucrose which reduces the sweetness without altering their inherent taste and flavor and improves the food texture because of its higher viscosity than sucrose. It further lowers down the freezing point of water than sucrose or high fructose syrup, so can be used to control the freezing points of frozen foods. These enzymes are also used for the preparation of viscous, stable starch solutions used the clarification of haze formed in beer or fruit juices, or for the pretreatment of animal feed to improve the digestibility.

7.5.2 Detergents

Amylases are the second type of enzymes used in the formulation of enzymatic detergents, and 90% of all liquid detergents contain these enzymes. These enzymes are used for degrading the residues of starchy foods. The needed characteristics are: activity at lower temperatures since washing of clothes at low temperatures protects the colors of fabrics and reduces energy consumption; alkaline pH, maintaining the necessary stability under detergent conditions; and the oxidative stability, one of the most important criteria for their use in detergents where the washing environment is very oxidizing.

Amylase from *Bacillus vallismortis* (Suganthi et al. 2015) was proved to possess a significant compatibility with the commercial laundry detergents and the results of washing performance test confirmed its effectiveness. Available data on the optimized culture conditions enables an easily adaptable setup of large scale production of the enzyme for use in detergent formulations.

7.5.3 Direct Fermentation of Starch to Ethanol

Conversion of biomass resources (especially starchy materials) to ethanol in large-scale processes is very perspective because it can be used as biofuel and starting material for various chemicals. For the ethanol production, starch is the most used substrate due to its low price and easily available raw material in most regions of the world. Other advantages of renewable biomass exploration for 1st generation biofuel production are reduced greenhouse gas emissions and alleviation of climate change. However, as it is derived from the edible fraction of food plants (corn, rapeseed, sugar beet) this could result in increasing the food price and food security. According to EASAC (2012) biomass cultivation for first generation biofuels put food, agriculture and natural ecosystems at risk. Additionally, the need in large quantity of α-amylase increases the cost of biofuel. Fuelzyme®—Verenium Corporation (San Diego, CA, USA) is based on α-amylase originated from *Thermococcus* sp. isolated from a deep-sea hydrothermal vent. Fuelzyme® is applied to mash liquefaction during ethanol production, releasing dextrins and oligosaccharides with better solubility and with low molecular weight. It operates in a pH range of 4.0–6.5 and temperatures above 110 °C (Callen et al. 2012).

7.6   Conclusion

Enzymatic degradation of starch is considered efficient, cost effective and an environmentally friendly process realized by amylolytic enzymes. As discussed here, α-amylases are versatile enzymes that are used widely in food industry, detergents, ethanol production etc., with a prospective tendency for applications in medicinal, clinical and analytical chemistry.

Considerable progress in last decade has been made in search of extremophilic producers of novel amylases due to the industrial requirements for operation activity at harsh industrial conditions, although the true diversity of extremophiles has not yet been fully explored. However, still an effective combination of amylolytic enzymes acting in similar reaction conditions is not available and much hope is assigned to the molecular biology methods to increase yield and improve the properties of enzymes.

Take Home Message

• Starch is composed of amylose and amylopectin, linked to another one through the glycosidic bond. Enzymes that act on starch and related polymers include α-amylase, glucoamylase, α-glucosidase, cyclodextrin glucanotransferase, cyclodextrinase, and pullulanase. Ca^{2+} is added to enhance the activity and/or stability of α-amylases. The amylolytic enzymes differ in their amino acid sequences, reaction mechanisms, catalytic activities and structural characteristics. Based on their mode of action, the enzymes are divided into two main categories: endoamylases (α-amylases, pullulanases, isoamylases) and exoamylases (β-amylase, glucoamylase).

• Thermostable amylases were also synthesized by some thermotolerant microorganisms, like alkali-and thermostable amylase from a polyextremophile *Amphibacillus* sp. NM-Ra2 and thermostable and acidophilic amylase from thermophilic *B. licheniformis* JAR-26. Alkaliphilic α-amylase was isolated from *B. licheniformis* strain AS08E growing and producing at extremely alkaline pH (12.5). Microbial amylases are used in starch processing industry, food industry, detergent industry. They are used for production of high fructose corn syrup and ethanol. A thermostable maltogenic amylase of *G. stearothermophilus* is used commercially in the bakery industry. Enzymes from thermoacidophiles suggest such promising properties, and therefore, they can be used in starch as well as in textile and fruit juice industries. These enzymes can be added to the dough of bread to degrade the starch in the flour into smaller dextrins, which results in improvements in the volume and texture of the product.

References

Adrio JL, Demain AL (2014) Microbial enzymes: tools for biotechnological processes. Biomolecules 4:117–139

Agüloglu S, Enez B, Özdemir S, Matpan-Bekler F (2014) Purification and characterization of thermostable α-amylase from thermophilic Anoxybacillus flavithermus. Carbohydr Polym 102:144–150

Bai Y, Huang H, Meng K et al (2012) Identification of an acidic α-amylase from Alicyclobacillus sp. A4 and assessment of its application in the starch industry. Food Chem 131:1473–1478

Ballschmiter M, Fütterer O, Liebl W (2006) Identification and characterization of a novel intracellular alkaline α-amylase from the hyperthermophilic bacterium Thermotoga maritima MSB8. Appl Eviron Microbiol 72:2206–2211

Bertoldo C, Antranikian G (2002) Starch-hydrolyzing enzymes from thermophilic archaea and bacteria. Curr Opin Chem Biol 6:151–160

Callen W, Richardson T, Frey G et al (2012) Amylases and methods for use in starch processing. US Patent 8,338,131, 25 Dec 2012

Chai YY, Rahman RNZRA, Illias RM et al (2012) Cloning and characterization of two new thermostable and alkalitolerant α-amylases from the Anoxybacillus species that produce high levels of maltose. J Ind Microbiol Biotechnol 39:731–741

EASAC (2012) The current status of biofuels in the European Union, their environmental impacts and future prospects. EASAC Policy Report 19. http://www.easac.eu

Egorova K, Antranikian G (2005) Industrial relevance of thermophilic Archaea. Curr Opin Microbiol 8:649–655

Emtenani S, Asoodeh A, Emtenani S (2015) Gene cloning and characterization of a thermostable organic-tolerant α-amylase from Bacillus subtilis DR8806. Int J Biol Macromol 72:290–298

Jabbour D, Sorger A, Sahm K et al (2013) A highly thermoactive and salt-tolerant α-amylase isolated from a pilot-plant biogas reactor. Appl Microbiol Biotechnol 97:2971–2978

Jiang T, Cai M, Huang M et al (2015) Characterization of a thermostable raw-starch hydrolyzing α-amylase from deep-sea thermophile Geobacillus sp. Protein Expr Purif 114:15–22

Jyoti J, Lal N, Lal R et al (2009) Production of thermostable and acidophilic amylase from thermophilic Bacillus licheniformis JAR-26. J Appl Biol Sci 3:7–12

Khodayari F, Cebeci Z, Ozcan BD et al (2014) Improvement of ezyme ativity of a novel native alkaline and thermophile Bacillus sp. CU-48, producing α-amylase and CMCase by mutagenesis. Int J Chem Nat Sci 2:97–103

Kim JW, Kim YH, Lee HS et al (2007) Molecular cloning and biochemical characterization of the first archaeal maltogenic amylase from the hyperthermophilic archaeon Thermoplasma volcanium GSS1. Biochim Biophys Acta 1774:661–669

Kumar S, Khare SK (2012) Purification and characterization of maltooligosaccharide-forming α-amylase from moderately halophilic Marinobacter sp. EMB8. Bioresour Technol 116:247–251

Li D, Park JT, Li X et al (2010) Overexpression and characterization of an extremely thermostable maltogenic amylase, with an optimal temperature of 100 degrees C, from the hyperthermophilic archaeon Staphylothermus marinus. N Biotechnol 27:300–307

Li X, Li D, Park KH (2013) An extremely thermostable amylopullulanase from Staphylothermus marinus displays both pullulan- and cyclodextrin-degrading activities. Appl Microbiol Biotechnol 97:5359–5369

Liebl W, Angelov A, Juergensen J et al (2014) Alternative hosts for functional (meta) genome analysis. Appl Microbiol Biotechnol 98:8099–8109

Lim JK, Lee HS, Kim YJ et al (2007) Critical factors to high thermostability of an alpha-amylase from hyperthermophilic archaeon Thermococcus onnurineus NA1. J Microbiol Biotechnol 17:1242–1248

Mesbah NM, Wiegel J (2014) Halophilic alkali-and thermostable amylase from a novel polyextremophilic Amphibacillus sp. NM-Ra2. Int J Biol Macromol 70:222–229

Park JT, Suwanto A, Tan I et al (2014) Molecular cloning and characterization of a thermostable α-amylase exhibiting an unusually high activity. Food Sci Biotechnol 23:125–132

Park KM, Jun SY, Choi KH et al (2010) Characterization of an exo-acting intracellular α-amylase from the hyperthermophilic bacterium Thermotoga neapolitana. Appl Microbiol Biotechnol 86:555–566

Qin Y, Huang Z, Liu Z (2014) A novel cold-active and salt-tolerant alpha-amylase from marine bacterium Zunongwangia profunda: Molecular cloning, heterologous expression and biochemical characterization. Extremophiles 18:271–281

Roy JK, Mukherjee AK (2013) Applications of a high maltose forming, thermo-stable α-amylase from an extremely alkalophilic Bacillus licheniformis strain AS08E in food and laundry detergent industries. Biochem Eng J 77:220–230

Shafiei M, Ziaee AA, Amoozegar MA (2010) Purification and biochemical characterization of a novel SDS and surfactant stable, raw starch digesting, and halophilic α-amylase from a moderately halophilic bacterium, Nesterenkonia sp. strain F. Process Biochem 45:694–699

Sharma A, Satyanarayana T (2013) Characteristics of a high maltose-forming, acid-stable, and Ca^{2+}-independent α-amylase of the acidophilic Bacillus acidicola. Appl Biochem Biotechnol 171:2053–2064

Sharma S, Khan FG, Qazi GN (2010) Molecular cloning and characterization of amylase from soil metagenomic library derived from Northwestern Himalayas. Appl Microbiol Biotechnol 86:1821–1828

Souza PM, Oliveira Magalhães P (2010) Application of microbial α-amylase in industry—a review. Braz J Microbiol 41:850–861

Srimathi S, Jayaraman G, Feller G et al (2007) Intrinsic halotolerance of the psychrophilic alpha-amylase from Pseudoalteromonas haloplanktis. Extremophiles 11:505–515

Suganthi C, Mageswari A, Karthikeyan S et al (2015) Insight on biochemical characteristics of thermotolerant amylase isolated from extremophile bacteria Bacillus vallismortis TD6 (HQ992818). Microbiology 84:210–218

Wang P, Wang P, Tian J et al (2016) A new strategy to express the extracellular α-amylase from Pyrococcus furiosus in Bacillus amyloliquefaciens. Sci Rep 6. doi:10.1038/srep22229

Yang H, Liu L, Li J et al (2011) Heterologous expression, biochemical characterization, and overproduction of alkaline αa-amylase from Bacillus alcalophilus in Bacillus subtilis. Microb Cell Fact 10:77

Chapter 8
Extremophilic Ligninolytic Enzymes

Ram Chandra, Vineet Kumar, and Sheelu Yadav

What Will You Learn from This Chapter?

Extremophilic microorganisms have developed a variety of physiological strategies that help them to survive on different ecological niche such as extreme temperature, pH, salt concentration and pressure. It has been demonstrated that these microorganisms produce extracellular isoenzyme capable to degrade the ligninocellulosic waste and other related compounds for their growth and survival. These are known as extremophilic ligninolytic enzyme. The extremophilic enzymes are considered superior than normal enzyme because they allow the performance of industrial processes even under adverse condition in which conventional proteins are completely denatured. The common extremophilic ligninolytic enzymes are manganese peroxidase (MnP), lignin peroxidase (LiP) and laccase. These enzymes predominantly have been reported in fungus but their occurrence and role for decolourisation and detoxification of various environmental pollutants also have been reported in bacteria and actinomycetes. Biochemically, MnP and LiP are glycosylated haem protein with molecular weight (MW) ranging from 38 to 62.5 kDa (MnP: 38–62.5 kDa; LiP: 38–46 kDa) while laccases are monomeric, dimeric and trimeric glycoprotein with MW range from 50 to 97 kDa. The optimum activity at pH range for MnP and LiP in fungus is 3.0–5.0 while in bacteria pH range for these enzymes ranges from pH 4.0 to 9.0. The optimum activity for laccase in

R. Chandra (✉)
Environmental Microbiology Division, Indian Institute of Toxicology Research, Post Box No. 80, Mahatma Gandhi Marg, Lucknow 226001, UP, India

Department of Environmental Microbiology, Babasaheb Bhimrao Ambedkar Central University, Vidya Vihar, Raebareli Road, Lucknow 226025, UP, India
e-mail: prof.chandrabbau@gmail.com; rc_microitrc@yahoo.co.in

V. Kumar • S. Yadav
Department of Environmental Microbiology, Babasaheb Bhimrao Ambedkar Central University, Vidya Vihar, Raebareli Road, Lucknow 226025, UP, India

© Springer International Publishing AG 2017
R.K. Sani, R.N. Krishnaraj (eds.), *Extremophilic Enzymatic Processing of Lignocellulosic Feedstocks to Bioenergy*, DOI 10.1007/978-3-319-54684-1_8

fungus and bacteria are noted pH 4.0–10.0. The extremophilic activity of these enzymes is regulated due to presence of various salt bridge between amino acids to maintain their stability for catalytic function. Furthermore, the oxidation mechanism of these ligninolytic enzymes have revealed that MnP and lacasse require specific mediator (e.g. GST, tween 80, ABTS, HBT) while LiP does not require any mediator for oxidation of phenolics and non-phenolics compounds. The major biotechnological applications of these enzymes are decolourisation and detoxification of various lignin and ligninolytic waste. It has also scope for pulp biobleaching and ethanol production.

8.1 Introduction

Generally life exist on the earth in the moderate environment means with neutral pH, temperature between 22 and 40 °C and normal atmospheric pressure (1.0 atm) with humidity, nutrient and salt condition. However, the deviation beyond moderate environmental condition may create the extreme environment with very low or high pH, temperature, pressure, salinity, humidity depending upon the geographical situation also. Thus, for the survival life has been evolved and adapted specialised mechanism for growth in harsh climate such organisms are known as extremophiles. There is variable environmental conditions world over for the microbial growth which different for physiology and biochemical properties for adaptation.

Low temperature (cold) environment is found in fresh and marine waters, polar and high alpine soils and water ecosystem. Oceans represent 71% of earth's surface and 90% by volume, which are at 5 °C or colder at altitudes >3000 m. As the altitude increases, the temperature decreases at the rate of 6.4 °C per km progressively and even temperatures below –40 °C have been recorded. Low temperatures are characteristic of mountains where snow or ice remains year-round. The temperature is cold, part of the year, on mountains where snow or ice melts. Thus, cold environment dominates the biosphere. According to Morita (2000), cold environments can be divided into two categories (1) psychrophilic (permanently cold) and (2) psychrotrophic (seasonally cold or where temperature fluxes into mesophilic range) environments. Although habitats with elevated temperatures are not as widespread as temperate or cold habitats, a variety of high temperature, natural and man-made habitats exist in environment. These include volcanic and geothermal areas with temperatures often greater than boiling, sun-heated litter and soil or sediments reaching 70 °C, and biological self-heated environments such as compost, hay, saw dust and coal refuse piles. In thermal springs, the temperature is above 60 °C, and it is kept constant by continual volcanic activity. Besides temperature, other environmental parameters such as pH, available energy sources, ionic strength and nutrients influenced the diversity of thermophilic microbial populations. The best known and well-studied geothermal areas are in North America (Yellowstone National Park), Iceland, New Zealand, Japan, Italy and the

Soviet Union. Hot water springs are situated throughout the length and breadth of India, at places with boiling water (e.g. Manikaran, Himachal Pradesh, India). Geothermal areas are characterized by high or low pH.

Fresh water alkaline hot springs and geysers with neutral/alkaline pH are located outside the volcanically active zones. Solfatara fields, with sulphur acidic soils, acidic hot springs and boiling mud pots, characterized other types of geothermal areas. These fields are located within active volcanic zones that are termed 'high temperature fields'. Because of elevated temperatures, little liquid water comes out to the surface, and the hot springs are often associated with steam holes called fumaroles. With increase in temperature, two major problems are encountered, keeping water in liquid state and managing the decrease in solubility of oxygen. Therefore, the microbes growing above boiling point of water have been isolated from hydrothermal vents, where hydrostatic pressure keeps the water in liquid state; however, majority of them are anaerobes. Thermal vent sites have recently been found in Indian Ocean. The primary areas with pH lower than 3.0 are those where relatively large amounts of sulphur or pyrite are exposed to oxygen. Both sulphur and pyrite are oxidized abiotically through an exothermic reaction where the former is oxidized to sulphuric acid, and the ferrous iron in the latter to ferric form. Both these processes occur abiotically, but are increased 10^6 times through the activity of acidophiles. Most of the acidic pyrite areas have been created by mining and are commonly formed around coal, lignite or sulphur mines. All such areas have very high sulphide concentrations and pH values as low as one (pH 1.0). These are very low in organic matter, and are quite toxic due to high concentrations of heavy metals. In all acidic niches, the acidity is mostly due to sulphuric acid. Due to spontaneous combustion, the refuse piles are self-heating and provide the high-temperature environment required to sustain thermophiles. The illuminated regions, such as mining outflows and tailings dams, support phototrophic algae.

In alkaline environments such as soils, increase in pH is due to microbial ammonification and sulphate reduction, and by water derived from leached silicate minerals. The pH of these environments fluctuates due to their limited buffering capacity and therefore, alkalitolerant microbes are more abundant in these habitats than alkaliphiles. The best studied and most stable alkaline environments are soda lakes and soda deserts (e.g. East African Rift valley, Indian Sambhar Lake). These are characterized by the presence of large amounts of Na_2CO_3 but are significantly depleted in Mg^{2+} and Ca^{2+} due to their precipitation as carbonates. The salinity ranges from 5% (w/v) to saturation (33%). Industrial processes including cement manufacture, mining, disposal of blast furnace slag, electroplating, food processing and paper and pulp manufacture produce man-made unstable alkaline environments. Environments with high hydrostatic pressures are typically found in deep sea and deep oil or sulphur wells. Almost all barophiles isolated to date have been recovered from the deep sea below a depth of approximately 2000 m. High-pressure condition is also met within soils where factors such as high temperature, high salinity and nutrient limitation may exert further stress on living species. Bacteria adapted to such an extreme environment are able to grow around or beyond 100 °C, and 200–400 bar of hydrostatic pressure.

The majority of extremophiles that have been identified to date belong to the domain of Archaea. However, many extremophiles from the eubacterial and eukaryotic domain have also been recently identified and characterised. Recent study suggested that the diversity of organisms in extreme environments is far greater than was initially suspected. Extremophiles are an important source of stable and valuable enzymes present in specific environment. Their enzymes, sometimes called "extremozymes". Natural extremozymes have been isolated from thermophiles, halophiles, psychrophiles, acidophiles and alkaliphiles. But the structural feature of the enzyme from acidophiles and alkaliphiles is not much known. Accordingly, biological system and enzyme can even function at temperature between −5 and 130 °C, pH 0–12, salt concentration of 3–35% and pressure up to 1000 bar. In many cases, microbial biocatalyst, especially of extremophiles, are superior to the traditional catalyst, because they allow the performance of industrial processes even under extreme conditions where conventional proteins are completely denatured. Extremophilic enzymes are able to compete for hydration via alterations especially to their surface through greater surface charges and increased molecular motion. The unique structural characteristics of the archaeal polar lipids, that is, the sn-glycerol-1-phosphate (G-1-P) backbone, ether linkages, and isoprenoid hydrocarbon chains, are in striking contrast to the bacterial characteristics of the sn-glycerol- 3-phosphate (G-3-P) backbone, ester linkages, and fatty acid chains. The chemical properties and physiological roles of archaeal lipids are often discuss in terms of the presence of the chemically stable ether bonds in thermophilic archaea. However, based on the archaeal lipids analyzed thus far, as shown by lipid component parts analysis, the mesophilic archaea possess essentially the same core lipid composition as that of the thermophilic archaea. The ether bonds therefore do not seem to be directly related to thermophily. The chemical stability of lipids and the heat tolerance of thermophilic organisms exhibit because of the ether bonds of archaeal lipids are for the most part not broken down under conditions in which ester linkages are completely methanolyzed (5% HCl/MeOH, 100 °C for 3 h), it is generally believed that the archaeal ether lipids are thermotolerant or heat resistant. This implies that thermophilic organisms are able to grow at high temperature due to the chemical stability of their membrane lipids. As a matter of fact, all the thermophilic archaea possess ether lipids, but not all of the organisms possessing the so-called "thermophilic". The properties have enabled some extremophilic enzymes to function in the presence of nonaqueous organic solvents, with this potential properties we can design useful catalysts. Especially lignocellulolytic, amylolytic, and other biomass processing extremozymes with unique properties are widely distributed in thermophilic prokaryotes and are of high potential for versatile industrial processes.

In environment, important extremophilic enzymes have been reported as ligninolytic enzyme which constitutes manganese peroxidase (MnP), lignin peroxidase (LiP) and laccases. These enzymes act on broad range of their substrate in normal to diverse conditions. The demand of these enzymes have increased in recent year due to their commercial prospect and industrial applications. Such enzymes have also proven their utility in the pollution abatement, especially in the treatment of industrial waste/wastewater containing hazardous compound like phenols, chlorolignin, synthetic dyes, and polyaromatic hydrocarbons (PAHs) as well as recalcitrant organic compounds structurally similar to lignin. Microorganisms with systems of thermostable enzymes decrease the possibility of microbial contamination in large scale industrial reactions of prolonged durations. The mechanisms for many thermotolerant enzymes have been reported due to their structural properties i.e. presence of Ca^{2+}, saturated fatty acid, α-helical structure etc. Therefore, the present chapter has been focused on important group of extremophilic ligninolytic enzymes which have tremendous commercial value for industrial application but its distribution, mode of action still has to be understood much more large scale for biotechnological application.

8.2 Manganese Peroxidase

MnP (EC 1.11.1.13) is a ligninolytic extracellular oxidoreductase enzyme belong to class II fungal haem containing peroxidases produced by almost all wood colonizing white root and several litter decomposing basidiomycetes during secondary metabolism in response to nitrogen or carbon starvation. It has also been produce by some native bacterial strains (Bharagava et al. 2009; Yadav et al. 2011). MnP was first discovered in the mid-1980s in white-rot fungus *Phanerochaete chrysosporium* by two international research teams (M. Gold's and R. Crawford's groups) and characterised as another key oxidative enzyme for lignin degradation (Paszczyński et al. 1985). After nearly simultaneous discovery, it has been reported in a large number of ligninolytic fungi including *Phlebia radiata*, *Pleurotus ostreatus* *Bjerkandera adusta*, *Dichomitus squalens*, *Trametes versicolor*, *Lentinus edodes* and so on. The presence of MnP has been also reported in *Aspergillus terrus* strain and *Penicillium oxalicum* isolates-1. MnP is classified in carbohydrate-active enzymes (CAZy) database in auxiliary activities 113 families. It oxidised Mn^{2+} to Mn^{3+} chelate with organic acid and then oxidised various phenolic as well as non-phenolic compounds including model compounds viz. veratryl alcohol (VA: 3,4-dimethoxybenzyl) and benzyl alcohol. During the degradation process MnP system generate highly-reactive and non-specific free radicals that cleave carbon–carbon and ether inter-unit bonds of various phenolics and non-phenolics compounds. Therefore, a wide range of substrate oxidizing capability renders it an interesting enzyme for biotechnological applications in several industries. Potential applications for MnP include biomechanical pulping, pulp biobleaching, dye decolourisation, bioremediation of some recalcitrant organopollutants (i.e., high MW chlorolignin, chlorophenols, polycyclic aromatic hydrocarbons, nitroaromatic compounds) and production of high value chemical from residual lignin from biorefineries and pulp and paper side stream. MnP has a high potential for penetrating deep into the soil fines and in nature it catalysed plant lignin depolymerisation. MnP has also been reported for the degradation/detoxification of triclosan, aflatoxin, nylon, β-carotene, as well as lignite originated from humic acid.

8.2.1   Molecular Structure

MnP is a glycosylated haem protein with MW ranging from 38 to 62.5 kDa, and averaging at 45 kDa. It occurs as a series of isoforms (isozymes) encoded by a family of closely related genes and the sequence of cDNA and genomic clones of three different *mnp* genes (*mnp1*, *mnp2*, and *mnp3*) from *P. crysosporium* have been determined. In *P. crysosporium* putative metal have been identified upstream of *mnp1* and *mnp2* that are also involved in transcription regulation of these genes. The expression of MnP in nitrogen limited culture of *P. crysosporium* is regulated at the level of gene transcription by hydrogen peroxide (H_2O_2) and various chemicals including ethanol, sodium arsenite, and 2,4-dichlorophenol as well as by Mn^{2+} and heat shock. Recently, 11 different isoform of MnP and their genes have been characterised in *Ceriporiopsis subvermispora*. The amounts produced and strengths of these enzymes are different for each type of white rot fungi resulting in different oxidative activities. However, the regulation of different MnPs isoform can be largely dependent on the inducing compound (e.g. Mn^{2+}, VA, tween and sodium malonate) and nutrients. Each isoforms of MnP contain 1 mol of iron per mol of protein and differ mostly in their isoelectric points (pI) which are usually rather acidic (pH 3.0–4.0), through less acidic and neutral isoforms have found in certain fungi. MnP differs from other peroxidases in the structure of its substrate binding site. Recent evolutionary studies showed that MnP evolved from fungal generic peroxidases (similar to plant peroxidases) by developing a Mn-binding site. MnP then gave rise to versatyl peroxidases (VPs) by incorporating an exposed catalytic tryptophan and finally to LiPs by loss of the VP Mn-oxidation site, with the presence of the exposed tryptophan being characteristic of both LiP and VP crystal structures, and the latter also conserving the above Mn-binding site.

The crystal structure of MnP (PDB Id: 3m5q) from *P. chrysosporium* has been crystallized and subsequently analysed at different refined resolution and this was the third peroxidase (after cytochrome c peroxidase and LiP) which crystal structure has been solved (Sundaramoorthy et al. 2010). The structure of MnP consisting of two domains with heme sandwiched in between. Electronic absorption, electroparamagnetic resonance (EPR) and resonance Raman spectral evidence suggested that the heme iron in native MnP is in high spin, pentacoordinate, ferric state with histidine coordinated as the fifth ligand. The protein molecule of MnP contains ten major helices and one minor helix. MnP having five rather than four disulfide bond present in LiP and VP. The additional disulfide bond Cys341–Cys348 is located near the C-terminus of the polypeptide chain aids in the formation of Mn^{2+} binding site and responsible for pushing the C-terminus segment away from the main body of enzyme. The molecular structure of MnP and its Mn^{2+} binding site are shown in Fig. 8.1.

The active site of MnP consist of proximal His173 ligand H-bonded to a conserved Asp242 residue which contribute to the low negative reduction potential of the iron, and stabilisation of the oxidation states, compound-I (MnP-I) and

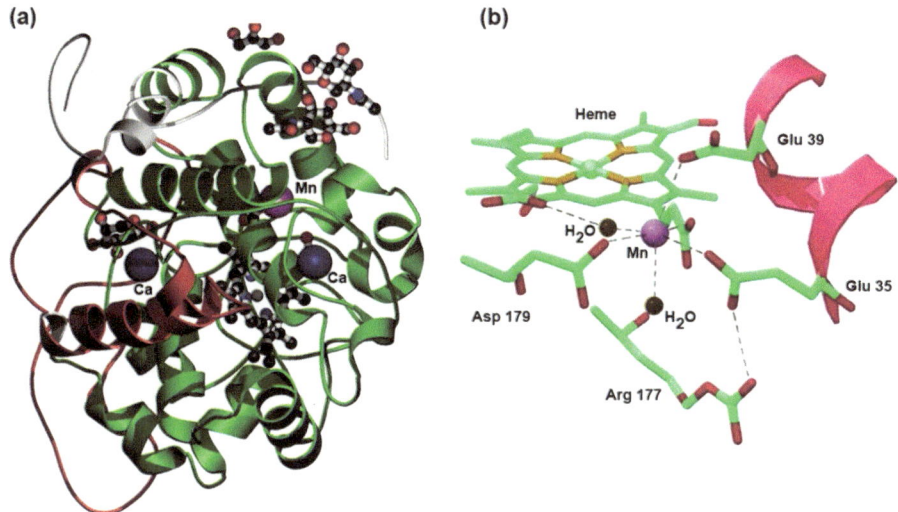

Fig. 8.1 Three-dimensional ribbon structure of *Phanerochaete chrysosprorium* MnP (**a**) MnP manganese binding site (**b**) (Sundaramoorthy et al. 2010, PDB entry 3m5q)

compound-II (MnP-II), and a distal side H_2O_2 binding pocket consisting as two conserved amino acid residues, His46 and Arg42. Arg42 implicated in stabilizing the MnP-I and MnP-II intermediate by forming a hydrogen bond with oxferryl oxygen. Crystal structure analysis of *P. crysosporium* MnP showed that Glu35, Glu39, and Asp179 are forming a Mn^{2+} binding site. This site has considerable flexibility to accommodate the binding of a wide variety of metal ions. The metal ligands, Glu35 and Glu39, move from their original Mn^{2+} binding conformations and this provides insights into the mechanism of MnP. Further, MnP crystal structure shows that the Mn^{2+} bind side chain of three amino acids, Glu35, Glu39, Asp179, one heme propionate, as well as two water molecules. The metal free high-resolution structures shows that Glu35 does not move there original position in MnP structure, Glu35 and Glu39 to adopt two conformations—"closed" conformations in the metal bound state and "open" conformations in the metal free state, possibly acting as a "gate", enabling a small carboxylic acid like oxalate or malonate to remove Mn^{3+} from the binding site. However, Cd^{2+} is a reversible competitive inhibitor of Mn^{2+} and bond to Mn^{2+} binding site on MnP, preventing oxidation of Mn^{2+}.

The substrate-bound MnP (Mn–MnP) contains about 357 amino acid residues, three sugar residues (GlcNac, GlcNac at Asn131 and a single mannose at Ser336), one iron (III) protoporphyrin IX prosthetic group, two calcium ions, a substrate Mn^{2+} ion, and 478 solvent molecules, including two glycerol molecules. High concentration of Ca^{2+} and Mg^{2+} enhance the activity of MnP. However, the substrate-free MnP model differs only in lacking the Mn^{2+} ion in the Mn binding site and in the number of solvent molecules, 549, which includes two glycerol molecules. A recent survey of over 30 fungal genomes provides evidence that MnP has three subfamilies (long, extralong

and short MnP) defined by the length of the C-terminus tail. The genome of
C. subvermispora is representative of the three MnP subfamilies.

8.2.2   Catalytic Cycle and Mode of Action

MnP catalyzes the oxidation of Mn^{2+} to Mn^{3+} in the presence of H_2O_2
(a cosubstrate). Mn^{2+} is a specific effector that induces MnP and represses LiP. In
the presence of 1 equiv H_2O_2, MnP forms MnP-I, a high valent oxo-Fe^{4+} porphyrin
based (Pi) free radical cation (step 1 in Fig. 8.2) which is in turn reduced by a bound
Mn^{2+} atom to form MnP-II, an oxo-Fe^{4+}porphyrin without the associated porphyrin
(Pi) free radical (step 2 in Fig. 8.2). However, in the absence of Mn^{2+} the addition of
2 equiv H_2O_2 yields MnP-II. The conversion of MnP-I to MnP-II can also be
achieved by addition of other electron donors, such as phenols and amines including *p*-cresol, guiacol, vanillyl alcohol, 4-hydroxy-3-methoxycinnamic acid,
isoeugenol, ascorbic acid, *o*-dianisidine, ferrocyanide and a variety of phenolic
compounds (Wariishi et al. 1988). MnP-II then oxidizes another Mn^{2+} ion, driving
the enzyme back to the ground state Fe^{3+} porphyrin (step 3 in Fig 8.2). In the
absence of substrate, the addition of excess H_2O_2 (250 equiv) drives MnP into
compound-III (MnP-III) which can be further oxidized until bleaching and irreversible inactivation (step 4 & 5 in Fig. 8.2).

The Mn^{3+} is a strong oxidizer (1.54V) and released from the MnP but it is quite
unstable in aqueous media. To overcome this drawback, white-rot fungi secrete
various organic acids such as oxalate or malate act as chelating agents enabling the

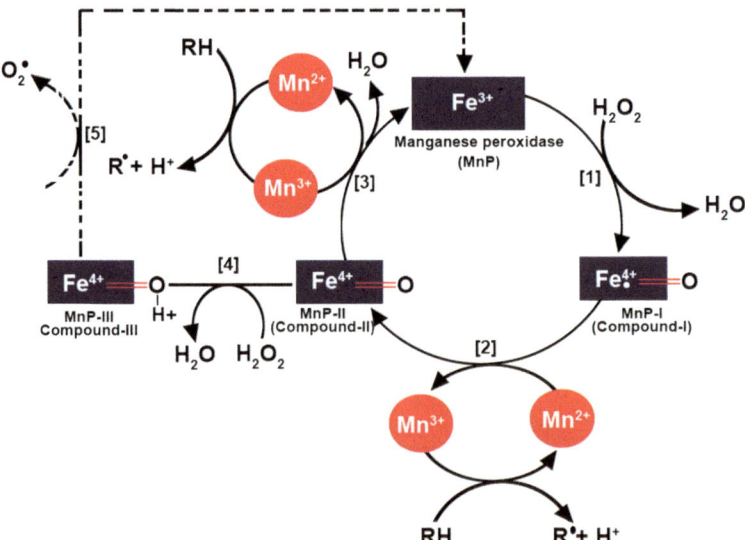

Fig. 8.2 Catalytic cycle of manganese peroxidase

formation of organic acid–Mn^{3+}complex. However, the addition of fumarate, malonate, tartrate, or lactate in the medium enhanced MnP production. The complex formation stabilizes Mn^{3+} so that bidentate ligated Mn^{3+} usually have redox potentials of around 0.7–0.9V and significantly lower oxidation capacities when compared to non-chelated Mn^{3+}. The redox potential of chelated Mn^{3+} depends on the chelator. The degradation of recalcitrant non-phenolic compounds has been limited with MnP generated Mn^{3+} chelates alone due to this lower oxidation power, but in the presence of some mediators or co-oxidants such as glutathione (GSH), polyoxyethylene sorbitan monoleate (tween 80), acetosyringone, methyl syringate, 3,5-dimethoxy-4-hydroxy-benzonitrile, linoleic acid, linolenic acids, it has effective in the oxidation of recalcitrant compounds. Mediators are easily oxidizable low MW compounds that can act as redox intermediates between the active site of the enzyme and a non-phenolic substrate. Mediators act also as electrons shuttles, providing the oxidation of recalcitrant complex substrates that do not enter the active site due to steric hindrances. It is therefore of primary importance to understand the nature of the reaction mechanism operating in the oxidation of a substrate by the oxidized mediator species derived from the corresponding mediator investigated. In the MnP-dependent oxidation of non-phenolic substrates, previous evidence suggests an electron-transfer (ET) mechanism with mediator syringl type phenols, towards substrates having a low oxidation potential. Alternatively, a radical hydrogen atom transfer (HAT) route may operate with ArOH type mediators, if weak C–H bonds are present in the substrate. A schematic pathway for the oxidation of substrate in presence/absences of mediator as shown in Fig. 8.3.

MnP catalyses the oxidation of Mn^{2+} to Mn^{3+} chelate Mn^{3+} to form stable complexes that diffuses freely and oxidized phenolic substrate (e.g. simple phenol, amines, dyes, phenolic lignin substructure and dimers) by one electron oxidation of the substrate, yielding phenoxy radical intermediate, which under undergoes rearrangement, bond cleavage, and non-enzymatic degradation to yield several breakdown products. In the presence of Mn^{2+}, malonate and H_2O_2, MnP from *P. chrysosporium* was found to calalyse C_α–C_β cleavage, alkyl cleavage and C_α-

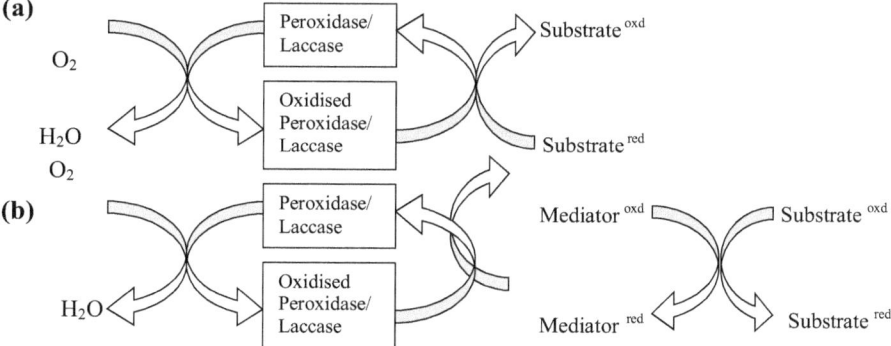

Fig. 8.3 Schematic representation of peroxidase/laccase-catalyzed redox cycles for oxidation of substrate in the absence (**a**) presence (**b**) of redox mediators

oxidation of phenolic arylglycerol β-aryl ether lignin model compounds (Tuor et al. 1992). In the absence of exogenous H_2O_2, MnP also has an oxidase activity against NADPH, GSH, dithiothreitol and dihydroxymaleic acid, forming H_2O_2 at the expense of oxygen. The oxidation of phenolic lignin model compound is shown in Fig. 8.4a.

The Mn^{3+} chelate only oxidising phenol portion of the lignin polymer under physiological conditions and cannot independently oxidised the non-phenolic parts. Therefore, alternative mechanism whereby MnP could oxidise non-phenolic compounds have been sought. For oxidation of non-phenolic compounds by Mn^{3+} involves the formation of reactive radical in the presence of mediators. It was found that in the presence of glutathione (GSH) MnP could oxidize non-phenolic lignin model compounds, veratryl, anisyl, and benzyl alcohol (Wariishi et al. 1989). They demonstrated that the Mn^{3+} formed oxidise thiol to thiyl radical that in turn abstracts a hydrogen from substrate (veratryl, anisyl, and benzyl alcohol) to forming a benzyl radical which react with another thiyl radical to yield an intermediate which decomposed to the benzaldehyde products (veratraldehyde, anisaldehyde and benzaldehyde).

Bao et al. (1994) reported that MnP from *P. chrysosporium* could oxidise a non-phenolic β-O-4 lignin model compounds in the presence of tween 80, an anionic surfactant made from an unsaturated fatty acid, oleic acid. They suggested that MnP oxidised the carbon–carbon double bond (C=C) in tween 80 to a peroxide which is known as lipid peroxidation and subsequently turned into a peroxy radical. As a result, the MnP-lipid system catalyse C_α–C_β cleavage, and β-aryl ether cleavage of non-phenolic diarylpropane and β-O-4 lignin, respectively. In the MnP-dependent peroxidation of unsaturated fatty acid, lipid free radicals also produce superoxide radicals O_2. The substrate oxidation mechanism involve benzyl hydrogen abstraction from benzyl carbon (C_α) via lipid peroxy radical followed by O_2 addition to form peroxy radical, and subsequent oxidative cleavage and non-enzymatic degradation as shown in Fig 8.4b. It was suggested that this process might enable the white rot fungi to accomplish the initial delignification of wood.

Moreover, MnP has been reported to oxidised various halogenated compounds by the oxidation mechanism is different to described above. Halide ions are oxidized by a unique peroxidase mechanism, compared with other electron donors. Most electron donors are oxidized by MnP-I via a single- electron mechanism with the intermediate formation of MnP-II (see Fig. 8.2) whereas halides are oxidized by MnP-1via a two-electron mechanism, yielding the native enzyme directly. A novel MnP from *P. chrysosporium* exhibits haloperoxidase activity at low pH. In the presence of H_2O_2 MnP oxidizes bromide and iodide to tribromide and triiodide at optimum pH 2.5 and 3.0, respectively.

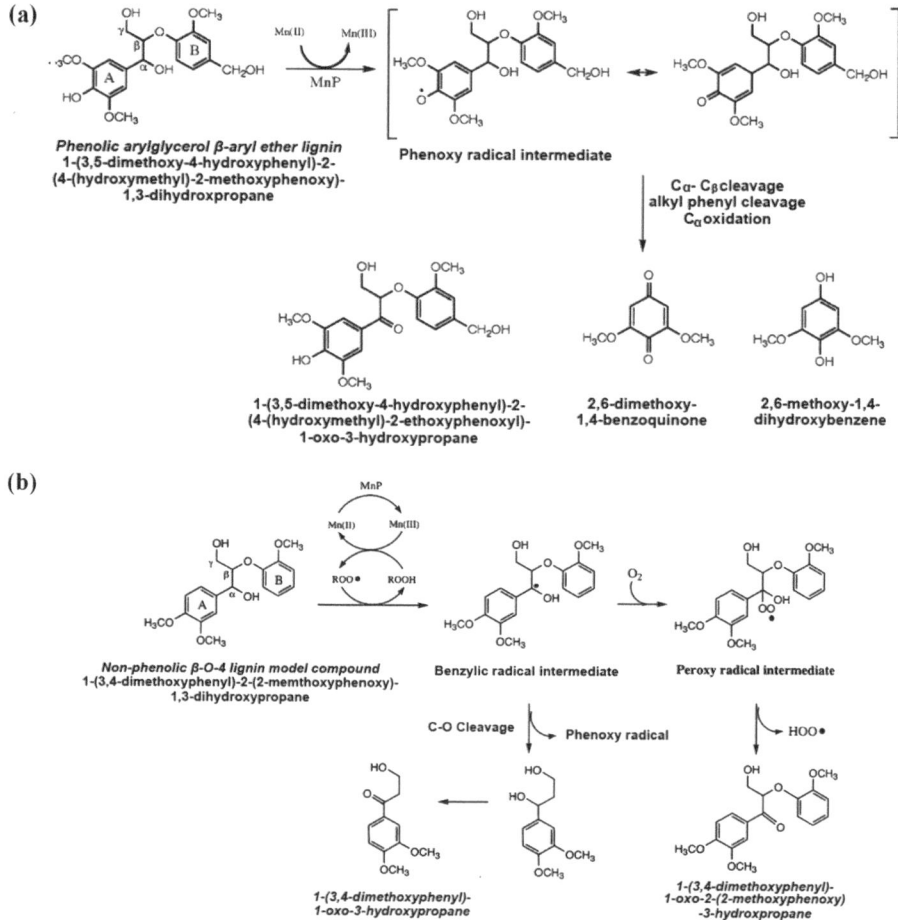

Fig. 8.4 Manganese peroxidase catalysed oxidation mechanism of phenolic aryglycerol β-aryl ether and (**a**) non-phenolic β-O-4 lignin model compound (**b**) [modified from Wong (2009), Tuor et al. (1992) and Bao et al. (1994)]

8.2.3   Common Substrate and Microorganisms

Microorganisms have the ability to interact, both chemically and physically, with substances leading to structural changes or complete degradation of the target molecule. A huge number of MnP producing fungi and bacteria genera possess the capability to degrade various organic substrates as a sole carbon, nitrogen and, phosphorus for their growth and metabolism in natural and controlled environment. A range of MnP producing fungi and bacteria and their substrate are listed in Table 8.1.

Table 8.1 Some important MnP producing microorganisms in extremophilic conditions and their substrate

Microorganisms	Substrate	pH	Mediator	Temp.
Fungi				
Irpex Lacteus CD2	Remazol brilliant violet 5R, direct red 5B, remazol brilliant blue R, indigo camine, methyl green	3.5–6.0	–	40–60
Phanerochaete sordita YK-624	Reactive red 120, bleaching of hard wood craft pulp	4.5	Tween 80	30
Bjerkandera sp. BOS55	Orange II	4.5	–	20–30
Phanerochete Chrysosporium BKM-F-1767	Poly R-478	4.5	–	
Coriolus hirsutus	Melanoidins	4.5	–	
Lactobacillus kefir	Sucrose-glutamic acid, sucrose- aspartic acid, glucose-glutamic acid	7.2–7.4	–	30
Phaneochete sordida YK-624	Aflatoxin B1	4.5	Tween 80	30
Nematoloma forwardii	[U-^{14}C]pentachlorophenol, [U-^{14}C] catechol, [U-^{14}C] tyrosine, [U-^{14}C] tryptophan, [4,5,9,10-^{14}C] pyrene, [U-^{14}C]2 amino-4,6-dinitrotoluene [^{14}C] pyrene, [^{14}C] anthracene, [^{14}C] benzo(a)pyrene, [^{14}C]benz(a)pyrene, [^{14}C] phenantherene, [^{14}C]- synthetic lignin (DHP)	4.5	–	30
Phanerochete crysosporium	Non-phenolic β-vanillyl alcohol; nonphenolic β-vanillyl dimer; Nonphenolic β-1 diarylpropane lignin model dimers	4.5	GSH Tween 80	30
Anthracophyllum discolor	Pyrene, anthracene, fluranthene, phenanthrene)	4.5	–	25
Ganoderma lucidum	Crescent, magna, textile effluent	4.5	–	25
Pleurotus ostreatus	2,2 -Bis- (4-hydroxyphenyl) propane (Bisphenol A)	4.5	–	25
Clitocybula dusenii b11	Lignite originated humic acid	4.0	GSH	37
White rot fungi IZU-154	Nylon-66			
Schizophyllum sp. F17	Congo red, orange G, orange IV	4.0–7.4	–	25
Trichophyton rubrum LSK-27	Plant derived lignin	4.5	–	30 or 40 °C
Penicillium oxalicum isolate 1	Plant derived lignin	4.5	–	37
Ceriporiopsis subvermispora	Plant derived lignin	2.0–5.0	–	25

(continued)

Table 8.1 (continued)

Microorganisms	Substrate	pH	Mediator	Temp.
White rot fungi IZU-154	Polyethylene		Tween 80	
Bacteria				
Bacillus sp. IITRM7	Sucrose aspartic acid maillard product	7.0	–	35
Raoultella planticola IITRM15	Sucrose aspartic acid maillard product	7.0		35
Enterobacter sakazakii IITRM16	Sucrose aspartic acid maillard product	7.0	–	35
Bacillus licheniformis (RNBS1)	Melanoidins	7.3	–	35
Bacillus sp. (RNBS3)	Melanoidins	7.3	–	35
Alcaligenes sp. (RNBS4)	Melanoidins	7.3	–	35

All values of temperature (Temp.) are given in °C

8.2.4 Screening of MnP Producing Microorganisms Its Substrate, Bioassay and Purification

MnP oxidised a wide range of phenolic and non-phenolic compounds as a common substrate in the presence of H_2O_2. There are various phenolic compounds (e.g. 2,2-azinobis (3-ethylthiazoline-6-sulfonate (ABTS), 2,6-dimethyloxyphenol (DMP), vanillylacetone, ferulic acid (4-hydroxy-3-methoxycinnamic acid), syringol, guaiacol, isoeugenol, *p*-methoxyphenol, syringaldazine, divanillylacetone, phenol red and coniferyl alcohol, [3-methyl-2-benzothiazolinone hydrazone (MBTH)], 3-(dimethylamino) benzoic acid (DMAB), *p*-cresol, *o*-dianisidine, catechol, hydroquinone) and non-phenol compound (e.g. vanillyl alcohol, VA and benzyl alcohol) in the presence used for in vitro MnP assay. The most commonly used substrate for MnP assay is guaiacol, a natural phenolic product first isolated from guaiac resin and the oxidation of lignin.

The isolated and purified microorganisms are screened for MnP enzyme activity by plate assay method using phenol red or guaiacol as substrate in minimal salt media agar plate where the screening of MnP is based on as colorless halo formation around the microbial growth. The assay plate containing composition (g/l) 3.0 peptone, 10.0 D-glucose, 0.6 KH_2PO_4, 0.001 K_2HPO_4, 0.4 $ZnSO_4$, 0.0005 $FeSO_4$, 0.05$MnSO_4$, 0.5 $MgSO_4$, H_2O_2, and 20.0 agar supplemented with 0.2% guaiacol. After incubation, MnP activity are visualised on plate by formation of reddish brown zone around the microbial growth due to the oxidative polymerization of guaiacol as shown in Fig 8.5a. While in another method MnP producing

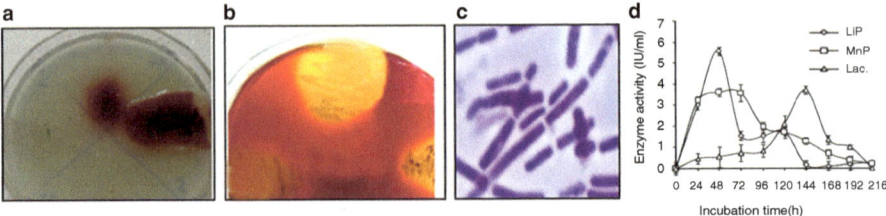

Fig. 8.5 Plate Showing growth of MnP producing microorganisms by using different substrate (**a**) guaiacol (**b**) phenol red (**c**) light microscopy (**d**) ligninolytic enzyme activity

bacteria has been screened on modified GPYM as well as in broth amended with different concentration of sucrose aspartic acid-maillar product (SAA-MP) using method describe by Yadav et al. (2011). In this method, MnP producing microbial culture are inoculated on GPYM agar plates containing D-glucose, 1.0; peptone, 0.1; K_2HPO_4, 0.1 and $MgSO_4 \cdot 7H_2O$, 0.05 with different concentration of SAA-MP (800–3600 mg/l) supplemented with 0.1% phenol red (w/v). After incubation the potential bacteria developed changing the deep orange colour of phenol to light yellow as shown in Fig 8.5b. Chandra and Singh (2012) studied the production of ligninolytic enzyme during pulp paper mill effluent degradation by bacterial consortium (Fig. 8.5c). They found that MnP activity are lower comparison to laccase but higher than to LiP as shown in Fig. 8.5d.

The bioassay of MnP activity is carried out by measuring optical density (OD) at 270 nm ($\varepsilon = 11.59 \, mM^{-1} \, cm^{-1}$) due to the formation of Mn^{3+}-malonate complex at 4.5 as shown in Fig. 8.6. This method is based on oxidation $MnSO_4$. The assay mixtures containing 50 mM sodium malonate buffer (pH 4.5), 0.5 mM $MnSO_4$, 0.1 mM H_2O_2 and 100 μl of enzyme solution. The reaction is initiated by adding H_2O_2 at 25 °C and Mn^{3+}-malonate complex measured spectrophotometrically at 270 nm. The MnP activity has also been measured by using ABTS, DMP or VA as the substrate under the conditions as described above and the oxidation has been followed by monitoring optical density at 414 nm ($\varepsilon = 36 \, mM^{-1} \, cm^{-1}$) for ABTS or at 470 nm ($\varepsilon = 49.6 \, mM^{-1} \, cm^{-1}$) for DMP or at 310 nm ($\varepsilon = 9.3 \, mM^{-1} \, cm^{-1}$) for VA, respectively. The MnP activity can also been determined spectrophotometrically by using phenol red as substrate at 610 nm. This activity assay is based on the oxidation of phenol red in the presence of H_2O_2 and Mn^{2+} and the oxidation product is measured by spectrophotometrically. In this method, a reaction mixture containing enzyme extract (700 μl), 0.2% phenol red, 2 mM sodium lactate (50 μl), 2.0 mM $MnSO_4$, 0.1% egg albumin, 2 mM H_2O_2 in 20 mM sodium succinate buffer at 4.5 pH. The oxidation product is measure by spectrophotometer recording the absorbance at 610 nm ($\varepsilon = 22 \, mM^{-1} \, cm^{-1}$). However, the manganese independent peroxidase activity is determined using 2.0 mM EDTA instead of MnSO4 solution. A another very sensitive bioassay method for measurement of MnP activity by using 3MBTH)/DMAB as a substrate has also been developed. In this method, the reaction mixture contained 0.07 mM MBTH, 0.99 mM DMAB, 0.3 mM $MnSO_4$, 0.05 mM H_2O_2 and the sample in a 100 mM succinic/lactic acid buffer (pH 4.5).

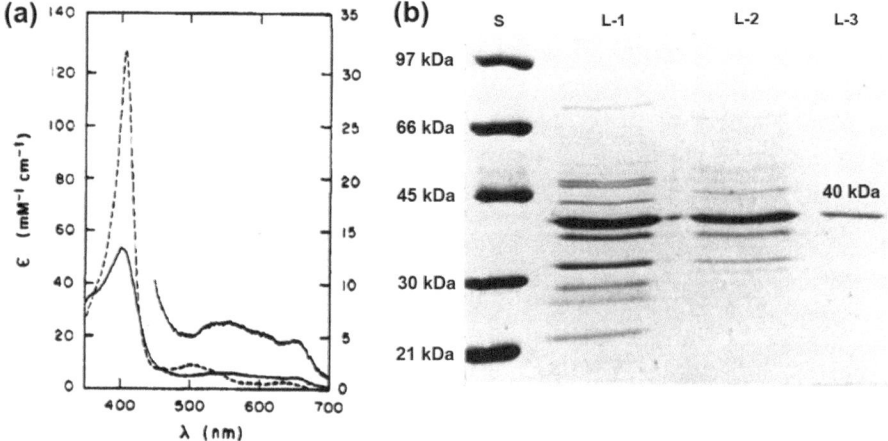

Fig. 8.6 UV-Visible spectrum of purified native MnP enzyme (*dashed line*) (**a**) Native PAGE for MnP (**b**) symbol S: Molecular weight standard marker; L-I crude MnP extract; L-2 partially purified MnP after ammonium sulfate precipitation; L-3 Gel filtration chromatography of purified MnP

This reaction mixture yields a deep purple color with a broad absorption band with a peak at 590 nm ($\varepsilon = 53.0 \ mM^{-1} \ cm^{-1}$). All activities are expressed in international unit (IU). IU defined as the amount of enzyme produced 1μmol of product per minute under the assay condition used.

Various enzyme/protein purification techniques are frequently employed purifying MnP from microbial culture. MnP has been purified from *Lentinula edodes* by applying the crude extract on cold acetone ($-20 \ ^\circ C$) and Sephadex G-100 column, respectively and characterized by denaturing sodium dodecyl sulfate-polyacrylamide gel electrophoresis (SDS-PAGE). They obtained the terminal specific activity of 5496 U mg^{-1} with a 6.76 fold. Further, MnP has also been isolated and purified from *Irpex lacteus* CD2 by using diethylaminoethyl cellulose (DEAE) sepharose column and characterised by SDS-PAGE using 10% polyacrylamide gel. They obtained MnP a terminal specific activity of 24.9 U/mg protein with 29.3-fold. In this method, the liquid culture of microorganisms first centrifuged at $5000 \times g$ for 20 min. Then the culture supernatant has concentrated by 80% ammonium sulfate at $4 \ ^\circ C$. The pellets have dissolved in sodium acetate buffer (20 mM, pH 4.8) then the enzymatic crude extract has dialyzed against the same buffer to remove ammonium sulfate and then applied to a DEAE sepharose column previously equilibrated with sodium acetate buffer (20 mM, pH 4.8). The MnP has eluted with a linear gradient of 0–1 M NaCl in the same buffer at a flow rate of 1 ml/min. The proteins in the eluted fractions are detected by recording the absorbance at 280 nm as shown in Fig 8.6a. Further, active fractions containing MnP activity is recorded after gel documentation of native-PAGE as shown in Fig 8.6b. The purified MnP has been verified by SDS-PAGE using 10% polyacrylamide gel. The MW of the purified MnP has been estimated in comparison to standard MW marker.

8.2.5   Effect of Environmental Parameters, Organic Solvent and Heavy Metal on MnP Activity

The ability of MnP to tolerate high temperature, different metal ions and organic solvents is very important for the efficient application of this enzyme in the biodegradation and detoxification of industrial waste. The activity and stability of MnP is strongly influenced by the pH, temperature, and time of incubation. The effect of temperature and buffer type on the stability of MnP has been previously investigated by various workers. Sutherland and Aust (1996) found that MnP from *P. crysosporium* was most stable at pH 5.5 and temperature at of below 37 °C. They also found that MnP become inactive at high temperature due to loss of Ca^{2+}, required for the stability and activity. The engineering of a disulfide bond (A48C and A63C) near the distal calcium binding site of MnP by double mutation showed the improvement in thermal stability as well as pH (pH 8.0) stability in comparison to native enzyme (MnP). The disulfide bond adjacent to the distal calcium ligand Asp47 and Asp64 stabilizes the recombinantly expresses MnP against the loss of calcium. The MnPs (G1 and G2) from *Ganoderma* sp. YK-505 exhibited 100% MnP activity after treatment for 60 min at 60 °C, but lower than 20% residual activity with 10 mM H_2O_2. Recently, a thermostable and H_2O_2 tolerant MnP isolated and purified from the culture medium of *Lenzites betulinus* named as L-MnP. The purified L-MnP has the highest H_2O_2 tolerance among MnPs reported so far. It retained more than 60% of the initial activity after thermal treatment at 60 °C for 60 min, and also retained more than 60% of the initial activity after exposure to 10 mM H_2O_2 for 5 min at 37 °C. Now a novel MnP (CD2-MnP) has been purified and characterised from the white-rot fungus *I. lacteus* CD2. The CD2-MnP has strong capability for tolerating different metal ions such as Ca^{2+}, Cd^{2+}, Co^{2+}, Mg^{2+}, Ni^{2+} and Zn^{2+} as well as organic solvents such as methanol, ethanol, DMSO, ethylene glycol, isopropyl alcohol, butanediol and glycerin. CD2-MnP exhibited high stability in pH range from 3.5 to 6.0 and optimal temperature was determined to be 70 °C. All these purified MnP are known as wild MnP (wMnP). wMnP are not well suited for industrial used, which often required particular substrate specificities and application conditions (including pH, temperature and reaction media) in addition to high production levels. Thermostable enzymes are typically tolerant to many other harsh conditions often required in industry, such as the presence of organic co-solvents, extreme pH, high salt concentrations, high pressures, etc. Therefore, the development of recombinant MnP (rMnP) for industrial application through protein engineering and heterologous expression is in process.

8.3 Lignin Peroxidase

LiP (EC 1.11.1.14) is an extracellular H_2O_2 dependent heme containing glycoprotein, produced by white-rot fungi and similar to the lignin-synthesizing plant peroxidases. It was first discovered in nitrogen and carbon limited cultures of *P. chrysosporium* and since then has become one of the most studied peroxidases (Glenn et al. 1983). It has also been reported to be produced by many white rot fungi including *Phlebia flavido-alba*, *Bjerkandera* sp. strain BOS55, *T. trogii*, *Phlebia tremellosa* and *P. chraceofulva*. Several LiP isozymes have also been detected in cultures of *P. chrysosporium*, *T. versicolor*, *B. adusta* and *Phlebia radiate* and so one. LiP possess high redox potential (700–1400 mV), low optimum pH 3.0 to 4.5, ability to catalyze the degradation of a wide number of aromatic substrates such VA, methoxybenzenes and also a variety of non-phenolic lignin model compounds as well as a range of organic compounds with a redox potential up to 1.4 V (versus normal hydrogen electrode) in the presence of H_2O_2.

LiPs can catalyse the oxidative cleavage of Cα–Cβ linkages, β-O-4 linkages, and other bonds present in lignin and its model compounds. The enzyme also catalyzes side-chain cleavages, benzyl alcohol oxidations, demethoxylation, ring-opening reactions and oxidative de chlorination. Moreover, bacteria are worthy of being studied for their ligninolytic potential due to their immense environmental adaptability and biochemical versatility. There is wide range of examples where bacteria like *Pseudomonas aeruginosa*, *Serretia marcescens*, *Nocardia*, *Arthrobacter*, *Flavobacterium*, *Micrococcus*, *Xanthomonas* sp. have been identified as lignocellulosic-degrading microorganisms. Therefore, identification of bacteria having lignin oxidizing enzymes would be of significant importance. The LiP activity is associated with primary growth of bacteria and thus the delignification process is presumed to be the result of primary metabolic activity and not dependent upon other factors such as stress to induce production.

8.3.1 Molecular Structure

LiP a monomeric glycoprotein of 38–46 kDa (and pI of 3.2–4.0) containing 1 mol of iron protoporphyrin IX per 1 mol of protein, catalyzes the H_2O_2-dependent oxidative depolymerization of lignin. LiP has a distinctive property of an unusually low pH optimum near pH 3.0 in extremophilic environment. In general, LiP which has MW 40 kDa contains 343 amino acids residues, 370 water molecules, a heme group, four carbohydrates, and two calcium ions. The crystal structure of LiP (PDB Id: 1LGA) isolated and purified from *P. chrysosporium* as shown in Fig. 8.7a. The heme is embedded in a crevice between the two domains, but is accessible from the solvent via two small channels. The secondary structure of enzyme molecule contains eight major and eight minor α-helices and two anti-parallel β-sheet, and it is organised in a proximal and a distal domain. The heme is embedded in a crevice

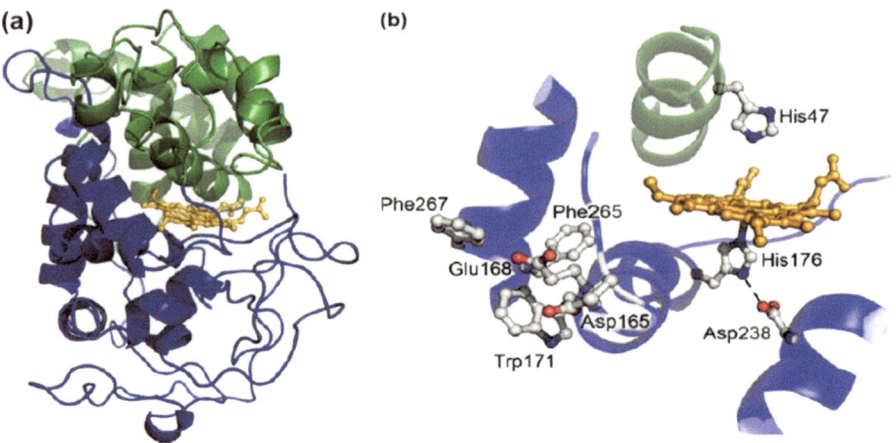

Fig. 8.7 Three-dimensional ribbon structure of *P. chrysosporium* lignin peroxidase (**a**) detail of the heme environment (**b**) (Poulos et al. 1993; PDB entry 1LGA)

between the domains: two small channels connect the prosthetic group to the solvent. The LiP contains eight Cys residues, all forming four disulfide bridges. There are two calcium-binding sites, one in each domain, with possible function of maintaining the topology of the active site. The heme is embedded in a crevice between the domains: two small channels connect the prosthetic group to the solvent. The heme iron is predominantly high spin, pentacoordinated with His176-N at the proximal side as the fifth ligand, and Wat339 H-bonded to the distal His47-N (Fig. 8.7b). The His is associated with the high redox potential of LiP. The enzyme's redox potential rises when the His has a reduced imidazol character. In addition, a greater distance between the His and the heminic group increases the redox potential of the enzyme. Another characteristic related with LiP's high redox potential is the invariant presence of a tryptophan residue (Trp171) in the enzymes surface. Trp171 seems to facilitate electronic transference to the enzyme from substrates that cannot access into the heminic oxidative group. This Trp171 residue has been suggested an important role in the binding and oxidation of VA, a fungal secondary metabolite produced by at the same time as LiP and oxidised by LiP. VA participates in the oxidation of different aromatic molecules.

8.3.2 Catalytic Cycle and Mode of Action

The catalytic cycle of LiP is similar to that of other peroxidases like MnP where ferric enzyme is first oxidized by H_2O_2 to generate the two-electron oxidized intermediate, compound-I (LiP-I). In this reaction 2-electron oxidation of ferric [Fe^{3+}]

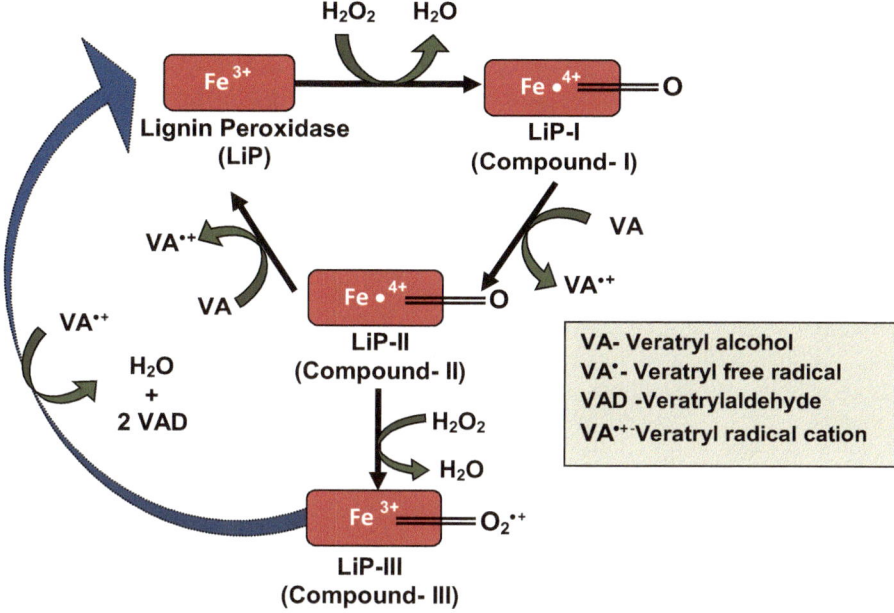

Fig. 8.8 Catalytic cycles for lignin peroxidase

LiP produces LiP-I intermediate, a oxoferryl iron porphyrin radical cation [Fe^{4+}=O] with the reduction of H_2O_2. Next, LiP-I is reduced by one electron donated by a substrate such as VA, yielding the 1-electron oxidized enzyme intermediate, compound-II [Fe^{4+}=O] (LiP-II), and a free radical product (VA$^{•+}$). VA$^{•+}$ acts as a redox mediator in the oxidation of lignin. VA$^{•+}$ is capable of mediating oxidation of secondary substrates typically not oxidized by LiP. The catalytic cycle is completed by the one-electron reduction of LiP-II by a second substrate molecule. But, in the absence of a reducing substrate, the enzyme can undergo a series of reactions with H_2O_2 to form compound-III (LiP-III), oxyperoxidase. LiP-III is stable, but prolonged incubation of enzyme with H_2O_2 in the absence of a reducing substrate such as VA can cause irreversible inactivation of the enzyme. In the presence of VA, however, LiP-II undergoes multiple turnovers without any detectable inactivation. Because VA is normally produced by white rot fungus, some workers conceptualized that VA protects the enzyme from the action of H_2O_2- dependent inactivation and participate as a redox mediator between the enzyme and substrates which cannot get inside the heminic center. Moreover, VA$^{•+}$ converts LiP-III to the native enzyme via the formation of veratryaldehyde and H_2O, potentially making more enzymes active for the oxidation of lignin. The catalytic cycle of LiP as shown in Fig. 8.8.

Some studies have observed that substrates are not oxidized by LiP such as anisyl alcohol and 4-methoxymandelic acid they are oxidized in the presence of VA. They proposed that the one-electron oxidized product of VA, the aryl cation radical, is able to mediate the oxidation of substrates typically not oxidized by the enzyme. The aryl cation radical is a diffusible species, capable of acting at a distance. In another studies concluded that the stimulation of 4-methoxymandelic acid and anisyl alcohol oxidation is due solely to the ability of VA to prevent inactivation of lignin peroxidase. They claimed that enzyme in the presence of anisyl alcohol and excess H_2O_2 leads to the formation of inactive LiP-III. LiP can be inhibited by cyanide and chloride. Chloride is expected to be a competitive inhibitor because it is demonstrated that chloride is not a substrate for LiP.

Like MnP, LiP is capable of oxidizing a wide variety of phenolic compounds including ring- and N-substituted anilines. It oxidise a wide range of aromatic compounds (guaiacol, vanillyl alcohol, catechol, syringic acid, acetosyringone, etc.) preferentially at a much faster rate compared to non-phenolic substrates. Using a lignin model dimer as the substrate, the cation radical decays with the spontaneous C_α–C_β fission of the alkyl side chain, with the products resembling those found when fungi degrade lignin. In the reduction of LiP-I and LiP-II, phenolic substrates are converted to phenoxy radicals. In the presence of oxygen, the phenoxy radical may react to form ring-cleavage products, or they may otherwise also lead to coupling and polymerization. It has been reported that LiP-catalyzed oxidation of the lignin model dimer compound 1,2-di (3,4-methoxyphenyl)-1,3-propanediol results in C_α–C_β cleavage to yield veratraldehyde. LiP-catalyzed oxidative reaction of phenolic compounds is typically associated with rapid decrease in enzyme activity. The decrease is likely caused by the accumulation of the inactive LiP-III during catalysis. Phenoxy radicals, unlike the non-phenolic VA discussed below, are unable to revert LiP-III to the native enzyme, although both substrates show similar rate constants for the reaction of LiP-I. The reactivity of phenolic compounds with LiP-I is much higher than that with LiP-II, and the rate constant decreases as the size of the substrate increases as demonstrated in the oxidation of oligomers of phenolic β-O-4 lignin model compounds.

LiP shows a very high redox potential (1.2 V at pH 3.0) compared to laccases (~0.8 V at pH 5.5), horseradish peroxidases (0.95 V at pH 6.3) and MnP (0.8 V at pH 4.5). This property enables LiP to catalyze the oxidation of non-phenolic aromatic compounds, even in the absence of a mediator. The oxidative reaction of non-phenolic diarylpropane and β-O-4 lignin model compounds of lignin involves initial formation of radical cation via 1e− oxidation, followed by side-chain cleavage, demethylation, intramolecular addition, and rearrangements. Oxidation of the A ring giving rise to C_α–C_β cleavage is the major route. In the mechanism, only the formation of the radical cation is enzyme catalyzed, and subsequent reactions of the substrate are nonenzymatic. Decay of the radical cation depends on the nature of the substituents on the aromatic ring. Electron-donating groups, such as alkoxy groups, on the aromatic ring favor the formation and stabilization of the aryl radical cation. LiP also oxidised the VA, a metabolic product at the same time as LiP produced by *P. chrysosporium*. The addition of VA is known to cause an increase in LiP activity and the rate of lignin mineralization. At pH 6.0, the second-order rate constant for LiP-catalyzed oxidation of VA

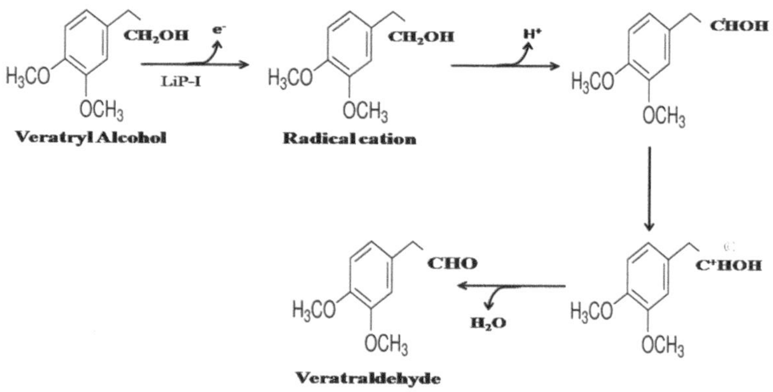

Fig. 8.9 Degradation mechanism of veratryl alcohol by lignin peroxidase

is similar to that of β-O-4 dimer (6.7×10^3 and 6.5×10^3 M^{-1} s^{-1}, respectively). The radical cation formed in the first reduction by LiP-I exist as a complex with LiP-II, which is catalytically active on a second VA molecule to form an aldehyde. The $VA^{\cdot+}$ generated in the reduction step decays by deprotonation at Ca, a typical reaction of alky aromatic radical cations, to form veratraldehyde as shown in Fig. 8.9. Under aerobic condition, however, additional oxidative pathways involving activated oxygen species occur leading to quinone formation and aromatic ring cleavage.

8.3.3 Common Substrate and Microorganisms

The association of ligninolytic enzymes with lignin breakdown arises because the enzyme can oxidize lignin-related aromatic compounds. Single-ring aromatic substrates are frequently used, including phenolic compounds i.e. guaiacol, vanillic acid, and syringic acid, and non-phenolic VA, and dimethylphenylenediamine. This type of substrates have been useful for characterization of catalytic cycles, particularly on the formation and reactions of the oxidized enzyme intermediates. Another group of substrates consists of lignin model dimers that are frequently used to investigate specific bond cleavages. These compounds are synthetic mimics of the common lignin substructures, such as diarylpropane and β-aryl ether dimer. The β-O-4 lignin model compounds are the most important type for elucidating lignin degradation, as arylglycerol β-aryl ether or β-O-4 bond is the most prevalent linkage type in lignin, accounting for about 50% of the interunit connections in gymnosperm and 60% in angiosperm. These model compounds can be synthesized either as phenolic or non-phenolic in nature. Phenolic subunits are present only about 10–20% in lignin. However, demethylation and ether cleavage reactions in enzyme-catalyzed degradation of phenolic compounds generate phenolic products,

Table 8.2 Different LiP producing microorganisms in extremophilic environment and their substrate

Microorganisms	Substrate	pH	Temp. (°C)
Fungi			
Phanerochaete chrysosporium	Azo dyes, penta chlorophenol, veratryl alcohol, anthracenes	3	39
Phanerochaete flavido-alba	Synthetic dehydropolymerized lignins (DHPs), veratryl alcohol	3–4	30
Bjerkandera sp. strain BOS55	2-chloro-1,4-dimethoxybenzene, veratryl alcohol	3	30
Trametes trogii	Lignin, veratryl alcohol	3	30
Phlebia ochraceofulva	Lignin, dimethyl succinate, veratryl alcohol	3	30
Phlebia tremellosa	Lignin, dimethyl succinate, veratryl alcohol	3	30
Bacteria			
Pseudomonas sp. SUK1	*n*-propanol	3	30
Pseudomonas aeruginosa	Lignin, bromophenol blue, veratryl alcohol		
Bacillus megaterium	Lignin, veratryl alcohol	7	37
Serretia marcescens	Lignin, bromophenol blue and veratryl alcohol	4	30
Bacillus subtilis	Alkaline lignin, veratryl alcohol	6	37
Arthrobacter globiformis	Lignin, veratryl alcohol	7	37
Actinomycetes			
Streptomyces viridosporus T7A	Vanillic acid, syringic acid	6	35
Streptomyces sp. AD001	2,4 dichlorophenol, 4-aminoantipyrene	7	35

which can in turn be the substrate for further breakdown. LiP producing microorganisms and their substrate are listed in Table 8.2.

8.3.4 Screening of LiP Producing Microorganisms, Its Substrate, Bioassay and Purification

The isolated and purified bacterial strain has been screened for LiP activity by plate assay method. The plate assay is generally performed using different substrate such as azure B (0.002%) or methylene blue (0.025%) in B & K agar medium plates containing dextrose 1%, peptone 0.5%, NaCl 0.5%, beef extract 0.3% and CuSO$_4$ (1 mM) (Chandra and Singh 2012). The disappearance of blue colour of the media confirmed the presence of LiP activity around the bacterial growth as shown in Fig. 8.10a (Chandra and Singh 2012). Moreover, LiP activity is determined spectrophotometrically during degradation of organic pollutants by recording the increase in absorbance at 310 nm through the oxidation of VA to veratryl aldehyde ($\epsilon_{310} = 9300$ M^{-1} cm^{-1}). The reaction mixture contained 100 mM sodium tartrate pH 3.0, 2 mM

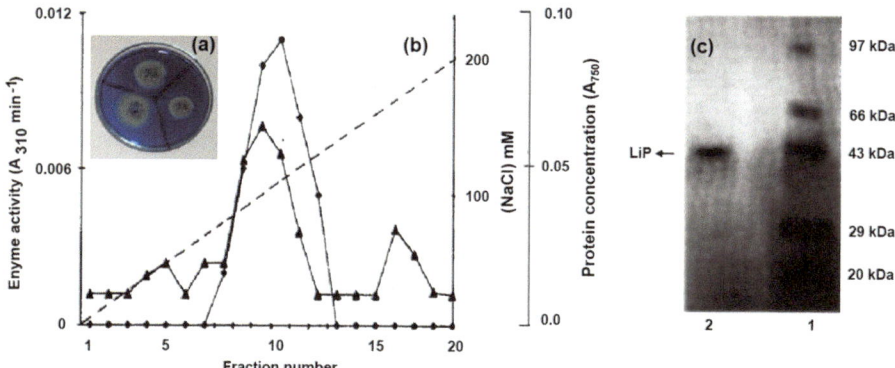

Fig. 8.10 (**a**) Plate assay of lignin peroxidase (**b**) elution profile from a DEAE column (*filled triangle*) activity profile (*filled circle*) protein at 750 nm; (straight line) NaCl gradient. Fractions of 5 mL were collected. (**c**) SDS-PAGE analysis of purified lignin peroxidase. Lane 1 contains the MW markers (from *top*): phosphorylase (97.4 kDa), bovine serum albumin (68 kDa), ovalbumin (43 kDa), carbonic anhydrase (29 kDa), soyabean trypsin inhibitor (20.1 kDa) and lysozyme (14.3 kDa). Lane 2 contains the purified lignin peroxidase. 50 mL was loaded (Yadav et al. 2009)

VA, LiP, and the reaction mixture was incubated at initiated by the addition of 0.5 ml H_2O_2 (final concentration 0.5 mM). The formation of the predominant product 2-chloro-1,4-benzoquinone from 2-chloro-1,4-dimethoxybenzene is measured at a wavelength of 255 nm using a molar extinction coefficient of 16,900 M^{-1} cm^{-1} (ten Have et al. 1998).

The LiP activity can also been determined spectrophotometrically by monitoring the oxidation of azure B as substrate in presence of H_2O_2 at 610 nm. The reaction mixture contained sodium tartrate buffer (50 mM, pH 3.0), azure B (32 µM), 500 µl of culture filtrate 500 µl of H_2O_2 (2 µM). OD is taken at 651 nm after 10 min. One IU of LiP activity is defined as activity of an enzyme that catalyzes the conversion of µ mole of substrate per minute. The purification of LiP enzyme has been reported by using the method described by Yadav et al. (2009). In this method, the partial purification of enzyme separated from exhausted medium is usually done by 70% ammonium sulphate saturation. The mixture is then stored in a cold room for 24 h to precipitate all the proteins and the precipitation is separated by centrifugation for 10 min. The supernatant is discarded and the remaining precipitate is further dissolved with 5 ml of 1M citrate phosphate buffer (pH 8.0) the concentrated enzyme mixture is subjected to dialysis. Further, the dialyzed enzyme is loaded onto a DEAE-cellulose column of size 1–16 cm, which is pre-equilibrated with the same phosphate buffer. The adsorbed enzyme is washed with 50 mL of the same buffer and is eluted by applying a linear gradient of NaCl (0–200 mM; 50 mL buffer 150 mL buffer containing 200 mM NaCl).The elution profile of the LiP activity from the DEAE cellulose column is given in Fig. 8.10b. The purpose of dialysis is to remove undesired small molecular weight molecules from a mixture in which the desired species of molecules are too large to travel across the membrane. Ordinarily this process is utilized during protein purification in which salting out procedure has been employed as the initial step with ammonium sulphate. After the protein is precipitated from the initial source, it is re-dissolved in buffer and then poured into a dialysis bag. The homogeneity of the enzyme preparation is checked by SDS-PAGE. The separating gel contains 12% acrylamide in 0.375M Tris-HCl buffer (pH 8.8) and the stacking gel remains 5% acrylamide in 0.063M Tris-HCl buffer (pH 6.8). Proteins are visualized by silver staining as shown in Fig 8.10c.

8.4   Laccase

Laccases (EC 1.10.3.2) are multi copper-containing polyphenol oxidases that are widely distributed in microorganisms, insects, and plants, showing a specific function in each of them. From this group, white rot fungi are the most studied laccases. It catalyze the oxidation of various aromatic compounds, particularly those with electron-donating groups such as phenols ($-OH$) and anilines ($-NH_2$), by using molecular oxygen as an electron acceptor. In nineteenth century, laccases was first isolated from exudates of the Japanese tree *Rhus vernicifera*. Laccases use molecular oxygen to oxidize a variety of aromatic and non-aromatic hydrogen donors via a mechanism involving radicals. These radicals can undergo further laccases catalyzed reaction and/or non-enzymatic reaction such as polymerization, and hydrogen abstraction. Therefore, laccase has also the ability to oxidize phenolic and non-phenolic substrates. The phenolic substrate oxidation by laccases result in formation of an aryloxyradicals an active species that is converted to a quinone in the second stage of the oxidation. Though, typical substrate of laccases known to be diphenol oxidase, monophenol e.g. sinapic acid or guaiacol can also oxidize polyamines, aminophenols, lignin, aryl diamine, inorganic ions and it may mitigate the toxicity of some polycyclic hydrocarbon. However, 2,2′-azino-bis (3-ethylbenzothiazoline-6-sulphonic acid) or ABTS is substrates which are most commonly used, does not form quinone and is not pH dependent. Laccases have been mostly isolated and characterized from plants and fungi, but only fungal laccases are used currently in biotechnological applications for the detoxification of complex industrial wastewater. Unfortunately, these enzymes usually work only efficiently under mild acidic conditions (pH 4.0–6.0) whereas the temperature range (30–55 °C) for catalytic activity is suboptimal. In contrast, little is known about bacterial laccases, which broad range substrate specificity for industrial application.

Most of study has been focused on white-rot fungus in which *Phlebia floridensis* showed higher thermostability at pH 4.5 and *T. versicolor* is useful for dye decolourization. Recently, in bacterial laccase polyphenol oxidase activity has been reported in an *Azospirillum lipoferum*, phenol oxidases, laccase were isolated from cell extracts of the soil bacterium *Pseudomonas putida* F6. Four strains of the bacterial genus *Streptomyces* (*S. cyaneus*, *S. ipomoea*, *S. griseus* and *S. psammoticus*) and the white-rot fungus *T. versicolor* were studied for their ability to produce active extracellular laccase in biologically treated wastewater with different carbon sources. Similarly, *Proteus mirabilis, Bacillus* sp., *Raoultella planticola* and *Enterobacter sakazakii* are used for degradation of persistent organic pollutants from biomethanated distillery spent wash, but it is still lacking for industrial application (Chandra and Singh 2012). Laccases have low substrate specificity make this interesting for degradation of several compounds with a phenolic structure because they are extracellular and inducible, do not need a cofactor, and have low specificity. Therefore, laccases have been employed in several areas such as bioremediation of aromatic recalcitrant compounds, treatment of effluents polluted with lignin, chemical synthesis, degradation of a wide number of textile dyes, and biomass pretreatment for biofuel production.

Due to the properties of their substrate, the enzymes participating in the breakdown of lignin and other related compounds should be exclusively extracellular. While this is without exception true for the LiP and MnP of white rot fungi and bacteria, the situation is not the same with laccases. Although most laccases studied so far are extracellular enzymes, while in wood-rotting fungi and some bacteria species are usually also found intracellular laccase activity. Blaich and Esser (1975) has been observed that most white-rot fungal species produced both extracellular and intracellular laccases with isoenzymes showing similar patterns of activity. *T. versicolor* was produced laccases both in extracellular and intracellular fractions when grown on glucose, wheat straw, and beech leaves (Schlosser et al. 1997). The intra and extracellular presence of laccase activity was also detected in *P. chrysosporium* and *Suillus granulates*. The fraction of laccase activity in *Neurospora crassa*, *Rigidoporus lignosus* and one of the laccase isoenzymes of *P. ostreatus* is also probably localized intracellularly or on the cell wall. The localization of laccase is probably connected with its physiological function and determines the range of substrates available to the enzyme. It is possible that the intracellular laccases of fungi as well as periplasmic bacterial laccases could participate in the transformation of low MW phenolic compounds in the cell. The cell wall and spores-associated laccases were linked to the possible formation of melanin and other protective cell wall compounds. However, laccase has been reported intracellularly as in *A. lipoferum*, *M. mediterranea* and in *B. subtilis*. The bacterial cells must have some strategy to cope with the intracellular presence of laccase and its toxic by-products. Rearrangement of the electron transport system has been hypothesized to be one of the ways in which the laccase-positive cells adapt to endogenous substituted quinones generated as products of laccase catalyzed reaction. Because, in fungi, extracellular localization of the enzymes helps them circumvent the problem of the reactive species, such as semiquinones and quinines that are generated by laccases while oxidizing aromatic substrates. These reactive species are powerful inhibitors of the electron transport system in both bacteria and mitochondria. The loss of cytochrome c oxidase activity and acquistion of resistance to quinone analogues has been demonstrated in a laccase-positive variant of *A. lipoferum*.

8.4.1 Molecular Structure

Laccases are monomeric, dimeric and tetrameric glycoproteins, generally having fewer saccharide compounds (10–25%) in fungus and bacteria than in plant enzymes. The carbohydrates, which are 10–45% of the total molecular mass, are covalently linked, and due to this property of enzymes show high stability. Mannose is one of the major components of the carbohydrates attached to laccases. The molecular weight of a laccase is determined to be in the range of 50–97 kDa from various experimental reports. The structure of laccase from *T. versicolor* containing approximately 500 amino acid residues organized in three β-barrel sequential domains. The three domains are distributed in a first domain with 150 initial amino acids, a second domain between the 150 and 300 residue, and a third domain from the 300 to 500 amino acid. The structure is stabilized by two disulfide bridges localized between domains I and II and between domains I and III. Figure 8.11a, b shows the three dimensional ribbon structure of bacterial lacasse from *Bacillus subtilis* XI based on the homology modelling of *B. subtilis* MB24 laccase PDB Id: 2x88A) (Guan et al. 2014) and fungal laccase from *T. versicolor*, respectively (PDB Id: 1N68) (Robert et al. 2003).

Laccases of three molecular forms of isozyme have been reported namely, Lac I, Lac II and Lac III. D'Souza-Ticlo et al. (2009) reported that the MW of Lac II is 56 kDa from *Cerrena unicolor*. In another study the MW of laccase has been reported ~97 kDa in *T. versicolor*. The biochemical properties of spore coat protein Cot A of *B. subtilis* has been reported to be similar to multicopper oxidases including laccases (Hullo et al. 2001). Further, the studies have revealed that Cot A contains all the structural features of a laccase including the reactive surface-exposed copper center (T1) and two buried copper centers (T2 and T3). The most thermostable laccases have been isolated from *Streptomyces lavendulae*, with half-life of 100 min at 70 °C, and in *B. subtilis*, for which Cot A reported a half-life of 112 min at 80 °C. The redox potential of any substrate also plays a very important role in laccase activity. The redox potential of bacterial laccases ranges from 0.4 to 0.5 V, but they are active and stable at high temperatures (66 h at 60 °C), at pH 7.0–9.0 and at high salt concentrations. The first gene and c-DNA sequences were recorded for laccases from the fungi *Neurospora crassa*, *Aspergillus nidulans*, *Coriolus hirsutus* and *Phlebia radiata*. Since, then the number of sequenced laccase genes has considerably increased. The sequences mostly encode polypeptides of approximately 500–600 amino acids in *Bacillus subtilis*. Cot A is 513 amino acids long, and its MW is 65 kDa typical eukaryotic signal peptide sequences of about 21 amino acids are found at the N-terminal of the protein sequences. In addition to the secretion signal sequence, laccase genes from *N. crassa*, *Podospora anserina*, *Myceliophthora thermophila* and *Coprinopsis cinerea* contain regions that code for N-terminal cleavable propeptides. These laccases also have C-terminal extensions of controversial function, i.e., the last amino acids from the predicted amino acid

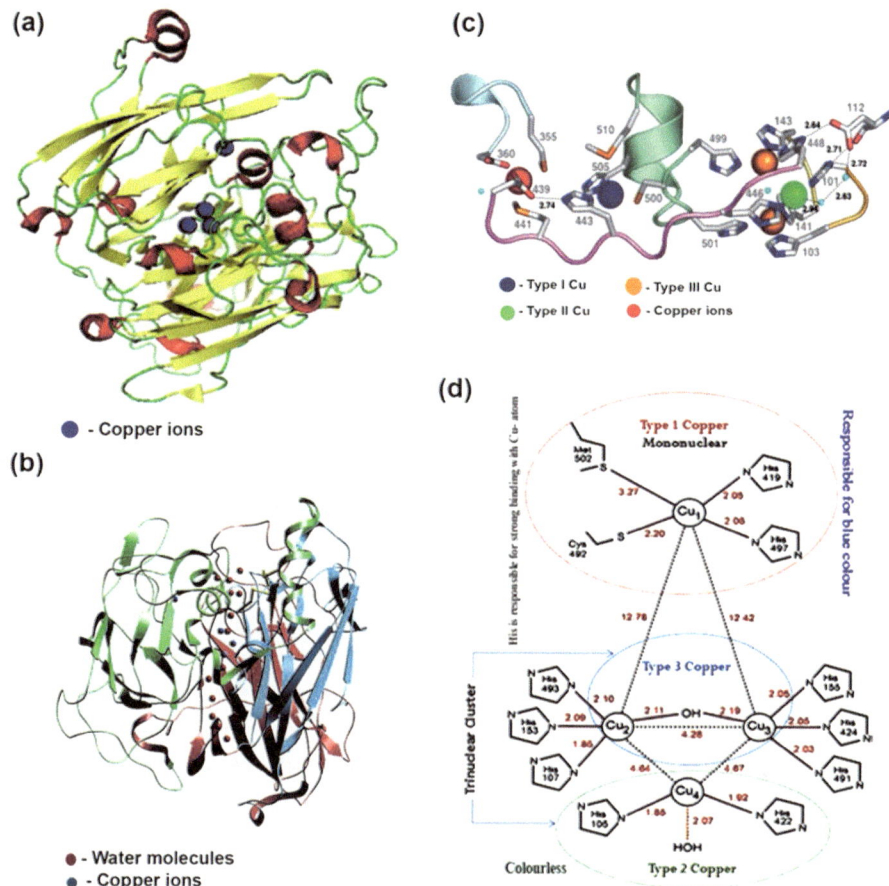

Fig. 8.11 Three-dimensional ribbon structure (**a**) *Bacillus subtilis* X1 laccase based on the *Bacillus subtilis* MB24 laccase crystal structure (PDB entry 2x88A) (Guan et al. 2014) (**b**) Laccase of *Trametes verssicola* showing the two channels leading to the T2/T3 cluster (PDB entry 1GYC) (Piontek et al. 2002) (**c**) Active site of CueO. Each Cu center is closely connected with each other. Asp112 is located behind the trinuclear Cu center, forming hydrogen bonds with the imidazoles coordinating to a T2Cu and a T3Cu directly and indirectly through a water molecule (PDB entry 1N68) (Robert et al. 2003) (**d**) The laccase active site showing the relative orientation of the Cu atoms including interatomic distances among all relevant ligands

sequence are not present in the mature protein. Figure 8.11c, d shows the active site of copper oxidase. The active sites of CueO (T1Cu, T2Cu, and T3Cu) in ribbon form are shown in different colours (blue, green and orange, respectively), and the figure also shows that Asp112 is located behind the tri-nuclear copper center.

8.4.2 Catalytic Cycle and Mode of Action

The catalytic mechanism of a laccase involves the reduction of the oxygen molecule, including the oxidation by one electron of a wide range of aromatic compounds, which include polyphenols, and methoxy-substituted monophenols and aromatic amines. The use of molecular oxygen as the oxidant and water as generated by-product are very common catalytic features. Because, laccases have a four copper (Cu) atom in their active site which participate in oxygen reduction and water production (Fig. 8.11d). The laccase's four Cu atoms are disseminated in three types of cores or places: type 1 Cu (T1Cu), type 2 Cu (T2Cu), and type 3 Cu (T3Cu). These cores are in two metallic active sites: the mononuclear location T1 and the trinuclear location T2/T3. It is believed that laccase catalysis involves the following mechanism. Initially reduction of T1Cu by reducing substrate. Further, Internal electron transfer from the T1Cu to T2Cu and T3Cu. Subsequently, reduction of oxygen to water at T2Cu and T3Cu site. The T1Cu gives the protein its blue colour absorbance at about 600 nm depends on the intensity of the Scys–CuII bond and the ligand field strength. T2Cu does not give any colour, but it is EPR detectable, and T3Cu contains a pair of atoms in binuclear conformation that gives a weak absorbance at 330 nm but is not detected by EPR. Spectroscopy combined with crystallography has provided a detailed description of the active site in a laccase. T2Cu and T3Cu form a trinuclear center, which is involved in the catalytic mechanism.

In the catalytic mechanism, an oxygen molecule binds to the trinuclear cluster of the asymmetric activation site, and it is postulated to restrict access of the oxidizing agent. During the steady state, laccase catalysis indicates that O_2 reduction takes place. The bonds of the natural substrate, lignin, are cleaved by laccase through C_α–C_β cleavage and aryl cleavage. Subsequently, the lignin degradation is caused by phenoxy radicals leading to the oxidation of the α-carbon or by the cleavage of the bond between the α-carbon and β-carbon. This oxidation results in oxygen-centered free radicals, which can be converted by a second enzyme catalyzed reaction to quinone. The quinone and the free radicals can then undergo polymerization. The T2Cu center coordinates with two His ligands and water as ligands. The Type 3 coppers are each 4-coordinate, having three His ligands and a bridging hydroxide (Fig. 8.11d). The reduction of oxygen by a laccase appears to occur in two $2e^-$ steps. The first is the rate determining step. In this, the T2/3Cu bridging mode is reduced by the first $2e^-$. The peroxide-level intermediate facilitates the second $2e^-$ reduction (from the T2Cu and T1Cu centers) in which the peroxide is directly coordinated to the reduced T2Cu, and the reduced T1Cu is coupled to the T3Cu by covalent Cys–His linkages.

The above mentioned information is described for blue copper oxidases, i.e. blue laccases, but some authors have reported a small group of laccases that lack the 600 nm band and hence the blue color; some of these non-blue laccases (dubbed "yellow" or "white") feature a high redox potential allowing them to oxidize non-phenolic compounds without any mediators. Both types of laccases (yellow and blue) have similar MW (70 kDa) and specific activities, and it is observed that the yellow laccases are obtained from the culture grown on solid state medium, while the blue forms were isolated from culture grown on liquid medium without lignin. The yellow laccases are formed by the modification of the blue forms with low molecular weight lignin decomposition products, and some non-blue laccases (yellow) have high redox potentials, allowing them to oxidize non-phenolic compounds without any mediator (Pozdnyakova et al. 2006). Therefore, it is assumed that the yellow form of a laccase is a result of binding of the aromatic product of lignin degradation to the blue laccase. It has been postulated that a yellow laccase might contain endogenous mediators derived from lignin, which perform the role of exogenous mediator in the reaction of non-phenolic compounds. Due to insufficient information about yellow laccases, more research is required in this field. In Fig. 8.3a the simplest case is the one in which the substrate molecules are oxidised to the corresponding radicals by direct interaction with the copper cluster. However, the substrates of interest cannot be oxidized directly by laccases, either because they are too large to penetrate into the enzyme active site or because they have a particularly high redox potential. By mimicking nature, it is possible to overcome this limitation with the addition of "redox mediators", which act as intermediate substrates for laccases, whose oxidized radical forms are able to interact with the bulky or high redox potential substrate targets Fig. 8.3b

Laccases catalyze the removal of one electron from the phenolic hydroxyl groups of phenolic lignin model compounds, such as vanillyl glycol, 4,6-di (t-butyl)guaiacol, and syringaldehyde, to form phenoxy radicals, which generally undergo polymerization via radical coupling. The reaction is also accompanied by demethylation, formation of quinone, resulting in ring cleavage. The degradation of phenolic β-1 lignin substructure models occurs via the formation of phenoxy radicals, which leads to C_α–C_β cleavage, C_α oxidation, alkyl–aryl cleavage, and aromatic ring cleavage. Laccase-catalyzed oxidation of phenols, anilines, and benzenethiols correlates with the redox potential difference between the TiCu site of the laccase and the substrate. The presence of electron withdrawing o- and p-substituents reduces the electron density at the phenoxy group, and thus it is more difficult to oxidize the phenolic substrate. Bulky substituents, which impose steric interference with substrate binding, cause a decrease in reactivity.

Laccase has been found to oxidize non-phenolic model compounds and β-1 lignin dimers in the presence of a mediator, indicating that the enzyme plays a significant role in the depolymerization of lignin and pulp delignification. The most studied mediators for laccases are ABTS, 1-hydroxybenzotriazole (HBT), and 3-hyroxyanthranilic acid (HAA). The oxidation is different for ABTS and HBT, involving a dication and a benzotriazolyl-1-oxide radical, respectively. Oxygen uptake by the laccase is faster with ABTS than with HBT, but the oxidation of non-phenolic substrate is comparable for both the mediators. The ABTS-mediated oxidation of a non-phenolic substrate proceeds via an electron transfer mechanism. In the oxidation process, the ABTS is first oxidized to a radical cation (ABTS˙$^+$), and then to a dication (ABTS^{2+}), with redox potentials of 472 mV (ABTS/ABTS˙+) and 885 mV (ABTS˙+/ABTS^{2+}), respectively. The dication is the active

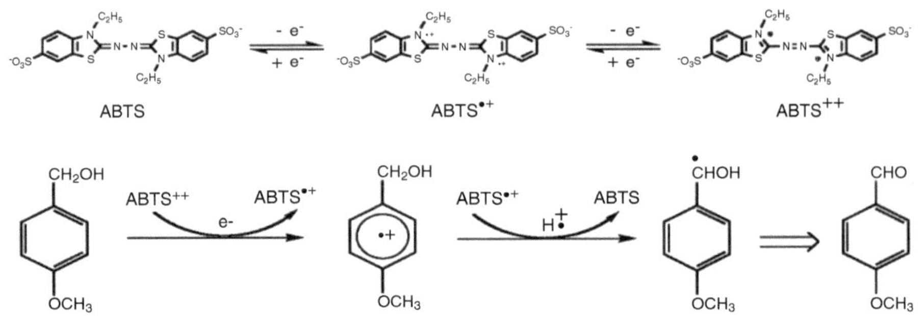

Fig. 8.12 Radical H-atom abstraction and electron transfer mechanism

intermediate, which is responsible for the oxidation of the non-phenolic substrate. The HBT⁻ mediated oxidation of non-phenolic substrate involves the initial oxidation of HBT to HBT⁺ by laccase, followed by an immediate deprotonation to form an N-oxy radical. The latter abstracts the benzylic H-atom from the substrate, converting it to a radical. The oxidation of VA by laccase-ABTS and by laccase-HBT (radical Habstraction mechanism) is presented in Fig. 8.12. HBT/HBT⁺, with an E_o value of 1.08 V, has a mediator efficiency with laccase that is higher than that of ABTS. The use of laccase/HBT for the bleaching of paper pulp and for the removal of lipophilic extractives has been described. Recently, two lignin-derived phenols, syringaldehyde and acetosyringone, have been shown to act as effective laccase mediators for the removal of lipophilic compounds from paper pulp.

The degradation of non-phenolic β-O-4 model compounds, which represent the major substructure in lignin, has been studied using laccase-mediator systems. Four types of reactions, β-ether cleavage, C_α–C_β cleavage, C_α-oxidation and aromatic ring cleavage, are catalyzed by the laccase–BHT (butylated hydroxytoluene) coupled system. In the oxidation of a non-phenolic β-O-4 lignin model dimer, 1-(4-ethoxy-3-methoxyphenyl)-1,3-dihydroxy-2-(2,6-dimethoxyphenoxy)propane, the coupled enzyme/HBT system catalyzes the 1e− oxidation of the substrate to form a β-aryl radical cation or benzylic (Ca) radical intermediates. The electron density of the aromatic ring affects the 1e− oxidation by the laccase/1-HBT couple. Substrates containing electron-donating groups favor aromatic ring cleavage products. The β-aryl radical cation is converted to the product via an aromatic ring cleavage, and the benzylic radical is cleaved at the C_α–C_β bond, similar to a Baeyer–Villiger reaction. The β-ether cleavage of the β-O-4 lignin substructure is caused by reaction with the Ca-peroxy radical intermediate produced from the benzylic radical. The rate of oxidation depends on the kcat of the laccase for the mediator and the stability of the enzyme to inactivation by the free radical of the mediator.

The ability of LiP, MnP and lacasses to degrade lignin has been studied in diverse industrial processes and bioremediation of contaminated soils and water, but this ability is non-identical between these three types of enzymes. This may be due to that enzyme-substrate interactions are different. The study of the interactive mechanisms involved in ligninolytic enzyme and lignin is indeed important in understanding enzyme reactions and may provide further insights to the development of biodegradation technologies. Ligninolytic enzymes are reported for degradation of lignin by direct interactions of ligninolytic enzymes in terms of a long-range electron transfer process. However, little is known about the effect of ligninolytic enzymes structures on the lignin biodegradation at the molecular level. The molecular docking approach can be used to model the interaction between a small molecule and a protein at the atomic level, which allow us to characterize the behavior of small molecules in the binding site of target proteins as well as to elucidate fundamental biochemical processes. Crystallographic water is a major challenge in molecular docking. These molecules are strongly bound to the receptor and observed across several crystallographic structures of a particular protein. In approximately 65% of the crystallographic protein-ligand complexes, at least one water molecule is involved in ligand-receptor recognition. The release of a crystallographic water molecule from its binding site is entropically favorable; however the process causes a simultaneous loss in enthalpy. To compensate for this enthalpy loss, a specific moiety of the ligand can be designed to mimic the interaction network of the displaced water through the formation of equivalent hydrogen bonds with the protein. Alternatively, structural water can be explicitly included in the docking experiments, allowing the formation of highly favorable hydrogen-bonding networks between the ligand and the target binding site. In this case, a variety of methods are available to evaluate which water molecules are strongly bound and, therefore, suitable for this purpose. Among these strategies one can highlight free energy perturbation calculations using Monte Carlo (MC) statistical mechanics simulations, which estimate the binding free energy for a given water molecule, allowing the discrimination between displaceable and strongly-bound structural water. MC methods generate poses of the ligand through bond rotation, rigid-body translation or rotation. The conformation obtained by this transformation is tested with an energy- based selection criterion. If it passes the criterion, it will be saved and further modified to generate next conformation. The main advantage of MC is that the change can be quite large allowing the ligand to cross the energy barriers on the potential energy surface, a point that isn't achieved easily by molecular dynamics based simulation methods.

8.4.3   Common Substrate and Microorganisms

Laccases have a broad range of substrate specificity. Due to this, microorganisms oxidize a wide range of environmental pollutants as a sole carbon or nitrogen source for their growth and metabolism. Laccases have been mostly isolated and characterized from plants and fungi, and only fungal laccases are currently used in biotechnological applications for the detoxification of complex industrial wastewater. Laccase producing microorganisms and their some common substrate are listed in Table 8.3.

Table 8.3 Some important laccase producing microorganisms in extremophilic environment and their substrate

Microorganisms	Substrate	pH	Temp. (°C)
Fungi			
Penicillium pinophilum TERI DB1	Distillery effluent decolourisation	8.5	25
Trametes versicolor	Non-phenolic lignin model dimer	5.0	25
Polyporus pinisitus	Direct red 28, Acid Blue 74	4–5	50
Myceliophthora thermophila	Direct red 28, Acid Blue 74	6.0	70–80
Trametes trogii	Nitrobenzene and anthracene	5.0	25
Trametes versicolor	Direct red 28, Acid Blue 74	5.0	65
Pleurotus sp.	2,4 Dichlorophenol, Benzo(a)pyrene [B(a)P]	5.0	28
Trichoderma atroviride	Catechol, o-cresol,	4.0–5.0	40–50
Phanerochaete crysosporium	Lignin	5.0	25
Daedalea flavida	Lignin	5.0	25
Pycnoporus coccineus	Anthracene, pyrene, fluoranthene, benzo[a] pyrene, phenanthrene	5.0	28
Phlebia sp.	Lignin	5.0	25
Phlebia floridensis	Lignin	5.0	25
Pseudochrobactrum glaciale (FJ581024)	Pulp paper mill effluent	5.0	25
Providencial rettgeri (GU193984)	Pulp paper mill effluent	5.0	25
Ganoderma lucidum	Antracene, benzo[a]pyrene, fluorine, acenapthene, acenaphthylene and benzo[a] anthracene	–	–
Bacteria			
Serratia marcescens (GU193982)	Black liquor	5.0	25
Citrobacter sp. (HQ873619)	Black liquor	5.0	25
Klebsiella pnumoniae (GU193983)	Black liquor	5.0	25
Staphylococcus saprophyticus	Brilliant blue, methyl orange, neutral red	3.0	32
Bacillus SF	Mordant black 9, mordant brown 96 & 15, acid blue 74	8.0	60
γ-Proteobacteria JB	Caramine	4–10	55
Bacillus subtilis	Syringaldazine, ABTS	3.0–7.0	75

Table 8.4 Natural and synthetic substrate of laccases

S. No.	Natural substrate	Synthetic substrate
1.	Acetosyringone	1- Hydroxylbenzotrizole (HBT)
2.	Syringaldehyde	N- Hydroxyphthalimide (HPI)
3.	Vanilin	Violuric acid (VLA)
4.	Acetovanillone	N- Hydroxylacetanlide (NHA)
5.	Sinapic acid	2,2,6,6-Tetramethylpiperidine- N-oxyl (TEMPO)
6.	Ferulic acid	Acetohydroxamic acid
7.	*p*-Coumaric acid	2,2,5,5- Tetramethylpyrrolidine-N-oxyl (PROXYL)
8.	Reduced glutathione	2,2'-Azinobis-(3-ethylbenzothia- zoline 6- sulfonic acid) (ABTS)
9.	Cystine	Guaicol
10.	Aniline	Methyl syringate
11.	4 hydroxybenzyl alcohol	

8.4.4 Screening of Laccase Producing Microorganisms, Its Substrate, Bioassay and Purification

Laccases can oxidize a wide range of molecules more than hundred different types of compound have been identified as substrate for laccase. There are various natural and synthetic substrates which are mentioned in Table 8.4, used for laccase assay. All substrates cannot be directly oxidized by laccases, either because of their large size which inhibit their penetration into the enzyme active site or because of their particular high redox potential. To overcome this hindrance, suitable chemical mediators are used which are oxidized by the laccase and their oxidized forms are able to interact with high redox potential substrate.

Although polyphenol oxidases copper proteins are able to oxidize aromatic compounds with molecular oxygen as the terminal electron acceptor. Polyphenol oxidases are associated with three types of activities:

(a) Catechol oxidase or *o*-diphenol: oxygen oxidoreductase (EC 1.10.3.1)
(b) Laccases or *p*-diphenol: oxygen oxidoreductase (EC 1.10.3.2)
(c) Cresolase or monophenols monooxygenase (EC 1.18.14.1)

Faure et al. (1995), compared commercial fungal laccase and catechol oxidase, purified from *Pyricularia oryzae* and *Agaricus bisporus*, respectively, with bacterial laccase from *A. lipoferum* by using several substrates and phenol oxidase inhibitors. Five classes of chemical compounds were investigated as substrates for laccase:

1. L-Tyrosine and several substituted monophenols such as *p*-coumaric and *o*-hydroxyphenylacetic or salicylic acids;

2. *o*-Diphenols (catechol, pyrogallol, guaiacol, and protocatechic, gallic, and caffeic acids), L-3, 4- dihydroxyphenylalanine and *o*-aminophenol, which could be oxidized by both laccase and catechol oxidase;

3. *p*-Diphenol and *p*-substituted aromatic compounds as typical *p*-phenol oxidase substrates such as hydroquinone, *p*-cresol, *p*-aminophenol and *p*-phenylenediamine;

4. *m*-Diphenols such as resorcinol, orcinol, 4-hexylresorcinol, and 5-pentadecylresorcinol;

5. Other laccase substrates such as syringaldazine, 1-naphthol, ABTS, and 4- and 5-hydroxyindoles. The range of substrates used by *A. lipoferum* laccase was similar to that used by *P. oryzae* laccase.

The screening methods of laccase producing microorganisms are similar to MnP as described previously but the minimal media does not contained H_2O_2. Bioassay of laccase activity is measured based on the oxidation of various substrate in presence of suitable buffer (i.e. sodium acetate) in acidic pH (5.0). The oxidation product is measured by spectrophotometer taking the absorbance at 420 nm or 450 nm depending upon substrate specificity. ABTS has been most commonly used substrate for laccase bioassay because it acts as cooxidant that can interact with laccase to accomplish electron transfer and it is chemically oxidized in two steps via $ABTS^+$ and $ABTS^{2+}$. Anisyl alcohol and benzyl alcohol can be better oxidized by $ABTS^{2+}$ than by $ABTS^+$. Laccase activity is assay through ABTS method. In this method reaction mixture contained 600 µL sodium acetate buffer (0.1 M, pH 5.0 at 27 °C), 300 µL ABTS (5 mM), 300 µL culture filtrate and 1400 µL distilled water. The mixture is then incubated for 2 min at 30 °C and the absorbance was measured immediately in 1-min intervals. One unit of laccase activity has been defined as activity of an enzyme that catalyzes the conversion of 1 mole of ABTS per minute. Laccase activity has been measured by another substrate guaiacol. The reaction mixture contained 10 mM sodium acetate buffer (pH 5.0), 2 mM guaiacol and 0.2 ml of culture supernatant was incubated at 25 °C for 2 h and the absorbance was read at 450 nm. The relative enzyme activity has been expressed as colorimetric units/ml (CU/ml).

In general plant lacasses are purified from sap or tissue extracts, whereas fungal lacasses are purified from culture are purified from culture (fermentation broth) of the selected organism. Various protein purification techniques are used for purification of laccase. Typical purification protocols involve ultrafiltration, ion-exchange, gel filtration and other electrophoretic and chromatographic techniques. Purification may be single or multi-step process. A single step lacasse purification from *N. crassa* are performed by using celite column chromatography and 54 fold purification with specific activity of 333 U/mg. Laccase from *T versicolor* are purified using ethanol precipitation, DEAE–sepharose, phenyl-sepharose and sephadex G-100 chromatography. *T. versicolor* 951022 excrete a single monomeric laccase showing a high specific activity of 91,443 U/mg for ABTS as substrate. Laccase from *T. versicolor* is purified with ion exchange chromatography followed by gel filtration with specificity activity of 101 U mL^{-1}

with 34.8 fold purification. Laccase from *Stereum ostrea* and obtained up to 70-fold purification from culture filtrate by a two step protocol-ammonium sulphate (80% w/v) and sephadex G-100 column chromatography.

8.4.5 Thermostability of Bacterial Laccases

Laccases are highly stable, industrially important enzymes that are capable of oxidizing a large range of substrates. Thermostability plays an important role in enzyme catalysis; several sequence and structural factors are involved in this phenomenon. Thermostable enzymes allow high process temperatures with higher associated reaction rates and less risk of microbial contamination. Some of the mechanisms/indicators of increased thermostability include: a more highly hydrophobic core, tighter packing (compactness), a deleted loop, greater rigidity (e.g. through increased proline content in the loop), higher secondary structure content, greater polar surface area, fewer thermolabile residues, increased H-bonding, higher pI, a disulfide bridge, more salt bridges and buried polar interactions. Moreover, the enhanced thermostability of a laccase from *Bacillus* sp. HR03 using site directed mutagenesis of the surface loop was achieved, in which glutamic acid (Glu188) was substituted with 2 hydrophilic(lysine and arginine) and 1 hydrophobic (alanine) residues. There are some bacterial species that show thermostability at different temperatures:

(a) Cot A from *B. subtilis* at 75 °C showed maximum activity and at 80 °C wild type Cot A has a life of 4 h, whereas recombinant Cot A from *E. coli* has a half-life of 2 h.
(b) *T. thermophilum* laccase has optimum activity at 92 °C with a half-life of 4 h at 80 °C and attains 60% of its activity after incubation for 10 min at 100 °C.

8.5 Industrial and Biotechnological Applications of MnP, LiP and Laccase

Ligninolytic enzymes are involved in the degradation of the complex and recalcitrant environmental pollutants. This group of enzymes is highly versatile in nature and they find application in a wide variety of industries. The biotechnological significance of these enzymes has led to a drastic increase in the demand for these enzymes in the recent time because of their oxidative ability toward a broad range of phenolic and non-phenolic compounds.

8.5.1 Industrial Waste Detoxification and Bioremediation

MnP, LiP and laccase have used to detoxify or remove various aromatic compounds found in industrial waste and contaminated soil and water. MnP have capability to mineralised various PAHs such as anthracene, benzo[a]pyrene, benz[a]anthracene, phenanthrene, [U-^{14}C]pentachlorophenol, [U-^{14}C]catechol, [U-^{14}C]tyrosine, [U-^{14}C]tryptophan, [4,5,9,10-14C]pyrene, [ring U-^{14}C]2-amino-4-6-dinitrotolune and 1.1.1-trichloro-2-2-bis-(4-chlorophenyl) ethane (DDT). MnP has also been reported for the elimination and detoxification of triclosan, an emerging persistent pollutant with ubiquitous presence in aquatic environment. Further, MnP detoxified afla-toxins B$_1$, lignite originated humic acid, nylon, polyethylene and amino carbonyl maillard products (melanoidins). LiP and MnP was effective in decolourizaing kraft pulp paper mill effluent. LiP was reported for mineralisation of PAHs like naph-thalene, phenanthrene, benzo[c]phenenthrene, benz[e]pyrene, benz[a]anthracene, pyrene, chrysene, anthracene, benzo[a] pyrene, perylene. Laccase has been reported for the degradation and decolourization of chlorophenol- and chlorolignin-containing black liquor and pulp paper mill wastewater. However, laccase has been reported to mineralised anthracene, benzo pyrene, fluorene and other 16 PAHs compounds which are listed by USEPA (United States Environmental Protection Agency) as priority pollutants of environment

8.5.2 Decolourisation of Textile Dye, Pulp Paper Mill and Distillery Effluent

MnP decolourise and degrade various types of synthetic dye; azo dye (reactive red 120, congo red, orange G and orange IV, remazol brilliant violet 5R, direct red 5B), anthraquinone dye (remazol brilliant blue R), indigo dye (Indigo Carmine) and triphenylmethane dye (Methyl Green) containing textile wastewater. LiP decolourised various dyes e.g. bromophenol blue, congo red, methylene blue, methyl green, methyl arrange, poly-R-478, poly S-119, poly T-128. Laccase have capability to decolourised a wide range of dyes e.g. cibanon red 2B-MD, cibanon golden yellow PK-MD, cibanon blue GFJ-MD, indanthrene direct black RBS, remazol brilliant blue R, congo red etc. Laccase catalysed textile dye bleaching may also be useful in finishing dyed cotton fabric. Under laccase catalysis, soluble dye precursors could be absorbed, oxidise and polymerised to give the desired tanning effect. LiP decolourised four distinct classes of dyes, they are- (a) tryphenylmethane(bromophenol blue), (b) heterocyclic dye (methylene blue and toluene blue O), (c) azo dye (congo red and methyl orange) and (d) polymeric dyes (Poly R-478, Poly S-119, AND Poly T-128).

8.5.3 Production of Ethanol and Value Added Products

Ligninolytic enzymes play a central role in lignin degradation. Due to high redox potential, MnP, LiP and laccase are of high industrial interest for delignification and production of ethanol and other cellulosed based low molecular weight chemicals such as vanillin, dimethoxy sulfoxide and phenol. Laccases have been used for the synthesis of several anticancer drugs such as actinocin, vinablastine and other pharmaceutical products. Laccase have been employed in several applications organic synthesis as the oxidation of functional group, the coupling of phenols and steroids, medical agent (anesthetics, anti-inflammatory, antibiotics and sedatives) and synthesis of complex natural products. MnP and LiP have potential to produce natural aromatic flavours compound e.g. vanillin, β-ionone, β-cyclocitral, dihydroactinidiolide, flavanoids and so one. Laccase can be applied to certain food processes that enhance or modify the colour appearance of food or beverages for the elimination of undesirable phenolic compounds, responsible for the browning haze formation and turbidity in clear fruit juice, beer and wine. The uses of laccase in baking process increase strength, stability and reduced thickness, thereby improving the machine ability of dough. Laccase have also been employed to sugar beet pectin gelation, baking and ascorbic acid determination.

8.5.4 Development of Biosensors

MnP and LiP are known as redox enzyme with efficient direct electron transfer (DET) properties with electrode. It may enable to use for development of biosensor based on DET, effective biofuels cell. However, laccase catalysis could be useful as biosensor for detecting oxygen and a wide variety of reducing substrate (phenols and anilines). A large number of biosensors containing laccase have been developed for immunoassay, glucose determination, aromatic amines and phenolic compounds determination. A carbon paste biosensor modified with a crude extract of the *P ostreatus* as a source of laccase source is proposed for catecholamine determination in pharmaceutical formulation.

8.5.5 Biomechanical Pulping and Pulp Biobleaching

MnP, LiP and laccase are the most important enzyme involved in biomechanical pulping and kraft pulp bleaching. In the laboratory scale consumption of refining energy in mechanical pulping was reduced with MnP pre-treatment. However, MnP degraded residual lignin of kraft pulp and enhanced the pulp bleaching effect. The laccases have also attracted considerable interest for pulp bio-bleaching. During lignin degradation, laccases are thought to act on small phenolic lignin fragments, in which the substrate reacts with the lignin polymer, resulting in its degradation. Laccase-catalyze textile dye bleaching may also be useful in finishing dyed cotton fabrics.

Take Home Message

- Lignin is amorphous complex polymer of phenylpropane units, which are cross-linked to each other with a variety of different chemical bonds. It confers rigidity and recalcitrant nature to the lignocellulosic biomass.
- The common extremophilic ligninolytic enzymes are manganese peroxidase (MnP), lignin peroxidase (LiP) and laccase. Such enzymes have also proven their utility in the pollution abatement, especially in the treatment of industrial waste/wastewater containing hazardous compound like phenols, chlorolignin synthetic dyes, and polyaromatic hydrocarbons (PAHs) as well as recalcitrant organic compounds structurally similar to lignin.
- Microorganisms with systems of thermostable enzymes decrease the possibility of microbial contamination in large scale industrial reactions of prolonged durations. The mechanisms for many thermotolerant enzymes have been reported due to their structural properties i.e. presence of Ca^{2+}, saturated fatty acid, α-helical structure etc.
- The ability of extremophilic ligninolytic enzymes to tolerate high temperature, different metal ions and organic solvents is very important for the efficient application of this enzyme in the biodegradation and detoxification of industrial waste.
- The engineering of a disulfide bond (A48C and A63C) near the distal calcium binding site of MnP by double mutation showed the improvement in thermal stability as well as pH (pH 8.0) stability in comparison to native enzyme (MnP). Some of the mechanisms/indicators of increased thermostability of laccases include: a more highly hydrophobic core, tighter packing (compactness), a deleted loop, greater rigidity (e.g. through increased proline content in the loop), higher secondary structure content, greater polar surface area, fewer thermolabile residues, increased H-bonding, higher pI, a disulfide bridge, more salt bridges and buried polar interactions.
- Moreover, the enhanced thermostability of a laccase from *Bacillus* sp. HR03 using site directed mutagenesis of the surface loop was achieved, in which glutamic acid (Glu188) was substituted with 2 hydrophilic (lysine and arginine) and 1 hydrophobic (alanine) residues.
- MnP, LiP and laccase have been used to detoxify or remove various aromatic compounds found in industrial waste and contaminated soil and water. It is used for the decolourisation of textile dye, pulp paper mill and distillery effluent. MnP, LiP and laccase are of high industrial interest for delignification and production of ethanol and other cellulose based low molecular weight chemicals such as vanillin, dimethoxy sulfoxide and phenol. Laccases have been used for the synthesis of several anticancer drugs such as actinocin, vinablastine and other pharmaceutical products. MnP, LiP and laccase are the most important enzymes involved in biomechanical pulping and kraft pulp bleaching. MnP and

LiP are known as redox enzyme with efficient direct electron transfer (DET) properties with electrode. It may enable to use for development of biosensor based on DET, effective biofuels cell.

References

Bao W, Fukushima Y, Jensen KA, Moen MA, Hammel KE (1994) Oxidative degradation of non-phenolic lignin during lipid peroxidation by fungal manganese peroxidase. FEBS Lett 354:297–300

Bharagava RN, Chandra R, Rai V (2009) Isolation and characterization of aerobic bacteria capable of the degradation of synthetic and natural melanoidins from distillery effluent. World J Microbiol Biotechnol 25:737–744

Blaich R, Esser K (1975) Function of enzymes in wood destroying fungi. 2. Multiple forms of laccase in white rot fungi. Arch Microbiol 103:271–277

Chandra R, Singh R (2012) Decolourisation and detoxification of rayon grade pulp paper mill effluent by mixed bacterial culture isolated from pulp paper mill effluent polluted site. Biochem Eng J 61:49–58

D'Souza-Ticlo D, Sharma D, Raghukumar C (2009) A thermostable metal-tolerant laccase with bioremediation potential from a marine-derived fungus. Mar Biotechnol 11(6):725–737

Faure D, Bouillant ML, Bally L (1995) Comparative study of substrates and inhibitors of *Azospirillum lipoferum* and *Pyricularia oryzae* laccases. Appl Environ Microbiol 61:1144–1146

Glenn JK, Morgan MA, Mayfield MB, Kuwahara M, Gold MH (1983) Anextracellular H2O2-requiring enzyme preparation involved in lignin biodegradation by the white-rot basidiomycete *Phanerochaete chrysosporium*. Biochem Biophys Res Commun 114:1077–1083

Guan ZB, Zhang N, Song CM, Zhou W, Zhou LZ, Zhao H, Xu CW, Cai YJ, Liao XR (2014) Molecular cloning, characterization, and dye-decolorizing ability of a temperature and pH-stable laccase from Bacillus subtilis X1. Appl Biochem Biotechnol 172:1147–1157

Hullo MF, Moszer I, Danchin A, Martin-Verstraete I (2001) CotA of Bacillus subtilis is a copper-dependent laccase. J Bacteriol 183(18):5426–5430

Morita RY (2000) Low temperature environments. Encycl Microbiol 3:93–98

Paszczyński A, Huynh VB, Crawford R (1985) Enzymatic activities of an extracellular, manganese-dependent peroxidase from *Phanerochaete chrysosporium*. FEMS Microbiol Lett 29(1–2):37–41

Piontek K, Antorini M, Choinowski T (2002) Crystal structure of a laccase from the fungus trametes versicolor at 1.90-Å resolution containing a full complement of coppers. J Biol Chem 277:37663–37669

Poulos TL, Edward SL, Wariishi H, Gold MH (1993) Crystallographic refinement of lignin peroxidase at 2 Ao. J Biol Chem 268:4429–4440

Pozdnyakova NN, Turkovskaya OV, Yudina EN, Rodakiewicz-Nowak Y (2006) Yellow laccase from the fungus *Pleurotus ostreatus* D1: Purification and characterization. Appl Biochem Microbiol 42(1):56–61

Robert SA, Wildner GF, Grass G, Weichse A, Amrus A, Rensing C, Montfort WRA (2003) Regulatory copper ion lies near the T1 copeer site in the muclticopper oxidase CueO. J Biol Chem 278:31958–31963

Schlosser D, Grey R, Fritsche W (1997) Patterns of ligninolytic enzymes in T. versicolor. Distribution of extra- and intracellular enzyme activities during cultivation on glucose, wheat straw and beech wood. Appl Microbiol Biotechnol 47:412–418

Sundaramoorthy M, Gold MH, Poulos TL (2010) Ultrahigh (0.93 Å) resolution structure of manganese peroxidase from *Phanerochaete chrysosporium*: implications for the catalytic mechanism. J Inorg Biochem 104(6):683–690

Sutherland GR, Aust SD (1996) The effects of calcium on the thermal stability and activity of manganese peroxidase. Arch Biochem Biophys 332(1):128–134

ten Have R, Hartmans S, Teunissen PJM, Field JA (1998) Purification and characterization of two lignin peroxidase isozymes produced by *Bjerkandera* sp. strain BOS55. FEBS Lett 422 (3):391–394

Tuor U, Wariishi H, Schoemaker HE, Gold MH (1992) Oxidation of phenolic aryglycerol β- aryl ether lignin model compounds by manganese peroxidase from Phanerochaete chrysosporium: Oxidative cleavage of an α-carbonyl model compound. Biochemistry 31:4986–4995

Wariishi H, Akileswaran L, Gold MH (1988) Manganese peroxidase from the basidiomycete *Phanerochaete chrysosporium*: spectral characterization of the oxidized states and the catalytic cycle. Biochemistry 27(14):5365–5370

Wariishi H, Valli K, Renganathan V, Gold MH (1989) Thiol-mediated oxidation of nonphenolic lignin model compounds by manganese peroxidase of Phanerochaete *chrysosporium*. J Biol Chem 264(24):14185–14191

Wong DW (2009) Structure and action mechanism of ligninolytic enzymes. Appl Biochem Biotechnol 157(2):174–209

Yadav M, Yadav P, Yadav KDS (2009) Purification and characterization of lignin peroxidase from *Loweporus lividus* MTCC-1178. Eng Life Sci 9(2):124–129

Yadav S, Chandra R, Rai V (2011) Characterization of potential MnP producing bacteria and its metabolic products during decolourisation of synthetic melanoidins due to biostimulatory effect of D-xylose at stationary phase. Process Biochem 46(9):1774–1784

Chapter 9
Extremophilic Pectinases

Prasada Babu Gundala and Paramageetham Chinthala

What Will You Learn From This Chapter?

Most of the harsh environments are filled with extremophiles. These extremophiles mostly adopted to such extreme conditions by secreting new macromolecules or enzymes with conformational change. Pectic enzymes are naturally, extremozymes which acts best at acidic conditions however, there are alkaline pectic enzymes also prevails and extensively used in fruit juice industry. Most of the pectinase producers belongs to fungal populations among them *Aspergillus* sp. are most notable pectinase produces. The action of pectin enzymes on pectins results in pectin oligomers which are good pre and probiotic agents. In addition, pectinases are extensively used in fruit juice extraction, textile processing, degumming of plant bast fibres, retting of plant fibres, waste water treatment, tea and coffee fermentation, paper and pulp industry, animal feed and in oil extraction.

9.1 Diverse Habitats and Their Extremities

Our planet 'earth' is a global ecosystem which supports life. An ecosystem includes all the organisms that live in a particular place, together with their physical environments. The familiar world we live is normal full of oxygen, never too cold nor too hot and protected by our atmosphere from most damaging radiations. The edges of /beyond this normal world is called '**extreme environment**'. These

P.B. Gundala (✉) • P. Chinthala
Department of Microbiology, Sri Venkateswara University, Tirupati 517502, India
e-mail: prag.babu@gmail.com; paramageetham@yahoo.com

© Springer International Publishing AG 2017
R.K. Sani, R.N. Krishnaraj (eds.), *Extremophilic Enzymatic Processing of Lignocellulosic Feedstocks to Bioenergy*, DOI 10.1007/978-3-319-54684-1_9

Fig. 9.1 World famous natural extreme environments. (**a**) Yellow stone National park, (**b**) Mono lake, (**c**) Octopus spring and (**d**) RioTinto

harsh environments gives challenging conditions for living organisms and this could be from ecosystem, climate, landscape or location. The harsh environments may be with high pH (alkaline), low pH (acidic), extremely hot, cold, hypersaline, under pressure, high radiation, without water, oxygen and light, presence of heavy metals, organic compounds, anthropogenically impacted habitats and astrobiology. There are some natural extreme environments such as" *Monolake*" Which is naturally hyper saline and alkaline ,"*Octopus spring*" which is alkaline hot spring with 95 °C, "Rio tinto" extremely acidic river which is full of heavy metals and yellow stone national park which possess unique geothermal properties (Fig. 9.1). Many extreme environments are usually dominated by the microbial communities. The organisms that thrive well in extreme environments are known as extremophiles (Satyanarayana et al. 2005). The microorganism that resides in extreme conditions not only tolerate specific extreme condition(s), but usually requires for survival and growth. The limits of the growth and reproduction of microbes are −12 °C to more than +100 °C, pH 0–13, hydrostatic pressure up to 1400 atm and salt concentrations of saturated brines.

In the earth cold environments occurs in fresh and marine waters, glaciers, polar and high alpine regions. Refrigerators and freezers are man made cold environments into permanently cold (psychrophilic) and seasonally cold (psychrotrophic) environments. Even though environments with high temperature are not wide spread as cold habitats, a variety of natural and man made high temperature habitats such as volcanic, geothermal areas, sun heated litter and soil, sediments, biological self heated environments such as compost, hay, coal, refuse piles and saw dust prevails. Most important geothermal areas are in North America, Iceland, Japan, Italy, Soviet union, India (Manikaran, Himachalpradesh). The presence of large amounts of sulfur or pyrite creates an environments with pH lower than 3. All such acidic environments are very low in organic matter and high in heavy metals. Like wise when large amounts of Na_2CO_3, NaCl (5–33%) present in environments creates alkaline environments. There are also alkaline environments created by industrial process like cement manufacturing mining, furnace slag, electroplating, food processing, pulp and paper manufacturing. Some environments like open oceans, seas, fresh water lakes, subsurface areas and desert soils are nutrient poor environments (oligotrophic). Deep sea, deep soil or sulfur well have high hydrostatic pressures (200–400 bars) and excert some stress. There are also natural hyper saline lakes, artificial solar lakes, where the salt concentration is about 3–3.5 mol/l NaCl. These hyper saline bodies may be acidic (Deep sea in the middle coast) or alkaline (Great salt lake, USA). When the water activity (a_w) is below 0.85 creates highly osmotic pressure environments. These are found in concentrated syrups. This extreme osmotic stress and low water activity results in dessication and creates xerophilic environment. Intense sources of radiations near nuclear reactions or by beams of sterilization also creates stress conditions. The environments with petroleum or synthetic organic solvents have high concentration of organic solvents.

9.2   Extremophiles

Microbes have been evolving for nearly 4.0 billion years and are capable of exploiting a vast range of energy sources and thriving in almost every habitats including extreme environments. These amazing creatures thriving in extreme environments are called extremophiles and are found in econiches with extremes of temperature (Thermophiles and Psychrophiles), pressure (Basophiles), alkalinity (alkaliphiles), acidity (Acidophiles), salinity (Halophiles), low nutrient conditions (Oligotrophs), extra dry conditions (Xerophiles), high levels of organic solvents (Organic solvent tolerant organisms), heavy metals (Metal tolerant organisms) and radiations (Radio resistant forms) (Table 9.1). These extremophiles have developed adaptations to survive in such extreme habitats, which include new mechanisms of energy transduction, regulating intracellular environment and metabolism, maintaining the structure and functioning of membranes and enzymes and so on.

Table 9.1 Distribution of extremeophiles in extreme environments

Environmental condition	Extremophilic group	Natural habitat	Extremes conditions	Typical microbes
Low temperature	Psychrophiles	Antarctic sea water	4 °C	*Bacillus* TA41
High temperature	Thermophiles	Geothermal marine sediments	100 °C	*Pyrococcus furiosus*
Low PH	Acidophiles	Acid mine drainage	pH 2.0; 75 °C	*Metallsphaera sedule*
High PH	Alkaliphiles	Sewage sludge	pH 10.1; 56 °C	*Clostridium paradoxum*
High salt	Halophiles	Hypersaline waters	4.5M NaCl	*Halobacterium halobacillus*
High pressure	Basophiles	Deep sea hydro thermal vent	250 atm; 85 °C	*Methanococcus janaschii*
Osmotic solvents	Osmophiles	Subsurface and sediments	$>0.85 \, a_w$	*Candida pelliculosa*
Dessication	Xerophiles	Soil, rocky areas.	–	*Chloroflexus aurantiacus*
Nutrient poor	Oligotrophs	Soils, subsurface environments	–	*Rhodococcus erythropolis N9T-4*
Ionizing radiations	Radiation resistant forms	Ionizing radiations	Gamma radiations	*Deinococcus radiodurans*
Presence of heavy metals	Metallo tolerents	Mine tailings	Presence of Cd^{++}. Co^{++}, Zn^{++}, Ni^{++}, Cu^{++}	*Alcaligenes eutrophicus*
Presence of organic solvents	Organic solvents resistant forms	Mud samples of Kyushu island Japan	More than 50% toluene	*Pseudomonas putida IA2000*

9.3 Extremozymes

Extremophiles are adopted at the molecular level to withstand these harsh conditions and it is true that enzymes also shows optimum functional activity at harsh condition. The enzymes from such extremophiles are called 'Extremozymes' with increased stability and activity at extremes. Due to these extreme stability extremozymes proffer innovative opportunities for biocatalysis and biotransformations. The most successful extremozyme 'Taq polymerase' from thermophilic bacteria *'Thermus aquaticus'* isolated by F.C. Lowyer of the California Institute of Technology, USA. A plenty of natural extremozymes have been isolated from thermophiles, halophiles, psychrophiles, acidophiles and alkaliphiles (Table 9.2). Natural enzymes from mesophiles differs from thermophilic and halophilic extremozymes by small number of amino acid replacements, without changing the overall protein confirmation or by possessing excess acid residues on the

Table 9.2 Pectin content in selected feedstocks

Source	Content (%wt)
Orange peel	30
Lemon nuggets	6
Lemon peel	32
Lemon pulp	25
Apple pulp	20.9
Sugar beet pulp	16.2
Peach	18
Mango	21
Pumpkin	22

enzyme surface. Further, using enzyme engineering, it is now possible to design extremozymes using site directed mutagenesis, random mutagenesis, covalent modification, multiple covalent immobilization or cross linking, protein-protein non covalent associations, antibody binding.

9.4 Pectins

9.4.1 Occurrence and Distribution of Pectic Substances

Pectins are high molecular polysaccharides that are formed in all land plants and many species of Algae. They form the major components of the middle lamella, a thin layer of adhesive extracellular material found between the primary walls of adjacent young plant cells. The pectins have multiple functions in plant growth and development. In plants the pectic substances also found in leaves, bark, roots, tubers, stalks and fruits of plants. However, large quantities of pectic substances are found in fruits, berries, flax seeds and root crops in particular sugar beet. Pectic substances contribute to a plant tissues turgor or increases a plant resistance to draught and help preserve fruits and vegetables under storage conditions. Pectins are abundant in citrus fruits (Orange, lemons, grape fruits) and in apples (Table 9.2). Commercially pectins are derived from citrus peel and apple pomace.

9.4.2 Structure and Their Classification

In nature pectin is the most complex macromolecule and heterogenous in both chemical structure and molecular weight. Naturally, pectins consists of three major polysaccharides with a back bone of galacturonic acid residues linked by $\alpha(1-4$ linkages) D-galacturonic acid residues usually referred as galacturonans. These are homogalacturonan, rhamnogalacturonan-I and rhamnogalacturonan-II (Fig 9.2).

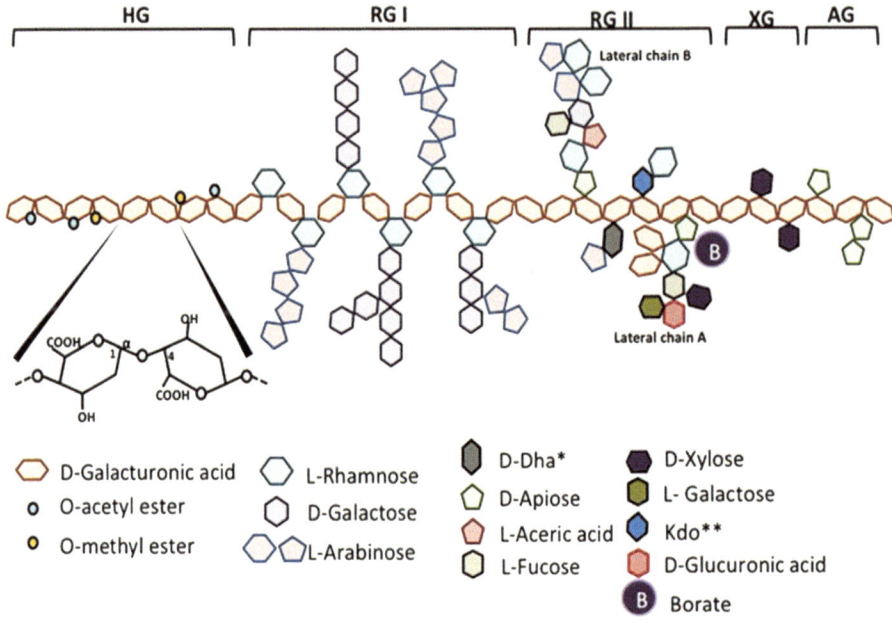

*D-Dha = 3-deoxy-D-lyxo-2-heptulosaric acid
**Kdo = 3-deoxy-D-manno-2-octulosonic acid

Fig. 9.2 Schematic representation of pectin structure. *AG* arabinogalactan, *HG* homogalacturonan, *RG* rhamnogalacturonan, *XG* xylogalacturonan)

Homogalacturonan is a linear chain α(1–4) D-galacturonic acid and residues with a variable degree of methylesterification at the carboxyl group. It could be O-acetylated at C-2 or C-3 depending on the source (Vincken et al. 2003). Rhamnogalacturonan-I consists of repeating units of the disaccharide α(1–2) L rhamnose α(1–4) D-Galacturonic acid. The galacturonic acid residues can be O-acetylated at C-2 or C-3 while 20–80% of the rhamnose residues can be substituted at C-4 or C-3 with neutral sugar side chains. The neutral sugar varies with plant sources and could be D-Galactose, L-arabinose and D-xylose, D-glucose, D-mannose, L- fucose and D-glucturonic acid. However, D-galactose, L-arabinose and D-xylose are most prevalent. Rhamnogalacteronan-II has a back bone of α (1–4) D-Galacturonic acid. The side chains attached to the back bone include 2-keto- 3- deoxy-D mono- octulosonic acid, 3-deoxy-D-lyxo-2-heptulo sonic acid, apiose and aceric acid (York et al. 1985). These rhamnogalacturonan-I and II domains are also called as hairy regions. In addition to these three domains. Voragen et al. (1995) described another three domains namely arabinogalactans, arabinans and xylogalacturonans which lack galacturonan back bone. However, commercial pectins are structurally less complex without neutral sugars. Commercial pectins consists mainly of a back bone of α-(1–4)–D-galacturonic acid with partial methyl esterification of the carboxyl groups. The composition and chemical structure of the elements of pectin varies with environmental conditions, plant source and plant development stage etc.

The American chemical society classified pectic substances into four main types based on the type of modification of back bone chain (Be Miller 1965): (a) Proto pectin, (b) Pectic acids, (c) Pectinic acids and (d) Pectins.

(a) **Proto pectins**: Water insoluble pectic substance present plant tissues. Upon restricted hydrolysis protopectin yields pectin or pectic acids

pectic acid (α - 1 , 4 - galacturonic acid)

$$\text{Protopectin} + H_2O \xrightarrow{\text{PPase}} \text{Pectin}$$
(Insoluble)   (Soluble)

(b) **Pectic acids:** These are galacturonans that contains negligble amounts of methoxyl groups. Normal or acid salts of pectic acids are called pectales.

(c) **Pectinic acid:** These are the galacturonans that contains methoxy groups (75%). Pectinotes are normal or acid salts of pectinic acids. Normal or acid salts of pectinic acids are reffered to as pectinates.

(d) **Pectin (Polymethyl galactenate):** It is a polymeric material in which 75% of the carboxyl groups of the galacturonate units are esterified with methanol. It is located in the cell wall and gives rigidity to the cell wall.

The polygalacturonic acid chain is partly esterified with methyl groups and the free acid groups may be partly or fully neutralised with sodium, potassium or ammonium ions. The ratio of esterified GalA groups to total GalA groups is termed as the DE. Based on the DE pectins are subdivided into two groups (a) High ester pectins (HM pectins) (>50 DE) and Low esterified pectins (LM) (<50 DE). These two groups of pectin gel by different mechanisms. To form gels HM-pectin requires a minimum amount of soluble solids and a pH within a narrow range, around 3.0. HM-pectin gels are thermally reversible. In general, HM-pectins are hot water soluble and often contain a dispersion agent such as dextrose to prevent lumping. LM-pectins produce gels independent of sugar content. They are also are not as sensitive to pH as the HM-pectins are. LM-pectins require the presence of a controlled amount of calcium or other divalent cations for gelation.

9.4.3 Degradation of Pectins

Pectins are decomposed instinctively by deesterification as well as by depolymerisation, the rate of this decomposition depends on pH, water activity,

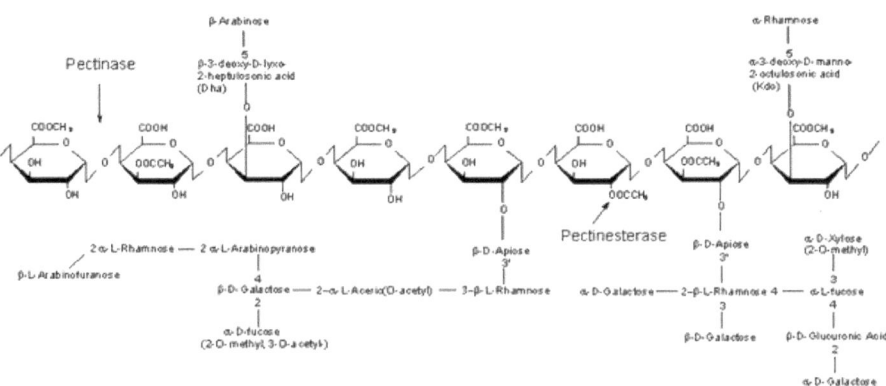

Fig. 9.3 Action sites of pectinase and pectinesterase on different pectin subunits

Table 9.3 Applications of pectic oligosaccharides and their source

Source	Importance
Citrus pectin	Stimulates apoptosis of human colonic adenocarcinoma cells
Carrots	Ability to block bacterial adherence of *E. Coli* to uroepithelial cells
Haw pectin	Reduces total cholesterol and triglycerides
Bergamot peal	Helps to improve probiotic bacteria and decrease pathogenic populations
Apple pectins	Decreases pathogenic bacteria
Sugar beet pectin	Different effects of DP_4 and DP_5 oligos and shows differential structure properties
Orange peal	Increases butyrate concentration along with increase in gut-bacteria
Sugar beet and Valencia oranges	Increased cancellation of acetate, Butyrate, propionate in the media

and temperature. Pectin depolymerization can be obtained by acid hydrolysis, hydrothermal processing, dynamic high pressure microfluidization (Chen et al. 2013) or photo chemical reaction in media containing TiO_2 and partial enzymatic hydrolysis (Concha and Zuniga 2012; Mandalari et al. 2006). However, enzymatic degradation has advantage over chemical hydrolysis and other physical methods. Enzymatic hydrolysis is target specific and produce specific useful oligosaccharides fragments (Fig 9.3). The enzymatic breakdown of pectins involves mainly two reactions i.e., hydrolysis (catalyzed by hydrolases) or β-elimination (catalyzed by lyases). High methoxyl pectins (HMP) or low methoxyl pectins (LMP) can be breakdown by endo polygalacturonases and yields 95% POS (Pectic oligosaccharides). A mixture of enzymes such as pectin—methyl esterases, endo polygalacturonases, endo arabinases and endo galactoses can also be employed to depolymerizing the pectins.

9.4.4   Industrial Importance of Pectin Degradation

Pectin depolymerization results in pectic oligosaccharides (POS). POS are a new class of prebiotics capable of extending a number of health promoting effects. Protection to cardiovascular system, colonic cells, stimulates apoptosis in human colonic adeno carcinoma, reduction of cell damage by heavy metals, anti obesity effects, anti infection, antitoxic, antibacterial and antioxidant properties. Further, they are most suitable for use in baby foods as they are not cytotoxic or mutagenic. Pectic polymer derivatives and oligosaccharides derived from pectins have positive effects on human health. These included immune regularly effects in the intestine, lowering the blood cholesterol level and slowing down and absorption of glucose in the serum of diabetic patients. Moreover, pectin oligosaccharides proved to be good probiotic components (Table 9.3).

To produce pectin oligomers pectin partial degradation or hydrolysis is necessary. Enzymatic degradation has several advantages over chemical hydrolysis since, enzymes acts on specific target and produce specific useful oligosaccharide fragments .To meet the demand of specific pectin derived oligosaccharides specialized mixture of Pectinolytic enzymes are required.

9.5   Pectic Enzymes/Pectin Degrading Enzymes

Enzymes are important biocatalyst for various industrial and biotechnological purposes and can work in many adverse condition compared to chemical catalysts. Microorganisms are preferred as a source of enzymes because of their short life span, high productivity rate, cost effective, and also free of harmful chemicals when compared to enzymes that are found in plant and animal source. Fifty percent of available enzymes are originated from fungi and yeast; 35% from bacteria, while the remaining 15% are either of plant or animal origin. The enzymes hydrolyzing the pectic substances are broadly known as pectic enzymes or pectinases. Pectinolytic enzymes or pectinases are a heterogeneous group of related enzymes that hydrolyze the pectic substances. So far, filamentous microorganisms are most widely used for pectinase production.

9.5.1   Nomenclature and Classification

Pectinases are a heterogeneous group of related enzymes therefore the classification of Pectinases is based on their preferred substrate (pectin, pectic acid or oligo D-galacturonate), the degradation mechanism (transelimination or hydrolysis) and the type of cleavage (random [endo] or terminal [exo]) (Kashyap et al. 2001).The three major groups are outlined in Fig 9.4.

Group-I-Protopectinases Protopectinases degrade the insoluble protopectin and give rise to highly polymerized soluble pectin. Pectinosinase is also synonymous with protopectinase (PPase). PPases are classified into two types, on the basis of their reaction mechanism (Sakamoto et al. 1994). A-type PPases react to the inner site, i.e. the polygalacturonic acid region of protopectin, whereas B-type PPases react to the outer site, i.e. on the polysaccharide chains that may connect the polygalacturonic acid chain and cell wall constituents. All three A-type PPases are similar in biological properties and have a similar molecular weight of 30 kDa. PPases-F is an acidic protein whereas PPase-L and -S are basic proteins. The enzymes have pectin releasing effects on protopectin from various origins. PPase-B, -C and -T have molecular weights of 45, 30, and 55 kDa respectively. PPase-B and -C have an isoelectric point (pI) of around 9.0 whereas PPase-T has a pI of 8.1

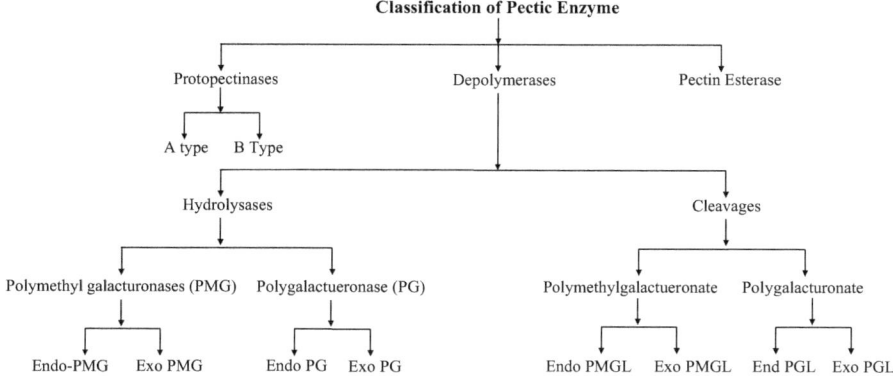

Fig. 9.4 Outlines of pectic enzyme classification

(Sakai, 1992). PPase-B, -C and -T act on protopectin from various citrus fruit peels and other plant tissues releasing pectin.

Group-II-Depolymerases Depolymerases catalyze the hydrolytic cleavage of α-(1–4) glycosidic bonds in the D-galacturonic acid moieties of the pectic substances. They exert their activity by hydrolyzing the glycosidilic linkages or by Cleavage. Depolymerases are submerged into hydrolases and cleavages

Sub group-I-Hydrolases The hydrolases include Poly methyl galacturonases (PMG) and Poly galacturonases (PG) and further classified into Exo and Endo-PMG and Exo PG or endo PG. Even though reports in literature on polymethylgalacturonases the existence of this enzyme is still in question (Sakai et al. 1993). Often, Polymethylgalacturonase preparations contaminated with PE can be mistaken for PMG containing preparations.

Polygalacturonases (PG) are the pectinolytic enzymes that catalyze the hydrolytic cleavage of the polygalacturonic acid chain with the introduction of water across the oxygen bridge. They are the most extensively studied among the family of pectinolytic enzymes. The PG involved in the hydrolysis of pectic substances are endo PG (E.C. 3.2.1.15) and exo PG (E.C. 3.2.1.67). Exo PG can be distinguished into two types: fungal exo PG which produce monogalacturonic acid as the main end product and the bacterial exo PG which produce digalacturonic acid as the main end product. Endo PG occurs in many organisms where as exo PG occurs less frequently.

Sub group-II-Cleavages Lyases (cleavages or transeliminases) perform non hydrolytic breakdown of pectates or pectinates, characterized by a trans eliminative split of the pectic polymer (Sakai et al. 1993). The lyases break the glycosidic linkages at C-4 and simultaneously eliminate hydrogen from C-5, producing a D 4:5 unsaturated products (Codner 2001) (Fig. 9.5). Lyases can be classified into the

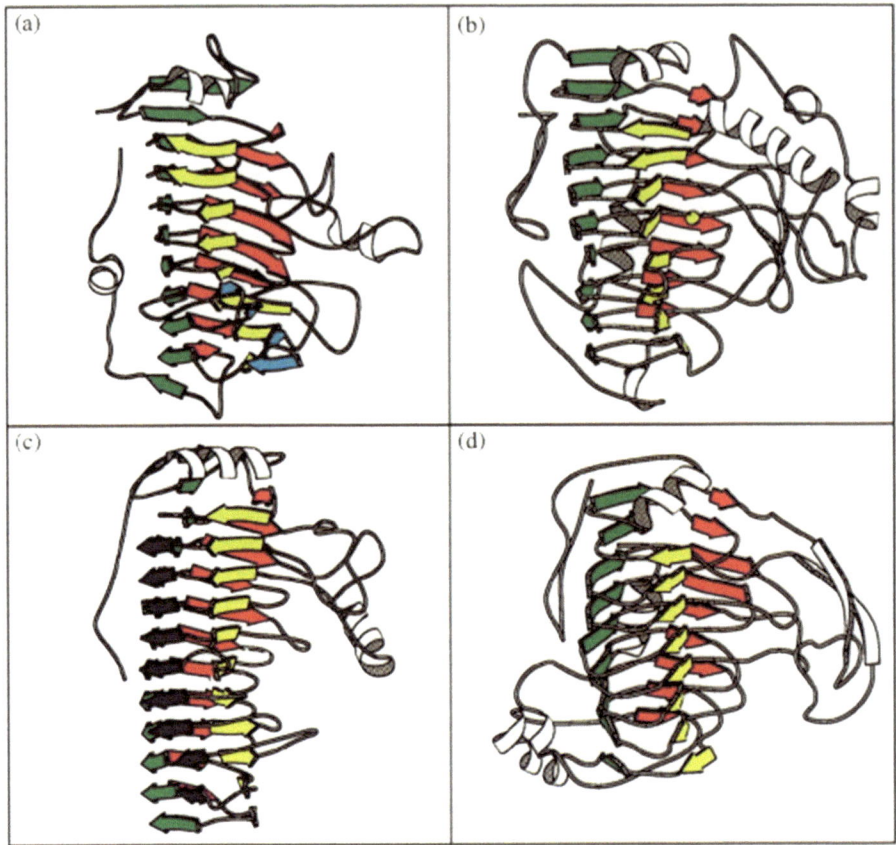

Fig. 9.5 Overall architectures of pectic enzyme. (**a**) Pectin methylesterase from *Erwinia chrysanthemi* (PDB code 1QJV), (**b**) Pectate lyase from *Bacillus subtilis* (1BN8), (**c**) Polygalacturonase from *Erwinia carotovora* (1BHE) and (**d**) Pectin lyase A from *Aspergillus niger* (1IDK). *Arrows* represent β-sheets and coils represent α-helices, parallel β-sheet 1 (PB1) is in *yellow*, PB2 in *green* and PB3 is in *red*. Polygalacturonase has an additional parallel β-sheet (PB1a), which is shown in *blue*. Pectin methylesterase has a small additional β-sheet shown in *light blue*

following types on the basis of the pattern of action and the substrate acted upon by them. These include

1. Endopolygalacturonate lyase (Endo PGL, E.C. 4.2.2.2);
2. Exopolygalacturonate lyase (Exo PGL, E.C. 4.2.2.9);
3. Endopolymethylgalacturonate lyase (Endo PMGL, E.C. 4.2.2.10);
4. Exopolymethylgalacturonate lyase (Exo PMGL).

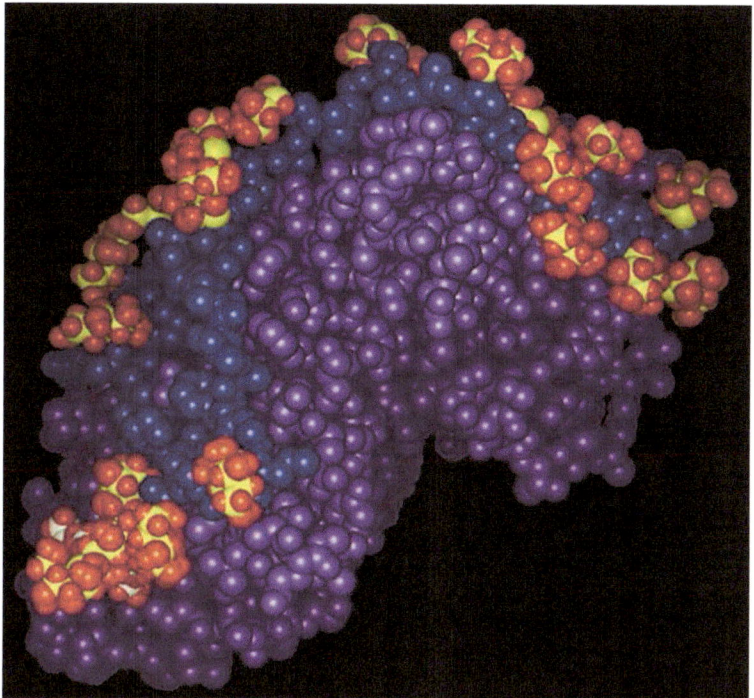

Fig. 9.6 A CPK space-filling model of RGase A, with the C-terminal tail shown in *blue* and the rest of the molecule shown in *purple*. All the O-linked glycosylation sites are located at the C-terminal tail. Mannose carbon atoms are shown in *yellow*, while the few *N*-acetylglucosamine carbons that can be seen are in *grey*. Carbohydrate oxygens are in *red*. The α(1–4)-linked mannoses are linked to threonine residues: 367, 370, 371, 372, 374, 376, 385, 386, 398, 405 and 408, and serine residues 368, 373, 380, 383, 400, 409 and 418. The N-linked glycosylation is found at Asn32 with the glycan tree: Manα1–6(Manα1–3) Manβ1–4GlcNAcβ1–4GlcNAc and at Asn299: Manβ1–4GlcNAcβ1–4GlcNAc

Group-III-Esterases Esterases catalyze the de-esterification of pectin by the removal of methoxy esters. Pectinesterases (PE) (E.C. 3.1.1.11) often referred as Pectin methylesterases, pectlylhydrolase, pectase, pectin methoxylase, pectin demethoxylase and pectolipase is a carboxylic acid esterase and belongs to the hydrolase group of enzymes (Whitaker 1984).

Recently, Schols et al. (1995) described a new group of Rhamnogalacturonan degrading enzymes. Until now four enzymes with a specific activity towards the RG I part of the pectin molecule have been reported. These are

(a) Rhamnogalacturonan hydrolase (RG hydrolase)
(b) Rhamnogalacturonan lyase (RG pectin lyase)
(c) Rhamnogalacturonan rhamnohydrolase (RG rhamnohydrolase)
(d) Rhamnogalacturonan galacturonohydrolase (RG galacturonohydrolase).

All enzymes are active towards the subunit of RG I, which contains strictly alternating galacturonic acid (GalA)—rhamnose (Rha) residues (Fig. 9.6).

RG-hydrolase and RG-lyase are endo-acting enzymes, the first one splits the GalA-Rha linkages while the second one cleaves Rha-GalA linkages leaving Δ4,5 unsaturated GalA at the non reducing end. RG-rhamnohydrolase and RG-galacturonohydrolase are exo-acting enzymes which release Rha or saturated GalA from the non reducing end of RG I respectively. RG-galacturonohydrolase in not able to split the unsaturated Gal A residue from the non reducing end (Mutter et al. 1998).

9.5.2 Pectinase Producing Organisms

Proto Pectinase Producers A-type PPases are produced by yeast and yeast-like fungi however, B-type PPases have been reported in *B.subtilis* IFO 12113, *B. subtilis* IFO 3134 (Sakai and Sakamoto, 1990) and *Trametes sp.* (Sakai et al. 1993). B-type PPases have also been produced by a wide range of *Bacillus sp.*

Depolymerases Producers Endo PG are generally distributed among fungi, bacteria and many types of yeast (Luh and Phaff 1951). They are also found in higher plants and some plant parasitic nematodes. They have been reported in many microorganisms, including *Aureobasidium pullulans, Rhizoctonia solani* Kuhn , Rhizopus stolonifer, *Fusarium moniliforme, Neurospora crassa, Thermomyces lanuginosus, Aspergillus sp. Mucor flavus*) and *Mucor circinelloides*. Endo PG have also been cloned and genetically studied in a large number of microbial species. In contrast, exo PG occur less frequently. They have been reported in *Bacteroides thetaiotamicron, Erwinia carotovora, Alternaria mali, Fusarium oxysporum , Ralstonia solanacearum Agrobacterium tumefaciens* and *Bacillus sp.*

Polygalacturonate lyases (Pectate lyases or PGLs) are produced by many bacteria and some pathogenic fungi with endo PGLs being more abundant than exo PGLs. PGLs have been isolated from bacteria and fungi associated with food spoilage and soft rot. They have been reported in *Colletotrichum lindemuthionum, Bacteroides thetaiotaomicron* and *Erwinia carotovora*. Very few reports on the production of polymethylgalacturonate lyases (pectin lyases or PMGLs) have been reported in literature. They have been reported to be produced from *Aspergillus japonicus, Pythium splendens, Pichia pinus, Aspergillus sp.* and *Thermoascus auratniacus.*

Pectin Esterase Producers PE is found in plants, plant pathogenic bacteria and fungi (Hasunuma et al. 2003). It has been reported in *Rhodotorula sp. Phytophthora infestans, Saccharomyces cerevisiae, Lachnospira pectinoschiza, Aspergillus niger, Lactobacillus lactis sp. cremoris, Penicillium frequentans, Aspergillus japonicus* and others. Rhamnogalacturonan degrading enzymes were reported from *A. aculeatus, A. niger, A. aculeatus* (Kofod et al. 1994).

9.5.3   Assay of Pectinases In Vitro

Protopectinase Assay. PPase activity can be assayed by measuring the amount of pectic substance liberated from protopectin by the carbazole sulphuric acid method or modified method. The pectin concentration is measured as D-galacturonic acid from its standard curve. One unit of PPase activity is defined as the enzyme that liberates pectic substance corresponding to 1 μmol of D-galacturonic acid per ml of reaction mixture under the assay conditions.

Depolymerase Assay. PG activity is determined on the basis of measuring, during the course of the reaction: (a) the rate of increase in number of reducing groups and (b) the decrease in viscosity of the substrate solution (Rexova-Benková and Marković 1976). The amount of reducing sugar can be readily measured by colorimetric methods like 3,5-dinitrosalicylate reagent method (Miller 1959) and the arsenomolybdate-copper reagent method (Collmer et al. 1988). One unit of enzyme activity is defined as the enzyme that releases 1 μmol/ml/min galacturonic acid under standard assay conditions. Viscosity reduction measurements have also found widespread use in determining the PG activity. The unit of enzyme activity is mostly selected as the amount of enzyme required for attaining a certain decrease of viscosity per unit time. However, this method has met with limited success. There is no direct correlation between viscosity reduction and the number of glycosidic bonds hydrolyzed. PG activity can also be determined by the cup-plate method. Cups are cut out from the solidified agar containing the substrate and are filled with the enzyme solution. After the lapse of a certain period of time, the zones of degraded substrate stained with iodine and quantified. PG isolated from different microbial sources differs markedly from each other with respect to their physico-chemical and biological properties and their mode of action. Calcium ions influence the activity of PG. However, in some cases the activity was inhibited. PG can be produced constitutively and in some cases they are inducible. Induction or stimulation is caused by the low concentration of pectins or oligo and monomeric fragments. Some PG are sensitive to catabolic repression, others inhibit *in vivo* mostly by a protein present in the host. Sometimes tannins or phenolic compounds present in the host tissue also inhibit PG activity. Among the PG obtained from different microbial sources, most have the optimal pH range of 3.5–5.5 and optimal temperature range of 30–50 °C. The molecular weight of these enzymes falls in the range of 30–60 kDa. *Aspergillus* and *Botritis sp.* produce endo PG of 85 kDa and 69 kDa respectively. Two endo PG (PG I and PG II), isolated from *A. niger* have an optimal pH range of 3.8–4.3 and 3.0–4.6 respectively (Singh and Rao 2002).

PE Assay PE activity is most readily followed by gel diffusion assay as described by Downie et al. (1998). Increased binding of ruthenium red to pectin, as the number of methyl esters attached to the pectin decreases, is used as the basis of the assay. The unit of activity in nano or picokatals is calculated based on the standard curve generated from the log-transformed commercial enzyme activity versus stained zone diameter. The sensitivity, specificity and simplicity of this PE assay are superior to all others. PE activity can also be measured by using a pH stat because ionization of the carboxyl group of the product releases a proton which causes a change in pH.

9.5.4 Applications of Pectic Enzymes

Over the years pectinases have been used in several conventional industrial processes such as textile, plant fiber processing, tea, coffee, oil extraction and treatment of industrial waste water containing pectinacious material. They have also been reported to work on the purification of viruses and in making of paper.

Fruit Juice Extraction The chief industrial application of pectinases is in fruit juice extraction and clarification. Pectin contributes to the fruit juice turbidity and viscosity. A mixture of pectinases and amylases are used to clarify fruit juices. Treatment of fruit pulps with pectinases also showed an increase in fruit juice volume from banana, grapes and apples. Pectinases in combination with other enzymes viz., cellulases, arabinases and xylanases have been used to increase the pressing efficiency of the fruits for juice extraction. Vacuum infusion of pectinases has a commercial application to soften the peel of citrus fruits, to replace hand cutting for the production of canned segments and to pickle processing where to avoid excessive softening during fermentation and storage.

Textile Processing and Bioscouring of Cotton Fibers Pectinases have been used to remove sizing agents from cotton in a safe and eco-friendly manner along with amylases, lipases, cellulases and hemicellulases with specific enzymes. Pectinases have been used in the Bioscouring is a novel process for removal of noncellulosic impurities from the fiber.

Degumming of Plant Bast Fibers Pectinases are used in degumming of textile fibres in combination with xylanases which is an ecofriendly and economical alternative to the chemical degumming.

Retting of Plant Fibers Pectinases have been used in retting of flax to separate the fibers and eliminate pectin.

Waste Water Treatment Food processing industries release waste water that contains pectin as by-product. Pretreatment of these waste waters with pectinolytic enzymes removes pectinaceous material and turn into suitable for decomposition by activated sludge treatment.

Tea and Coffee Fermentation Treatment with Pectinases accelerates tea fermentation and also destroys the foam forming property. They are also used in coffee fermentation to remove mucilaginous coat from coffee beans.

Paper and Pulp Industry During papermaking pectinase can be employed to depolymerise pectin and then to lower the peroxide bleaching.

Animal Feed Pectinases are used in the enzyme cocktail. Pectinases are used in the production of animal feeds to reduce the feed viscosity thereby to increase the absorption of nutrients.

Purification of Plant Viruses When the virions are restricted to phloem, to purify the virion particles alkaline pectinases and cellulases can be used.

Oil Extraction Citrus oils such as lemon oil can be extracted with pectinases. They destroy the emulsifying properties of pectin which interferes with the collection of oils from citrus peel extracts.

Improvement of Chromaticity and Stability of Red Wines To improve the visual characteristics red wine Pectinolytic enzymes are added to during processing. Enzymatically treated red wines also showed greater stability.

9.6 Extremophilic Pectinases

The stable increase in the number of newly isolated extremophilic microorganisms and the discovery of their enzymes by academic and industrial institutions underlines the enormous potential of extremophiles for application in future biotechnological processes. Enzymes from extremophilic microorganisms offer versatile tools for sustainable developments in a variety of industrial application because they show important environmental benefits due to their biodegradability, specific stability under extreme conditions, improved use of raw materials and decreased amount of waste products. Even though key advances have been made in the last decade, our knowledge on physiology, metabolism, enzymology and genetics of extremophilic microorganisms and their related enzymes is still limited. New techniques, such as genomics, metanogenomics, DNA evolution and gene shuffling, will lead to the production of enzymes that are highly specific for unbounded industrial applications. Due to the unusual properties of extremozymes from extremophiles, there is urgent need to optimize the already existing processes or to develop new sustainable technologies.

The notion that extremophiles are capable of surviving under non-standard conditions in non-conventional environments has led to the assumption that the properties of their enzymes have been optimized for these conditions. Depending upon the pH requirement for optimum enzymatic activity, pectinase enzyme is classified into acidic and alkaline pectinase. Acidic pectinases are useful in extraction, clarification and liquefaction of fruit juices (Kaur et al. 2004) and wines. Whereas, alkaline pectinases are widely used in the fabric industry, pulp and paper industry and in improving the quality of black tea.

Table 9.4 Acidic pectinases and their properties

Type of pectinase	Producer	Optimum pH	Properties
PG	*Saccharomyces pastorionus*	4.2	Mol. wt. 43 kDa pI: 5.4 Km. 0.62 mg/ml
Endo PG	*Saccharomyces pastorianus*	4.2	Mol. 143 kDa Km 0.59 µg/µl
Endo PG	*Cryptococcus albidus ver albitus*	3.75	Mol. wt. 41 kDa Km. 0.57 µg/µl pI. 8.1

9.6.1 Acidic Pectinases

Acidophiles are the organisms that thrive in acidic environments with pH less than 4.0.To survive under extreme pH they develop specific metabolic properties, genetic features and structural and functional characteristics of their macromolecules that are helpful in maintaining pH and distinguishing them from neutrophilic counter parts. The acidic pectinases are stable at low pH and have extensive applications in the extraction and clarification of both sparkling clear juices (apple, pear, grapes and wine) and cloudy (lemon, orange, pineapple and mango) juices and maceration of plant tissues. Additionally, acidic pectinases are useful in the isolation of protoplasts and saccharification of biomass. Most commonly these are isolated from fungal sources, especially from *Aspergillus niger* because fungi are potent producers of pectic enzymes and the optimal pH of fungal enzymes is very close to the pH of many fruit juices, which range from pH 3.0 to 5.5 (Table 9.4)

9.6.2 Alkaline Pectinases

Alkaline pectinases have been used in many industrial and biotechnological processes, such as textile and plant fiber processing, coffee and tea fermentation, oil extraction, treatment of industrial wastewater containing pectinacious material, purification of plant viruses and paper making. Especially in textile, cotton bioscouring with the alkaline pectinases would not affect the cellulose backbone and thus avoid fiber damage without pollution to environment in contrast to drastic alkaline conditions conventionally used (Klug-Santner et al. 2006). Now, alkaline pectinases have proved to be the most effective and suitable enzymes for cotton bioscouring (Wang et al. 2007). Up to now, some microorganisms have been studied for producing alkaline pectinases. The alkaline pectinases are produced predominantly from the genus *Bacillus sp.* and *Pseudomonas sp.* Although a few alkaline pectinases from various microbes have also been purified and characterized by Kobayashi et al. (2001), there is still a demand for the alkaline pectinases with

Table 9.5 Alkaline pectinases and their properties

Type of PG	Producer	Optimum pH
PGL	*Bacillus sp RK9*	10.0
PG	*Bacillus sp NT33*	10.5
PG	*Bacillus polymyxa*	8.4–9.4
PATE	*Bacillus pumilus*	8.0–8.5
PAL (Pectatelyaes)	*Amucola sp*	10.25
PATE	*Xanthomonas carperstrus*	9.5
PG	*Bacillus No. P-4-N*	10–10.5
PATE	*Bacillus stereothermophillus*	9.0
Pectin lyase	*Pencillum italicum CECT22941*	8.0
Pectin lyase	*Bacillus DT7*	8.0
PAL	*Bacillus subtilis*	8.5

high enzymatic activities and stable properties at alkaline conditions for a wide application (Klug-Santner et al. 2006).

The first paper on alkaline endopolygalacturonase produced by alkaliphilic *Bacillus* sp. strain P-4-N was published in 1972. The optimum pH for enzyme action was 10.0 for pectic acid. Fogarty and coworkers then reported that *Bacillus* sp. strain RK9 produced endopolygalacturonate lyase. The optimum pH for the enzyme activity toward acid-soluble pectic acid was 10.0. Subsequently, several papers on potential applications of alkaline pectinase have been published. The first application of alkaline pectinase-producing bacteria in the retting of Mitsumata bast. The pectic lyase (pH optimum 9.5) produced by the alkaliphilic *Bacillus* sp. strain GIR 277 has been used in improving the production of a type of Japanese paper. A new retting process produced a high-quality, non woody paper that was stronger than the paper produced by the conventional method. Cao et al. isolated four alkaliphilic bacteria, NT-2, NT-6, NT-33, and NT-82, producing pectinase and xylanase and strain NT-33, had an excellent capacity for degumming ramie fibers (Table 9.5).

9.6.3 Thermostable Pectinases

Thermostable enzymes and fungi have been topics for much research during the last two decades, but the interest in thermophiles. Thermal stability and activity of pectinases are of great significance in biotechnological processes. They offer robust catalyst alternatives, such as able to withstand to harsh conditions of industrial processing. Further, high temperatures often promote better enzyme penetration and cell wall disorganization of the raw materials (Paes and O'Donohue 2006). Primary reasons to choose thermo stable enzymes in bioprocessing is of course the intrinsic thermo stability which implies possibilities for prolonged storage (at room temperature), increased tolerance to organic solvents, reduced risk of

Table 9.6 Thermostable, Alkaline pectinases and their producers

Microorganisms	Type of pectinases	Optimum pH	Optimum temperature (°C)
Bacillu sp. NT-33	PG	10.5	75
Bacillus P-4-N	PG	10–10.5	65
Bacillus polymyxa	PG	8.4–9.4	45
Bacillus sp. PK-9	PGL	10.0	75
Bacillus subtilis	PAL	8.5	60–65
Pseudomonas syringe	PAL	8.0	40
Bacillus pumulus	PATE	8.0–8.5	45
Bacillus stearothermophiles	PATE	9.0	70
Bacillus sp DT-7	Pectate lyase	9.0	70

contamination as well as low activity losses during processing (when staying below the Tm of the enzyme) even at the elevated temperatures often used in raw material pre treatments. Discovery and use of thermo stable enzymes in combination with recombinant production have erased some of the first identified hinders (e.g. limited access and substrate specificity) for use in industrial biocatalysis. In industrial applications with thermophiles and thermo stable enzymes, isolated enzymes are today dominating over microorganisms (Turner et al. 2007). In these fields, finding new enzymes has special interest to improve the efficiency of the production systems. The high cost of the production is perhaps the major constraint in commercialization of new sources of enzymes. Though, using high yielding strains, optimal fermentation conditions and efficient enzyme recovery procedures can reduce the cost. In addition, technical constraint includes supply of cheap and pure raw materials and difficulties in achieving high operational stabilities, particularly to temperature and pH. Therefore, the understanding of various physiological and genetic aspects of pectinase is required for producing thermo stable and acid stable strains of pectinolytic fungi. (Table 9.6).

9.6.4 Cold-Active Pectinases

A diverse range of microbes have been discovered in cold environments and include representatives of the Bacteria, Eucarya and Archaea .Most microorganisms isolated from cold environments are either psychrotolerant (also termed psychrotrophic) or psychrophilic. Psychrotolerant organisms grow well at temperatures close to the freezing point of water, but have fastest growth rates above 20 °C, whereas psychrophilic organisms grow fastest at a temperature of 15 °C or lower, but are unable to grow above 20 °C. Irrespective of how they may be defined, 'psychro' microorganisms are cold-adapted and exhibit properties distinctly different from other thermal classes (e.g. thermophiles).The flexible structures of enzymes from psychrophiles (cold-adapted enzymes) compensates for the low kinetic energy present in cold environments. Because of their inherent flexible structure, cold-adapted enzymes show a reduction in activation enthalpy ($\Delta H\#$) and a more negative activation entropy ($\Delta S\#$) compared with mesophilic and thermophilic homologues. As a consequence, when temperature is decreased the reaction rate of enzymes from psychrophiles tends to decrease more slowly compared with equivalent enzymes from thermophiles. This balance of thermodynamic activation parameters is translated into relatively high catalytic activity (kcat) at low temperatures and a concomitant low structural stability compared with enzymes from mesophiles or thermophiles. The gain in enzymatic activity would be enormous if the reduction in $\Delta H\#$ was not accompanied by a concomitant decrease in $\Delta S\#$. For example, a decrease in $\Delta H\#$ of 20 kJ/mol would result in ~50,000-fold increase in kcat at 15 °C at constant $\Delta S\#$. However, in practice such a vast increase in activity is not observed as a result of enthalpy-entropy compensation http://onlinelibrary.wiley.com/doi/10.1111/j.1751-7915.2011. 00258.x/full-b62. This is reflected in the activity-stability-flexibility characteristics of many thermally adapted enzymes. The compositional and structural features that confer high flexibility to thermolabile cold-adapted enzymes are generally opposite to that of more rigid and stable mesophilic and thermophilic homologues. For example, psychrophilic enzymes tend to possess various combinations of the following features: decreased core hydrophobicity, increased surface hydrophobicity, lower arginine/lysine ratio, weaker inter-domain and inter-subunit interactions, more and longer loops, decreased secondary structure content, more glycine residues, less proline residues in loops, more proline residues in α-helices, less and weaker metal-binding sites, a reduced number of disulfide bridges, fewer electrostatic interactions (H-bonds, salt-bridges, cation–pi interactions, aromatic–aromatic interactions), reduced oligomerization and an increase in conformational entropy of the unfolded state. Genomic comparisons of psychrophiles vs. thermophiles have also revealed that distinct biases in amino acid composition is a trademark of thermal adaptation. In certain enzymes such as a zinc metalloprotease from an Arctic sea ice bacterium, the whole structure of the enzyme appears to be uniformly flexible (global flexibility) as a result of an overall decrease in H-bonding. However, in other enzymes flexibility has been shown to be localized in the structures surrounding or comprising the active site. Recently, the Psychrophilic and Pectinolytic Yeasts (PPY), which are able to degrade pectin compounds at low temperature were isolated, in order to develop the cold-active pectinases applied to food industries. The isolated PPY strains were identified to *Cystofilobasidium capitatum*, *C. larimarini*, *Cryptococcus cylindricus*, *C. macerans*, *C. aquaticus* and *Mrakia frigida* and the extracellular fraction of these strains exhibited pectin methylesterase (PME), pectin lyase (PNL) and polygalacturonase (PG) activities at 5 °C. Further, cold-active pectinases can help to reduce viscosity and clarify fruit juices at low temperatures.

9.7 Commercial Production/Status of Extremophilic Pectinases

Cost effective bulk production is required to commercialize any enzyme for industrial applications. Some researchers reported that Pectinases produced by solid state fermentation (SSF) should more stable higher stability for pH and temperature variations and were less affected by catabolic repression than Pectinases produced by SmF (Submerged fermentation). The Pectinases produced through SSF not only reduced the production costs but also the method is less polluting. So far different agricultural and agro industrial residues are used as substrates such as wheat bran, soybean, Cranberry and straw berry pomace, Coffee pulp and Coffee, Husk, Cocoa, Orange bagasse, sugarcane bagasse and wheat bran, apple pomace and from sugarbeet pulp. Due to their multifarious applications several companies are involved in commercial production of Microbial pectinases (Table 9.7).

Due to the wide range of applications of pectinases in the food industry the industrial production of pectinases has drawn world-wide attention. Molds are primarily used for the production of pectinases on a commercial scale. Suitable organisms include strains of *A. niger, A. wnetii, A. oryzae* and *Rhizopus sp.* Research for additional pectinase producers is hampered by the fact that only a limited number of microorganisms are approved for application in the food industry. Commercial production may be enhanced by a selection of more productive mutants which are not subject to catabolic repression or synthesize large quantities of enzyme without the necessity of an inducer. Although highly productive bacterial strains producing polygalacturonate lyase are known, still pectinases are not produced commercially from them. There are three different industrial methods used to produce microbial enzymes; the surface-bran culture (Koji) method, the deep-tank (submerged) process and the two-stage submerged process. According to Rombouts and Pilnik (1978) most pectinases are still produced by the surface method carried out in rotating drums. Although in general the submerged process is more widely used because of its easier control. A crucial factor in pectinase production is the composition of the medium. Details about such media are

Table 9.7 List of commercial suppliers of the pectinase enzyme[a]

Product	Supplier
Pan enzyme	C.H. Boehringer sohn; west Germany
Ultrazyme	Ciba-Geigy. A.G. Switzerland
Pectolase	Grinsteelvaekatt, Denmark
Solase	Kikkoman shoyuco, Japan
Pectinex	Schweizerische ferment, A.G. Switzerland
Rapidase: clarizyme	Society Rapidase, S.A: France
Klerzyme	Wallerstein, Co; USA
Pectinol: Rohament	Rohm, Gmb H, West Germany

[a]Adapted from Kashyap et al. (2001)

considered to be strictly confidential and are not released by the manufacturers. In general the medium will be a mixture of carbohydrates (glucose, molasses, and starch hydrolysates), N sources (NH4+ salts and yeast extract) and minerals. If the enzyme is not produced constitutively an inducer also has to be added. For reasons of economy pectin is not used much in production media but it is substituted by dried sugar cossettes, citrus peel or apple pomace. Control of pH is also very important. The highest enzyme production is achieved when the pH value drops from an initial value of about 4.5 to a more or less constant value of 3.5 during the course of the fermentation which usually takes 3–6 days. At extreme pH (7) a marked inactivation occurs. At the end of the fermentation the enzymes are extracted from the semisolid medium and mycelium. The dilute enzyme solution is concentrated and the enzymes are then precipitated with organic solvents or inorganic salts. Following precipitation the enzyme cake is centrifuged or filtered and then dried at low temperatures or spray dried (Sakai et al. 1993). Subsequently, it is ground to a particular particle size and used to prepare commercial enzyme formulations. Some preparations are sold as liquid concentrates. Pectinases are produced by a number of companies in Europe (Novo- Nordisk, Miles Kali-Chemie and Swiss Ferment Co.),the United States (Miles laboratories and Rohm and Haas Co.) and in Japan (Kikkoman Shoyu Co.) (Table 9.7).

9.8 Future Prospects of Extremophilic Pectinases

Owing to the importance of extremophilic pectinases there is an urgent need to develop pectinases with desirable physic chemical characteristics and low cost production have been the focus of much research. Enzyme producing companies constantly improves the products for more widespread use. The exploitation of knowledge of various factors/amino acids residues/non covalent interactions that is hydrogen bonding, electrostatic interactions, hydrophobic and vander vaals forces contributing towards the stability of an enzyme could be beneficial for producing the enzymes with higher stability through protein/enzyme engineering techniques. More over a better understanding on the correlation of pectic enzyme makeup and hydrolysis of pectins results in development of novel strategies to produce tailor made pectinase enzymes for the production of commercial pectin oligosaccharides.

Take Home Message

• Pectin is the most complex macromolecule and a heteropolysaccharide which is found in the cell wall of plants and fruits. It is composed of heterogenous in both chemical structure and molecular weight.
• Pectins consists of three major polysaccharides with a back bone of galacturonic acid residues linked by α (1–4 linkages). These are homogalacturonan, rhamnogalacturonan-I and rhamnogalacturonan-II. Homogalacturonan is a linear chain α(1–4)

D-galacturonic acid and residues with a variable degree of methylesterification at the carboxyl group. Rhamnogalacturonan-I consists of repeating units of the disaccharide α(1–2) L rhamnose α(1–4) D-Galacturonic acid. Rhamnogalacteronan-II has a back bone of α (1–4) D-Galacturonic acid. These rhamnogalacturonan-I and II domains are also called as hairy regions. Commercial pectins consists mainly of a back bone of α- (1–4)–D-galacturonic acid with partial methyl esterification of the carboxyl groups. Commercial pectins are structurally less complex without neutral sugars.

- The composition and chemical structure of the elements of pectin varies with environmental conditions, plant source and plant development stage etc. Based on the type of modification of back bone chain, the American chemical society classified pectic substances into four main types: Proto pectin, Pectic acids. Pectinic acids and Pectins.
- Pectinase producing organisms includes Proto Pectinase producers (yeast, yeast-like fungi, *B.subtilis* IFO 12113, *B. subtilis* IFO 3134 and *Trametes sp.*), Depolymerases Producers (*Aureobasidium pullulans, Rhizoctonia solani* Kuhn , Rhizopus stolonifer, *Fusarium moniliforme, Neurospora crassa, Thermomyces lanuginosus, Aspergillus sp. Mucor flavus,* and *Mucor circinelloides*), Pectin Esterase Producers (*Rhodotorula sp. Phytophthora infestans , Saccharomyces cerevisiae , Lachnospira pectinoschiza, Aspergillus niger, Lactobacillus lactis sp. cremoris, Penicillium frequentans , Aspergillus japonicas*).
- The assay for pectinase includes protopectinase assay and depolymerase assay. Pectic enzymes has wide range of applications including fruit juice extraction, textile processing and bioscouring of cotton fibers, degumming of plant bast fibers, retting of plant fibers, waste water treatment, tea and coffee fermentation, paper making and pulp bleaching, production of animal feeds, oil extraction, improvement of chromaticity and stability of red wines. The acidic pectinases are stable at low pH and have extensive applications in the extraction and clarification of both sparkling clear juices (apple, pear, grapes and wine) and cloudy (lemon, orange, pineapple and mango) juices and maceration of plant tissues. Additionally, acidic pectinases are useful in the isolation of protoplasts and saccharification of biomass. Alkaline pectinases have proved to be the most effective and suitable enzymes for cotton bioscouring. Cold-active pectinases can help to reduce viscosity and clarify fruit juices at low temperatures.

References

Be Miller JN (1965) Acid hydrolysis and other hydrolytic reactions of starch. In: Whistler RL, Paschall EF (eds) Starch chemistry and technology. Academic, New York, p 495
Chen J, Liang RH, Liu W, Li T, Liu CM, Wu S (2013) Pectic-oligosaccharides prepared by dynamic high-pressure microfluidization and their in vitro fermentation properties. Carbohydr Polym 91(1):175–182

Codner RC (2001) Pectinolytic and cellulolytic enzymes in the microbial modification of plant tissues. J Appl Bacteriol 84:147–160

Collmer A, Ried JL, Mount MS (1988) Assay methods for pectic enzymes. Methods Enzymol 161:329–335

Concha J, Zuniga ME (2012) Enzymatic depolymerization of sugar beet pulp: production and characterization of pectin and pectic-oligosaccharides as a potential source for functional carbohydrates. J Chem Eng 192:29–36

Downie B, Dirk LMA, Hadfield KA, Wilkins TA, Bennet AB, Bradford KJ (1998) A gel diffusion assay for quantification of pectinmethylesterase activity. Anal Biochem 264:149–157

Hasunuma T, Fukusaki EI, Kobayashi A (2003) Methanol production is enhanced by expression of an *Aspergillus niger* pectin methylesterase in tobacco cells. J Biotechnol 106:45–52

Kashyap DR, Vohra PK, Chopra S, Tewari R (2001) Applications of pectinases in the commercial sector: a review. Bioresour Technol 77:215–227

Kaur G, Kumar S, Satyanarayana T (2004) Production, characterization and application of a thermostable polygalacturonase of a thermophilic mold *Sporotrichum thermophile* Apinis. Bioresour Technol 94:239–243

Klug-Santner BG, Schnitzhofer W, Vrsanská M, Weber J, Agrawal PB, Nierstrasz VA, Guebitz GM (2006) Purification and characterization of a new bio scouring pectate lyase from *Bacillus pumilus* BK2. J Biotechnol 121:390–401

Kobayashi T, Higaki N, Yajima N, Suzumatsu A, Haghihara H, Kawai S, Ito S (2001) Purification and properties of a galacturonic acid releasing exopolygalacturonase from a strain of Bacillus. Biosci Biotechnol Biochem 65:842–847

Kofod LV, Kauppinen S, Christgau S, Andersen LN, Heldt-Hansen HP, Dorreich K, Dalboge H (1994) Cloning and characterization of two structurally and functionally divergent rhamnogalacturonases from *Aspergillus aculeatus*. J Biol Chem 269:29182–29189

Luh BS, Phaff HJ (1951) Studies on polygalacturonase of certain yeasts. Arch Biochem Biophys 33:212–227

Mandalari G, Bennett RN, Kirby AR, Lo Curto RB, Bisignano G, Waldron KW et al (2006) Enzymatic hydrolysis of flavonoids and pectic oligosaccharides from bergamot (Citrus bergamia Risso) peel. J Agric Food Chem 54(21):8307e8313

Miller GL (1959) Use of dinitrosalicyic acid reagent for determination of reducing sugars. Anal Chem 31:426–428

Mutter M, Beldman G, Pitson SM, Schols HA, Voragen AGJ (1998) Rhamnogalacturonan α-D galactopyranosylurono hydrolase an enzyme that specifically removes the terminal non reducing galacturonosyl residue in rhamnogalacturonan regions of pectin. Plant Physiol 117:153–163

Paes G, O'Donohue MJ (2006) Engineering increased thermostability in the thermostable GH-11 xylanase from *Thermobacillus xylanilyticus*. J Biotechnol 125:338–350

Rexova-Benková L, Marković O (1976) Pectic enzymes. Adv Carbohydr Chem 33:323–385

Rombouts FM, Pilnik W (1978) Enzymes in fruit and vegetable juice technology. Process Biochem 13:9–13

Sakai T (1992) Degradation of pectins. In: Winkelmann G (ed) Microbial degradation of natural products. Weinheim, VCH, pp 57–81

Sakai T, Sakamoto T (1990) Purification and some properties of a proto pectin solubilizing enzyme that has potent activity on sugar beet protopectin. Agric Biol Chem 53:1213–1223

Sakai T, Sakamoto T, Hallaert J, Vandamme EJ (1993) Pectin, pectinase and protopectinase: production, properties and applications. Adv Appl Microbiol 39:231–294

Sakamoto T, Hours RA, Sakai T (1994) Purification, characterization and production of two pectic-transeliminases with protopectinase activity from *Bacillus subtilis*. Biosci Biotechnol Biochem 58:353–358

Satyanarayana T, Raghukumar C, Shivaji S (2005) Extremophilic microbes: diversity and perspective. Curr Sci 89:78–90

Schols HA, Bakx EJ, Schipper D, AGJ V (1995) A xylogalacturonan subunit present in the modified hairy regions of apple pectin. Carbohydr Res 279:265–279

Singh SA, Rao AGA (2002) A simple fractionation protocol for, and a comprehensive study of the molecular properties of two major endopolygalacturonases from *Aspergillus niger*. Biotechnol Appl Biochem 35:115–123

Turner P, Mamo G, Karlsson EN (2007) Potential and utilization of thermophiles and thermostable enzymes in biorefining. Microb Cell Fact 6:9

Vincken J-P, Schols HA, Oomen RJ, McCann MC, Ulvskov P, Voragen AG (2003) If homogalacturonan were a side chain of rhamnogalacturonan I. Implications for cell wall architecture. Plant Physiol 132:1781–1789

Wang CF, Wang JN, Sheng ZM (2007) Solid-phase synthesis of carbon-encapsulated magnetic nanoparticles. J Phys Chem C 111(17):6303–6307

Whitaker JR (1984) Pectic substances, pectic enzymes and haze formation in fruit juices. Enzyme Microb Technol 6:341–349

York WS, Darvill AG, McNeil M, Stevenson TT, Albersheim P (1985) Isolation and characterization of plant cell walls and cell wall constituents. Methods Enzymol 118:3–40

Voragen AGJ, Pilnik W, Thibault JF, Axelos MAV, Renard CMGC (1995) Pectins. In: Stephan AM (ed) Food polysaccharides and their aplications. Dekker, New York, pp 287–339

Chapter 10
An Overview on Extremophilic Esterases

Roberto González-González, Pablo Fuciños, and María Luisa Rúa

What Will You Learn from This Chapter? Extremophiles, organisms that have evolved to exist in a variety of extreme environments, fall into a number of different groups, including thermophiles and hyperthermophiles, halophiles, psychrophiles or piezophiles. Extremophilic microorganisms have the potential to produce valuable enzymes able to function under conditions in which usually the enzymes from non-extremophilic members could not. Many novel enzymes have been isolated from these microorganisms to date; amongst all of them, hydrolases, and particularly esterases, are experiencing a growing demand. These lipolytic enzymes, having applications in food, dairy, detergent, biofuel and pharmaceutical industries, are promising catalysts that may lead towards more efficient and environmentally friendly processes.

This chapter summarizes the properties and features of esterases from the main extremophilic groups. As for the interest of hyperthermophilic esterases as biocatalysts, there are several advantages when performing industrial processes at high temperature, such as the increased solubility of polymeric substrates, reduction of viscosity, increased bioavailability, faster reaction rate, and the decreased risk of mesophilic microbial contamination. Psychrophilic esterases attract great attention because of their high catalytic efficiency at low temperature. This property not only saves energy, but it also represents a relevant advantage for processes involving heat-sensitive compounds. Additionally, their inherent high flexibility, compared to mesophilic esterases, allows for reactions under extremely low water conditions, in

R. González-González • M.L. Rúa (✉)
Department of Food and Analytical Chemistry, University of Vigo, Campus of Ourense, As Lagoas, 32004 Ourense, Spain
e-mail: mlrua@uvigo.es

P. Fuciños
International Iberian Nanotechnology Laboratory (INL), Avenida Mestre José Veiga, 4715-330 Braga, Portugal

© Springer International Publishing AG 2017
R.K. Sani, R.N. Krishnaraj (eds.), *Extremophilic Enzymatic Processing of Lignocellulosic Feedstocks to Bioenergy*, DOI 10.1007/978-3-319-54684-1_10

which the higher rigidity of mesophilic esterases limits the conversion yields. Halophilic esterases are stable and active at high salt concentrations, offering important opportunities in food processing, environmental bioremediation or biosynthetic processes.

10.1   Introduction

Carboxylesterase (EC 3.1.1.1, carboxylester hydrolases) and lipase (EC 3.1.1.3, triacylglycerol hydrolases) belong to a family of carboxylic-ester hydrolases that catalyze the hydrolysis of ester bonds found throughout the three phylogenetic domains of life. They catalyze the stereospecific hydrolysis, transesterification, and conversion of a variety of amines and primary and secondary alcohols. This property together with the fact that they do not require cofactors and are usually stable and active in the presence of organic solvents explain numerous biotechnological applications in areas such as food and medical biotechnology, organic chemical synthesis, paper manufacturing or biodiesel production (Bornscheuer 2002).

It is accepted that carboxylesterases differ from lipases in their preference toward water-soluble short-chain acylglycerols ($\leq$10 carbon atoms) and lack the interfacial activation. The majority of carboxylesterases belong to the α/β-hydrolase fold superfamily of enzymes according to conserved motifs (mainly a central core of anti-parallel beta-sheets surrounded by α-helices) and structural and biological properties (Ollis et al. 1992). However, other hydrolases have the α/β-hydrolase fold like serine carboxypeptidases, haloalkane dehalogenase, oxido-reductase or acetylcholinesterase.

Prokaryote-derived lipolytic enzymes have been classified into eight families (I–VIII) based on conserved sequence motifs and fundamental biological properties. Enzymes within family I are true lipases while those from families II–VIII are carboxylesterases (Arpigny and Jaeger 1999). Although this classification is widely used as a reference, certain modifications and extensions have been proposed (up to six new families have been incorporated) in order to accommodate new families discovered through metagenomics (López-López et al. 2014). The length of the protein sequences varies and also conserved sequence motifs vary among the different families. The most characteristic is a pentapeptide sequence motif GXSXG that is made up of four amino acids in addition to the catalytic serine; X may represent different amino acids and different conserved pentapeptide sequences in the existing families.

The discovery of new extremophilic microorganisms and their enzymes have a great impact on the field of biocatalysis. Extremophiles, that have evolved to exist in a variety of extreme environments, fall into a number of different classes including thermophiles, halophiles, acidophiles, alkaliphiles, psychrophiles, and barophiles (piezophiles). Polyextremophiles are those that can survive in more

Table 10.1 Ecology and classification of extremophiles (Pakchung et al. 2006)

Extremophile	Habitat	Genus
Thermophile	High temperature	
	Moderate thermophiles (45–65 °C)	*Pseudomonas, Bacillus, Geobacillus*
	Thermophiles (65–85 °C)	*Bacillus, Clostridium, Geobacillus, Thermotoga, Thermus, Aquifex,*
	Hyperthermophiles (>85 °C)	*Sulfolobus, Pyrolobus, Thermophilum*
Psycrophile	Low temperature (<15 °C)	*Psychrobacter, Alteromonas*
Alkalophile	High pH (pH >9)	*Bacillus, Pseudoalteromonas*
Acidophile	Low pH (pH < 2–3)	*Sulpholobus, Picrophilus*
Halophile	High salt concentration (2–5 M NaCl)	*Halobacterium, Haloferax, Halococcus*
Piezophile	High pressure (up 130 MPa)	*Shewanella, Moritella, Pyrococcus*
Metalophile	High metal concentration	*Ralstonia*
Radiophile	High radiation levels	*Deinococcus, Thermococcus*
Microaerophile	Growth in <21% O_2	*Campylobacter*

than one of these extreme conditions. Classification and some examples of extremophile genus are shown in Table 10.1.

The vast majority of extremophilic organisms belong to the prokaryotes, *Archaea* and *Bacteria* domains. These microorganisms produce unique biocatalysts that function under conditions in which non-extremophiles microorganisms could not survive. Due to their extreme stability (high thermostability, stability in organic solvents, high resistance to denaturing reagents and extreme pHs, high doses of radiation or tolerance to high levels of heavy metals,...) required in detergent, biofuel or pharmaceutical industries, enzymes from extremophiles (extremozymes) offer new opportunities for several industrial applications. This book chapter focuses on members of the carboxylic ester hydrolases (EC. 3.1.1.1).

In the recent years, there has been a great number of reports showing the identification of novel biocatalysts using metagenome-based technologies some of which show remarkable functional properties that are potentially useful for biotechnological applications. Basic steps of accessing non-cultivated microorganisms have been outlined earlier (Handelsman 2004). Up to now, more than 200 different esterases (and their biochemical characteristics) from the α/β-hydrolase superfamily were identified by metagenomic methods from a number of environments such us soils, compost, bioreactors, marine water or freshwater samples (Martínez-Martínez et al. 2013).

10.2 (Hyper)thermophilic Esterases

10.2.1 Biotechnological Interest

There are several advantages in performing industrial bioconversions at high temperatures, such as the improving of solubility and better accessibility of the enzyme to the substrate. Thus, in some cases, the use of polluting reagents to solubilize reactants and products can be avoided (Lagarde et al. 2002). Higher reaction rates can also be reached due to the reduction of viscosity and increased diffusion rate. Furthermore, the risk of microbial contamination with mesophiles is considerably diminished at high temperatures (Gomes and Steiner 2004).

10.2.2 Thermostability and Stability Against Chemicals

Thermal stability is one of the most valuable characteristics in the search for novel esterases, and extreme thermophiles have been shown to be an interesting source of them and other stable enzymes. Two types of protein thermostability are of relevance for industrial purposes: (i) thermodynamic stability which refers to the capability of an enzyme to be functional under denaturing conditions such as high temperature, presence of an organic solvent, detergents or extreme pH values and (ii) long-term stability where an enzyme does not lose its activity for prolonged incubation (Sharma et al. 2012). Structural alignment and comparative modelling have been applied in order to identify mechanisms related to thermal stability and resistance to chemical denaturation of several esterases, having being solved many three-dimensional structures to date. It has been observed that sequence and structural features that contribute to the stability of (hyper)thermophilic enzymes include changes in amino acid composition, higher hydrophobic interactions, increased number of ion pairs and salt bridges, decrease of solvent-exposed surface and oligomerisation (Levisson et al. 2009a, b; Byun et al. 2007).

Thus, apart from thermo-resistance, thermozymes frequently present an unusual resistance in the presence of several chemical and physical denaturing agents, which make them suitable for harsh industrial conditions where mesophilic enzymes could not function. In addition, their high stability combined with high retained activity levels at mesophilic temperatures might be valuable in order to cut down costs in many industrial processes: energy saving with the combination of lingering useful life of the biocatalysts. Intracellular enzymes from these microorganisms are adapted to operate optimally at (or near) the optimum growth temperature for the organism, being it 60 °C for a moderate thermophile or even 95 °C in the case of hyperthermophiles. As for extracellular thermozymes, they may be stable to temperatures considerably higher than the optimum growth temperature (Cowan and Fernandez-Lafuente 2011). As a biotechnological practical advantage of thermostability, when cloning and gene expression in a mesophilic host cell, purification of the protein can be rapidly performed to a high degree by simply heating the cell-free extract (e.g., 85 °C, 15 min). Almost all of the proteins from the host cell precipitate when this treatment is applied, while the recombinant thermo-stable protein remains in a soluble form (Atomi and Imanaka 2004). Most industrial processes in which esterases are used as biocatalysts are carried out at temperatures above 45 °C. Specifically, those involving treatment of fats are typically performed at temperatures up to 70 °C (Fuciños et al. 2011).

10.2.3   Finding Lipolitic Activity-Producing (Hyper) thermophiles and Their Carboxylesterases Characteristics

As shown in Table 10.2, several esterase-producing (hyper)thermophiles have been studied (Levisson et al. 2009a, b) and many highly thermotolerant esterases have been reported either isolated from wild-type strains (e.g. esterase E34Tt from *Thermus thermophilus* HB27, with a half-life of 135 min at 85 °C, an optimal temperature >80 °C and optimal pH of 8.1) (Fuciños et al. 2011) or expressed in heterologous hosts, such as the case of an esterase from *Aeropyrum pernix* K1 expressed in *Escherichia coli*, with an optimal temperature of 90 °C and a half-life over 160 h at 90 °C (Gao et al. 2003). As another remarkable cloning example, a highly thermostable recombinant esterase from *Pyrococcus furiosus* was described. The resulting enzyme displayed an optimal temperature of 100 °C and a half-life of 50 min at 126 °C (Ikeda and Clark 1998).

A big number of thermophilic carboxylesterases prefer medium chain (acyl chain length of around 6) p-nitrophenyl substrates (Table 10.2). Several enzymes from hyperthermophiles have also been tested for activity toward esters with various alcoholic moieties other than the standard p-nitrophenyl or 4-methylumbelliferyl esters. Other thermophilic esterases have been characterized for their ability to resolve racemic mixtures. For example, the kinetic resolution of the esterase Est3 from *Sulfolobus solfataricus* P2 was investigated using (R,S)-ketoprofen methyl ester. The esterase hydrolyzed the (R)-ester of racemic ketoprofen methylester and displayed an enantiomeric excess of 80% with a conversion rate of 20% in 32 h (Levisson et al. 2009a, b; Atomi and Imanaka 2004; Kim and Lee 2004).

Regarding to metagenomics approaches, current molecular biology techniques, direct genome shotgun sequencing, and molecular phylogenetic studies using metagenomes are nowadays making possible to build total environmental DNA libraries. They contain the genomes of unculturable microorganisms, which is allowing us to access to a big number of unknown (hyper)thermophilic and other enzymes with new and unique properties. Therefore, metagenomic libraries from thermal environments have been enormously useful over the last few years as for the aim of screening novel thermostable enzymes, including esterases. As an

Table 10.2 Characteristics of esterases from (hyper)thermophilic microorganisms

Microorganism	Enzyme name	MW (kDa)	Enzyme properties	Enzyme stability	Remarks	References
Aeropyrum pernix K1	Esterase	63	Best substrate: p-nitrophenyl C8 Optimal T: 90 °C Optimal pH: 8	Half-life over 160 h at 90 °C	Esterase and acylamino acid-releasing enzyme activities. Thermostability was protein concentration-dependent	(Gao et al. 2003)
Archaeoglobus fulgidus	Esterase AFEST	35.5	Best substrate: p-nitrophenyl C6 Optimal T: 80 °C Optimal pH: 6.5–7.5	Half-life: 26 min at 95 °C	80% of activity at 110 °C	(Manco et al. 2000)
Pyrobaculum calidifontis VA1	Esterase Est	34	Best substrate: p-nitrophenyl C6 Optimal T: 90 °C Optimal pH: 7.0	Half-life: 56 min at 110 °C	16% of retained activity at 30 °C. Active and stable in the presence of 80% water-miscible organic solvents	(Hotta et al. 2002)
Pyrococcus furiosus	Esterase	–	Best substrate: 4-methylumbelliferyl C2 Optimal T: 100 °C Optimal pH: 7.6	Half-life: 50 min at 126 °C	Half-life of 34 hours at 100 °C	(Ikeda and Clark 1998)
Sulfolobus solfataricus P1	Esterase	34	Best substrate: p-nitrophenyl C6 Optimal: 85 °C Optimal pH: 8	41% of remaining activity after 5 days at 80 °C	Detergent and organic solvent resistant (90% methanol, ethanol, 2-propanol, acetone). Activated by dimethyl sulfoxide	(Park et al. 2006)
Thermotoga maritime	Esterase (EstA)	267 (hexamer)	Best substrate: p-nitrophenyl C8 Optimal T: >95 °C Optimal pH: 8.5	Half-life: 1.5 h at 100 °C	Structure revealed the presence of an N-terminal immunoglobulin (Ig)-like domain	(Levisson et al. 2009a, b)
Thermus thermophilus HB27	Esterase E34Tt	34	Best substrate: p-nitrophenyl C10 Optimal T: >80 °C Optimal pH: 8.1	Half-life: 135 min at 85 °C	Detergent essential to solubilise the enzyme from cell membranes and for maintaining activity and stability	(Fuciños et al. 2011)

Sulfolobus solfataricus P2	Esterase (SsoPEst)	58.4	Best substrate: p-nitrophenyl C8 Optimal T: 80 °C Optimal pH: 5.5	Half-life: 1 h at 80 °C	Activity significantly inhibited by PMSF (phenylmethylsulfonyl fluoride)	(Shang et al. 2010)
Picrophilus torridus	Esterase EstA	66	Best substrate: p-nitrophenyl C2 Optimal T: 70 °C Optimal pH: 6.5	Half-life: 21 h at 90 °C	Remarkable preservation of activity in the presence of detergents, urea, and commonly used organic solvents	(Hess et al. 2008)
From metagenomic library	Esterase	29	Best substrate: p-nitrophenyl C5 Optimal T: 70 °C Optimal pH: 9	Half-life: 30 min at 80 °C	Stable at 70 °C for at least 120 min	(Tirawongsaroj et al. 2008)

example of a novel thermophilic esterase found by means of metagenomics, it is worth to mention the case of EstE1, mined from a screening of four independent metagenomic libraries of thermal areas of Indonesia. It displays a typical thermophilic profile: extremely stable at 80 °C in the absence of any stabilizer, with a high optimal temperature of 95 °C. Its activity at lower temperatures is remarkably high: 20% and 30% of its optimal activity is retained at 30 and 40 °C, respectively (Rhee et al. 2005).

10.2.4 Immobilization of Thermophilic Esterases

Enhancing of protein stability, modifications in enzyme functional properties or even recovery of specific proteins from complex mixtures, have been widely described and achieved through the development of effective methods for immobilizing. For example, the immobilization of large multi-subunit proteins with multiple covalent linkages (multipoint immobilization), results quite interesting for stabilizing proteins where the dissociation of their subunits is the initial step in enzyme inactivation. A combination of targeted chemistries, for both the support and the protein, sometimes also combined with chemical or genetic engineering, has enormously contributed to these purposes (Cowan and Fernandez-Lafuente 2011). As for immobilization studies of thermophilic esterases, many of them have been reported from genus such as *Geobacillus* or *Anoxybacillus*. As an example, the case study of an esterase from *Bacillus stearothermophilus* immobilized by multipoint covalent attachment to glyoxyl agarose. Optimized immobilization conditions increased the stability of the esterase preparations by factors up to 600-fold when compared to single-point covalent derivatives; retention of the initial activity after multipoint covalent attachment was 65%. Multipoint covalently linked esterase derivatives retained more than 70% of the initial activity after 1 week in 50% dimethyl sulfide or dimethylformamide at 30 °C. This structural stabilization was also evident under various denaturing conditions (e.g., in the presence of high concentrations of sodium chloride and organic solvents). As some examples of the importance of immobilizing, the use of enzymes as catalysts in the industrial biosynthesis of esters, fine chemicals or in order to protect hydroxyl groups of sugars, usually involves prior immobilization so as to improve the reactor performance and to facilitate the recovery of both reaction products and enzyme (Cowan and Fernandez-Lafuente 2011; Fernandez-Lafuente et al. 1995).

10.3 Psychrophilic Esterases

Psychrophilic microorganisms are adapted to live at temperatures below 5 °C. In fact, for some of them, a low temperature environment (<12 °C) is not only optimal but mandatory for sustained cell growth (Tutino et al. 2010). Psychrophiles have been isolated from natural terrestrial and aquatic environments such as Antarctic regions, glaciers, Arctic tundra soil, ocean depths, surfaces of plants and animals living in such cold environments or even in super-cooled, high-altitude cloud droplets, and refrigerated appliances (Gomes and Steiner 2004; Maiangwa et al. 2015).

Psychrophilic microorganisms comprise not only genus from Gram-negative (e.g. *Moraxella, Moritella, Polaribacter, Polaromonas, Pseudoalteromonas, Pseudomonas, Psychrobacter, Psychroflexus*) and Gram-positive bacteria (e.g. *Arthrobacter, Bacillus, Micrococcus*), but also archaea (e.g. *Halorubrum, Methanococcoides, Methanogenium*) and eukaryotes (e.g. *Candida, Cladosporium, Cryptococcus, Penicillium*) that have developed adaptive mechanisms to overcome the difficulties of living at reduced temperature (Gomes and Steiner 2004). For instance, most biological processes in mesophiles show little or no activity at low temperature, with drops on the reaction rate ranging from 16 to 80-fold when the temperature is reduced from 37 to 0 °C (Tutino et al. 2010). In contrast, psychrophiles display unique characteristics in their membranes and enzymes, which allow them to efficiently perform metabolic functions at low temperature, in some cases far below 0 °C (Gomes and Steiner 2004; Maiangwa et al. 2015).

Psychrophilic microorganisms have developed very different strategies of adaptation to cold environments, although psychrophilic enzymes do possess some common features enabling high catalytic efficiency at low temperatures (0–20 °C). Compared to enzymes from mesophiles, cold-adapted enzymes display a higher structural flexibility, particularly around the active site, for easier accommodation of substrates, and lower energy of activation. Such a high molecular flexibility at low temperature requires weakening the intramolecular forces that contribute to maintain the three-dimensional structure of the protein, so that the conformational changes necessary for reaching the activation state have a lower energy cost (Maiangwa et al. 2015).

Comparative studies using psychrophilic and mesophilic homologues showed that the increased flexibility of psychrophilic enzymes may be a result of (i) having fewer ionic interactions and hydrogen bonds; (ii) a decrease in compactness of the hydrophobic core (iii) a higher number of hydrophobic side chains exposed to the solvent; (iv) longer and more hydrophilic surface loops; (v) fewer proline and arginine residues; and (vi) a higher number of glycine residues (Tutino et al. 2010). As a consequence of these adaptive modifications, very frequently, enzymes from psychrophiles are also more thermolabile than those from mesophiles or thermophiles. A cold-adapted enzyme can be active (even optimally) at temperatures well above the temperature of the environment where the source microorganism was isolated. However, the loss of activity is often severe even at moderate temperatures near or below 50 °C, which is generally recognized as a consequence of a lack of selective pressure for thermostability in environments where the temperature is permanently cold (Gerday et al. 1997).

The above mentioned properties make psychrophilic enzymes very attractive for many industrial applications. Low temperature operation is desired either for energy saving purposes or due to substrate or product stability. Psychrophilic enzymes are also an added value in processes that require an easy enzyme inactivation by moderate heating. Particularly, psychrophilic esterases are receiving great attention due to their potential use in pharmaceutical, biotransformation and fine chemical industries, detergent and food industries or for the in situ bioremediation of fat-contaminated cold environments.

The use of psychrophilic esterases for ester synthesis in organic media represents a clear benefit. In organic media, the activity of mesophilic or thermophilic esterases is usually impaired by an excess of rigidity due to the low water content. Increasing the water content improves the flexibility of the enzyme, however, as the water content increases, the equilibrium shifts towards hydrolysis reducing the esterification yield. Thanks to their inherent flexibility, cold adapted esterases allow the use of lower water content in the reaction medium, while maintaining an optimal conversion yield (Tutino et al. 2010).

In the detergent industry, esterases are used for the removal of fatty stains, which decompose into more hydrophilic substances that are easily washed out. The use of psychrophilic esterases in formulations for cold washing of fatty stains would reduce the energy consumption, minimize the wear and tear of fabrics, and also allow for reducing the content of other undesirable chemicals in detergents (Maiangwa et al. 2015).

In the food sector, esterases have relevant applications in industries such as cheese manufacturing or flavour synthesis. In these areas, the use of cold-active esterases offers significant advantages for processes that need to be performed at low temperature to avoid the spoilage of food ingredients caused by undesirable side-reaction that can occur at high temperatures. A good example is the use of a lipolytic enzyme from Pseudomonas P38 for the synthesis of flavour esters. The cold-adapted enzyme catalyses the synthesis of butyl caprylate from butanol and caprylic acid in n-heptane at low temperatures. The biotransformation can be performed optimally (75% yield) between 15–20 °C (Joseph et al. 2008). This approach is also very convenient for the food industry because the obtained compounds can be labelled as 'natural', therefore representing a better alternative to the traditional chemical synthesis (Tutino et al. 2010).

A major challenge with psychrophilic esterases is their availability and cost of production. Few psychrophilic esterases have been studied, and even less have reached the market. The majority of the microbial diversity from cold environments is unculturable, and in those cases in which the microorganisms were successfully cultivated, the purification (needed for fine-chemical synthesis or pharmaceutical industry) was extremely difficult because of the esterases bonding to lipopolysaccharides also produced by the psychrophilic microorganisms (Maiangwa et al. 2015).

Their over-expression in heterologous hosts was also attempted. However, frequently the cold-adapted enzymes are unstable under the temperature conditions required for the expression within mesophilic bacteria (near 37 °C) (Joseph et al. 2008). Nonetheless, in the last few years, culture-independent metagenomic approaches have been applied to the discovery of novel psychrophilic esterases. Also, methods for recombinant production of cold-adapted esterases in

Table 10.3 Characteristics of cold-active esterases isolated from psychrophilic microorganisms

Organism	Esterase	MW	Enzyme properties	Enzyme stability	References
Alcanivorax dieselolei B-5(T)	EstB (expressed in *E. coli*)	45.1 kDa	Best substrate: p-nitrophenyl C6 Optimal pH: 8.5 Optimal T: 20 °C (95% max at 0 °C)	T: stable at 40 °C. Inactivated after 3 h at 55 °C	(Zhang et al. 2014)
Uncultured, from a biogas slurry metagenomic library	Est01 (expressed in *E. coli*)	44.8 kDa	Best substrate: p-nitrophenyl C4 Optimal pH: 8.0 Optimal T: 20 °C (43% max activity at 10 °C)	pH: stable at pH 9.0 T: 20% initial activity after 1 h at 40 °C. Inactivated after 20 min 50 °C	(Cheng et al. 2014)
Psychrobacter sp. Ant300	PsyEst (expressed in *E. coli*)	43 kDa	Best substrate: p-nitrophenyl C6 Optimal pH: 7.0–9. Optimal T: 35 °C (active at 5 °C)	T: Inactivated at 4 °C	(Kulakova et al. 2004)
Rhodotorula mucilaginosa	Esterase	86 kDa	Best substrate: NR Optimal pH: 7.5 Optimal T: 45 °C (20% max activity at 0 °C)	pH: unstable above pH 9 T: unstable above 50 °C	(Zimmer et al. 2006)
Uncultured, from an activated sludge metagenomic library	Lipo1 (expressed in *E. coli*)	35.6 kDa	Best substrate: p-nitrophenyl C4 Optimal pH: 7.5 Optimal T: 10 °C	pH: 70% initial activity after 24 h at 28 °C in the pH range of 6.5–8.5 T: stable at 10 °C. Inactivated above 30 °C	(Roh and Villatte 2008)

(continued)

Table 10.3 (continued)

Organism	Esterase	MW	Enzyme properties	Enzyme stability	References
Pseudoalteromonas arctica	EstO (expressed in *E. coli*)	44 kDa	Best substrate: p-nitrophenyl C4 Optimal pH: 7.5 Optimal T: 25 °C (50% activity at 0 °C)	pH: Unstable below pH 5.0 and above pH 10 T: 50% max activity after 5 h at 40 °C	(Al Khudary et al. 2010)
Streptomyces coelicolor A3(2)	EstC (expressed in *E. coli*)	35 kDa	Best substrate: p-nitrophenyl C4–C8 Optimal pH:8.5–9.0 Optimal T:35 °C	pH: retains full residual activity between pH 6–11 T: deactivated after 1 h at 40 °C	(Brault et al. 2012)
Psychrobacter pacificensis	Est10 (expressed in *E. coli*)	24.6 kDa	Best substrate: p-nitrophenyl C4 Optimal pH: 7.5 Optimal T: 25 °C (50% activity at 0 °C)	T: stable at room temperature. 80% initial activity after 2 h at 40 °C	(Wu et al. 2013)
Photobacterium sp. MA1-3	MA1-3 (expressed in *E. coli*)	35 kDa	Best substrate: p-nitrophenyl C4 Optimal pH: 8.0 Optimal T: 30 °C (45% max activity at 5 °C)	T: stable in the range 5–40 °C. Inactivated above 50 °C	(Kim et al. 2013)

NR: not reported

psychrophilic hosts are being developed. These strategies will offer high enzyme productions at competitive prices for the industry (Maiangwa et al. 2015).

Table 10.3 summarizes the properties of some psychrophilic esterases recently isolated using either classical microbiology techniques or molecular biology based approaches.

10.4   Halophilic Esterases

From the Greek roots *hals*, meaning salt, and *phil*, meaning loving or friendly with, halophily indicates that salt is required for function. The molecular ecology of extremely halophilic archaea and bacteria has been widely described. Halophiles like *Halobacterium, Haloferax, Haloarcula, Halococcus, Natronobacterium* and *Natronococcus* belong to the archaea group, while *Salinibacter ruber* is a bacterium. Halophilic organisms have been grouped into five categories according to salt tolerance: non-halophiles (<0.2 M salt), slight halophiles (0.2–0.5 M salt), moderate halophiles (0.5–2.5 M salt), borderline extreme halophiles (1.5–3.0 M salt), and extreme halophiles (2.5–5.2 M salt). Halophiles have developed two different adaptive strategies to cope with the osmotic pressure induced by the high NaCl concentration of the normal environments they inhabit (Schreck and Grunden 2014).

Some extremely halophilic bacteria accumulate salts such as sodium or potassium chloride (NaCl or KCl), up to concentrations that are isotonic with the environment (the "salting in" strategy). Enzymes from these halophiles have adapted to this environmental pressure by acquiring a relatively large number of negatively charged amino acid residues (Asp, Glu) on their surfaces to prevent precipitation. These organisms also prefer potassium as counter-ion as it has a much lower water-binding rate, which helps to maximize the available amount of water for the enzymes (Schreck and Grunden 2014).

This adaptation might confer additional stability to low water content environments and actually a number of enzymes from halophilic organisms stable and active at low water activity (as low as 0.75) have been reported (Ghanem et al. 2000). This makes identifying and using enzymes from this group of halophiles a clear approach for developing biotechnological processes based on synthesis reactions in non-aqueous media.

In contrast, moderate halophiles accumulate in the cytoplasm high amounts of specific organic solutes such as ectoine or hydroxyectoine, which function as osmoprotectants, maintaining low the intracellular salt concentration and stabilising biological structures without interfering with the normal metabolism of the cell (the compatible solute strategy) (Dalmaso et al. 2015). Intracellular enzymes from this group of halophiles do not normally show any particular adaptation to high salt concentrations but their cell walls, transporters, and periplasmic proteins typically are modified for high salt environments.

As widely described, halophilic esterases constitute an important group of biocatalysts with different biotechnological applications. Features and some examples of different properties of carboxylesterases from several halophiles are shown in Table 10.4.

Apart from being intrinsically stable and active at high salt concentrations, halophilic enzymes offer important opportunities in biotechnological applications, such as food processing, environmental bioremediation or any enzymatic transformation where is required the presence of organic solvents or low water activity.

Table 10.4 Characteristics of carboxylesterases from halophilic microorganisms

Organism	Esterase	MW (kDa)	Enzyme properties	Enzyme stability	References
Haloarcula marismortui	Esterase	–	Best substrate: p-nitrophenyl C5 Optimal pH: 7.5 Optimal T: 45 °C Optimal [NaCl] or [salt]: 4 M	pH: NR T: Loss of activity above 75 °C [NaCl] or [salt]: 50% activity lost without salt	(Camacho et al. 2009; Müller-Santos et al. 2009)
Haloarcula marismortui	Expressed in *E. coli*	34 (from sequence)	Best substrate: vinyl butyrate Optimal pH: 8.5 Optimal T: 37 °C Optimal [NaCl] or [salt]: 2 M NaCl/ 3.0 M KCl. Complete lack of activity without salt	pH: Stable between 5 and 9 T: 100% lost activity at 60 °C [NaCl] or [salt]: 100% loss without salt. Unfolded in salt-free medium	(Rao et al. 2009; Lv et al. 2010)
Haloarcula marismortui	LipC enzyme overexpressed in *E.coli* BL21	34 (from sequence)	Best substrate: p-nitrophenyl C2 Optimal pH: 9.5 Optimal T: 45 °C Optimal [NaCl] or [salt]: 3.4 M NaCl/3.0 M KCl	pH: NR T: stable at room T at 3.4 M NaCl concentration. As T increases, salt addition accelerates destabilization [NaCl] or [salt]: Circular dichroism revealed the maximal retention of the α-helical structure at the salt concentration matching the optimal activity	(Müller-Santos et al. 2009)
Thalassobacillus sp. strain DF-E4	Extracellular carboxylesterase	45	Best substrate: p-nitrophenyl C4 Optimal pH: 8.5 Optimal T: 40 °C Optimal [NaCl] or [salt]: 0.5 M NaCl	pH: Stable between 6.0 and 9.5 T: Stable at 45 °C for 1h [NaCl] or [salt]: Stable up to 4M NaCl; retained 90% activity after 12 h incubation	(Lv et al. 2010)
Halobacillus trueperi whb27	Esterase	35	Best substrate: Optimal pH: 10.0 Optimal T: 42 °C Optimal [NaCl] or [salt]: 2.5 M NaCl	–	(Yan et al. 2014)

Bacillus cereus strain *AGP-03*	Extracellular	41 kDa Dimeric (25 kDa and 16 kDa)	Best substrate: p-nitrophenyl C4 (V_{max} = 1654 U mg^{-1} protein; K_m = 52.46 μM) Optimal pH: 8.5 Optimal T: 55 °C Optimal [NaCl] or [salt]: 4.5% NaCl	pH: Stable between 5.5 and 10 T: Stable from 10 °C—75 °C (activity retained from 100 to 72%) [NaCl] or [salt]: Maximum activity at 4.5% NaCl. 100% stable between 4.5–11% NaCl	(Ghati and Paul 2015)
Marinobacter lipolyticus SM19	LipBL expressed in *E.coli* DH5α	45.3 kDa	Best substrate: p-nitrophenyl C6; tricaproin Optimal pH: 7.0 (p-nitrophenyl C4) Optimal T: 80 °C Optimal [NaCl] or [salt]: Highest activity in absence of NaCl	pH: NR T: 5% decrease in activity after 1h/45 °C; 80% decrease after 2h/50 °C [NaCl] or [salt]: NR	(Pérez et al. 2011)
Marinobacter lipolyticus	LipBL expressed in *E.coli* BL21	45.3 kDa	Best substrate: p-nitrophenyl C4 Optimal pH: 8 (p-nitrophenyl C6) Optimal T: NR Optimal [NaCl] or [salt]: Highest activity in absence of NaCl	pH: Stable between 5.5 and 10 T: 70% activity retained after incubation at 30 °C/1 h. 70% activity lost after 1h at 50 °C [NaCl] or [salt]: NR	(Pérez et al. 2012)

NR: not reported

That is because there is a relation among salt and organic solvent tolerances very often observed for halophilic enzymes because salt presence has the effect of reducing water activity (Schreck and Grunden 2014).

The recombinant esterase PE8 (219 aa, 23.19 kDa), from the marine bacterium *Pelagibacterium halotolerans* B2^T, is an alkaline esterase with an optimal pH of 9.5 and an optimal temperature of 45 °C toward *p*-nitrophenyl acetate (Wei et al. 2013). PE8 exhibited activity and enantioselectivity in the synthesis of methyl (*R*)-3-(4-fluorophenyl)glutarate ((*R*)-3-MFG), a pharmaceutically important precursor in the synthesis of the widely used antidepressant (−)paroxetine hydrochloride, from the prochiral dimethyl 3-(4-fluorophenyl)glutarate (3-DFG). (R)-3-MFG was obtained in 71.6% ee and 73.2% yield after 36 h reaction under optimized conditions (0.6 M phosphate buffer (pH 8.0) containing 17.5% 1,4-dioxane under 30 °C).

10.5 Piezophiles: Approaches Related to Lipolytic and Other Enzymes

Piezophiles have been broadly defined as those microorganisms that display optimal growth rates at high pressures. They have been observed among several prokaryotic genera such as *Shewanella, Colwellia, Moritella, Methanococcus, Psycromonas, Photobacterium, Pyrococcus* or *Thermus*, being Earth's oceans, with an average pressure of 38 MPa, their main home. Despite the fact that a vast portion of the world biosphere is a high-pressure environment, deep-sea piezophilic microorganisms are much less known than other microorganisms. The presence of hydrothermal vents in the sea floor has allowed knowing habitats where high pressure and high temperature conditions exist. Many enzymes that are stable at high pressures have been isolated from a wide variety of extremophilic microorganisms with optimal growth conditions above one atmosphere. Thus, marine organisms inhabit environments where they might be exposed to a temperature range of 1–300 °C and pressures that range from 0.1–110 MPa (Gomes and Steiner 2004; Dalmaso et al. 2015; Abe and Horikoshi 2001, Yano and Poulos 2003). It has not been fully elucidated whether piezophilic adaptation requires the modification of a few genes, metabolic pathways, or a more profound reorganization of the genome. Adaptation mechanisms against high pressure include reduction of cell division, modification of membrane and transport proteins and accumulation of osmolytes, which stabilize the proteins (Dalmaso et al. 2015).

Enzymes that can operate at high pressures and temperature have great advantages in biotechnological applications. Enzymatic reactions that have a negative change in activation volume ($\Delta V < 0$) are favored by increasing pressure; as an example, α-chymotrypsin catalyzes the hydrolysis of an anilide when $\Delta V < 0$ at increased pressure and the hydrolysis of an ester when $\Delta V > 0$ (Gomes and Steiner 2004). Despite the fact that pressure does not represent a major selective factor for protein structure and function in piezophiles and that their proteins do not need specific pressure-related adaptations (Yano and Poulos 2003), there are some examples of thermophilic protein stabilization by high pressure (Hei and Clark 1994). It has also been described the induction of stabilization and activation of several lipolytic enzymes and other hydrolases through high pressure mechanisms (Eisenmenger and Reyes-De-Corcuera 2009).

As already described (Abe and Horikoshi 2001; Yano and Poulos 2003), in spite of there being many potential biotechnological applications of piezophiles and piezoenzymes, there are not many known practical applications of piezophiles or piezophilic enzymes in the market. As a representative application, it is worth to mention the use of piezophiles in the formation of gels and starch granules (Demirjian et al. 2001). The main reason of the lack of known applications is the fact that it is not easy to cultivate piezophiles under their environment high-pressure conditions using current technology. Thus, investigation through metagenomics and many other prospections about the properties of these enzymes and other cellular components of piezophiles need to be continued (Gomes and Steiner 2004).

10.6   Polyextremophilic Esterases

Most of the microorganisms inhabiting extreme environments must thrive under multiple extreme conditions. For instance, the ocean depths combine high pressure and cold (in the ocean mud) or high temperatures (in the proximities of hydrothermal vents); hot springs frequently combine elevate temperatures and extreme pH values; and hypersaline areas can combine extreme osmotic pressure and either high (e.g. the Dead Sea in Israel) or cold temperature (e.g. brines within the sea ice in Polar regions). Excellent reviews on the poly-extremophilic microbial diversity of these environments have been reported (Dalmaso et al. 2015; Capece et al. 2013).

Polyextremophilic enzymes produced by these microorganisms display several unique properties for their applications in the food, detergent, chemical, or paper industries (Dalmaso et al. 2015). In the case of esterases, simultaneous resistance towards cold temperature and alkaline pH values (see Table 10.3 for examples) is greatly appreciated for cold-washing detergent formulations, and resistance towards extremes of pH and high concentration of organic solvents is suitable for applications in synthetic chemistry (see Table 10.2 for examples).

10.7   Relevance of Esterases to "Lignocellulosic Feedstocks"

Lignocellulose is the most abundant carbohydrate source in nature and it promises as an ideal renewable energy source. Second bioethanol generation is obtained mainly from lignocellulosic materials, which are not a food source in contrast to first bioethanol generation which derives from starch-based feedstocks (Cragg et al. 2015).

Lignocellulosic feedstocks are mainly composed of cellulose, hemicellulose and lignin. Hemicellulose is a complex branched polymer composed of different polysaccharides, such as pentoses (e.g., xylan and araban), hexoses (mainly mannan, glucan, glucomannan, and galactan), and pectin. Their backbones have branches composed of monomers such as D-galactose, D-xylose, L- arabinose, and D-glucuronic acid, generating various types of hemicelluloses that are strongly dependent on the tissue and plant species (Van Den Brink and De Vries 2011). All types of hemicelluloses are partially esterified with acetic acid. Acetylation increases their solubility but also prevents hydrolysis of glycosidic linkages of hemicelluloses by the corresponding hydrolases hampering the enzymatic saccharification to produce fermentable sugars (Cuervo-Soto et al. 2015).

Hemicellulose encases cellulose, i.e. a linear polymer of D-glucose subunits linked by β-1,4-glycosidic bonds. Both hemicellulose and cellulose are cross-linked in the plant cell walls with the hydrophobic network of lignin, the third most abundant component in lignocellulosic materials, which makes the chemical or enzymatic hydrolysis of cellulose and depolymerization of hemicellulose difficult. The highly organized crystalline structure of cellulose presents an obstacle to its hydrolysis. Lignin itself is almost completely resistant against microbial (and enzymatic) attack. Pretreatment steps (mechanical, chemical, or combinations of both) are always needed in order to deconstruct those recalcitrant structures and make them more accessible to enzymes (Bornscheuer et al. 2014).

On the other hand, enzymatic degradation of lignocellulosic materials needs the combined and synergistic action of several enzymes.

For cellulose to be degraded, three main activities are involved: chain-end-cleaving cellobiohydrolases, internally chain-cleaving endoglucanases, and ß-glucosidases, which hydrolyze soluble short-chain glucooligosaccharides to glucose.

Complete hydrolysis of the hemicellulose fraction requires two groups of enzymes: (1) endo- xylanase and β-xylosidase, which cleave the xylan main chain; and (2) accessory enzymes, which remove the side chains and break crosslinks between xylan and other plant polymers. The latter group consists of α-L-arabino- furanosidase, α-glucuronidase and a number of carbohydrate esterases (CE) (Wong 2006).

CE show a great diversity in substrate specificity and structure and are currently classified in the Carbohydrate-Active Enzymes (CAZymes) database (Cantarel et al. 2009), in 16 different CE families (CAZy; http://www.cazy.org). CE catalyze the de-O or de-N-acylation of substituted saccharides, acting on two classes of substrates: those in which the sugar plays the role of the "acid", such as pectin methyl esters and those in which the sugar behaves as the alcohol, such as in acetylated xylan. A number of possible reaction mechanisms may be involved: the most common is a Ser-His-Asp catalytic triad catalyzed deacetylation analogous to the action of classical lipase and serine proteases but other mechanisms such as a Zn^{2+} catalyzed deacetylation prevails in some families (Lombard et al. 2014). In terms of substrate specificity, the 16 CE families represent acetylxylan esterases, acetyl esterases, chitin deacetylases, peptidoglycan deacetylases, feruloyl esterases, pectin acetyl esterases, pectin methylesterases, glucuronoyl esterases and enzymes catalyzing N-deacetylation of low molecular mass amino sugar derivatives (Biely 2012). Feruloyl esterases (EC 3.1.1.73) do not fit into the established CE families and have been separately classified based on sequence similarities (Udatha et al. 2011). Among the accessory enzymes, feruloyl esterases play a key role in enhancing the accessibility of enzymes to and subsequent hydrolysis of hemicellulose fibers by removing the ferulic acid side chains and crosslinks. Ferulic acids are covalently linked to polysaccharides, including glucuronoarabinoxylans, xyloglucans, and pectins, through ester linkages. Ferulic acid is a cinnamic acid with the chemical name (3-methoxy- 4-hydroxy)-3-phenyl-2-propenoic acid, or 3-methoxy-4-hydroxy-cinnamic acid (Wong 2006).

Feruloyl esterases belong to a subclass of carboxylic esterases (EC 3.1.1). A wide range of bacteria and fungi has been reported to secrete these enzymes, which are highly inducible, depending on the growth substrates (Wong 2006). Feruloyl esterases have been extensively covered in recent excellent reviews (Wong 2006; Udatha et al. 2011; Aurilia et al. 2008; Faulds 2010; Fazary and Ju 2007; Koseki et al. 2009).

Thermostable enzymes that hydrolyze lignocellulose to its component sugars have significant advantages for improving the conversion rate of biomass over their mesophilic counterparts. Temperature might be advantageous to accelerate the lignocellulose deconstruction and facilitate the simultaneous saccharification prior or during fermentation of the resulting sugars to bioethanol. In an excellent review (Blumer-Schuette et al. 2014), the possibilities of the thermophilic deconstruction of lignocellulosic materials are extensively analysed, providing great information on thermophilic microorganisms and their thermozymes to degrade plant biomass.

10.8 Conclusions and Perspectives

Extremozymes have a great economic potential in many industrial processes and, in particular, esterases are one of the most important groups of biocatalysts for biotechnological applications.

In the recent years, there has been a great increase in the number of reports that, through metagenomic techniques, have accessed to the genomes of unculturable prokaryotes. This technology has led to the identification and characterization of a vast number of biocatalysts active under a myriad of conditions that ultimately reflect the primordial environment from which they were isolated. Esterases are a good example of that and they are well represented within microbial communities operating in most environments on the Earth.

Extremozyme libraries are commonly constructed using mesophilic hosts mainly due to the deepest knowledge of the expression systems and easy cultivation, but often low-level protein expression is achieved compromising any basic and biotechnological application where larger amounts of the recombinant protein would be needed. There has been a consistent effort in the scientific community to develop genetic systems for robust protein expression, mainly using archaeal genera of thermophilic organisms, but still more work is needed.

Finally, although some advances have been made, studies on the physiology, metabolism, enzymology and genetics of extremophiles are still limited. Therefore, it is expected that this area will experience an important growth, pursuing a better understanding of the application of esterases and other hydrolases from extremophiles.

Take Home Message:

- Carboxylesterase and lipase belong to a family of carboxylic-ester hydrolases that catalyze the hydrolysis of ester bonds. They catalyze the stereospecific hydrolysis, transesterification, and conversion of a variety of amines and primary and secondary alcohols. Sequence and structural features that contribute to the stability of (hyper)thermophilic enzymes include changes in amino acid composition, higher hydrophobic interactions, increased number of ion pairs and salt bridges, decrease of solvent-exposed surface and oligomerisation.
- The use of psychrophilic esterases in formulations for cold washing of fatty stains would reduce the energy consumption, minimize the wear and tear of fabrics and also allow for reducing the content of other undesirable chemicals in detergents.
- The use of cold-active esterases offers significant advantages for processes that need to be performed at low temperature to avoid the spoilage of food ingredients caused by undesirable side-reaction that can occur at high temperatures.
- Carboxylesterase can be produced from different extremophilic sources. They are produced from moderate thermophiles (45–65 °C) such as *Pseudomonas, Bacillus, Geobacillus*, Thermophiles (65–85 °C) such as *Bacillus, Clostridium, Geobacillus, Thermotoga, Thermus, Aquifex, Sulfolobus, Pyrolobus, Thermophilum*, Hyperthermophiles (> 85 °C) such as *Sulfolobus, Pyrolobus, Thermophilum*, Psychrophiles (low temperature < 15 °C) such as *Psychrobacter, Alteromonas*, Alkalophiles (pH > 9) such as *Bacillus, Pseudoalteromonas*, Acidophiles (pH< 2–3) such as *Sulpholobus, Picrophilus*, Halophiles (high salt concentration, 2–5 M NaCl) such as *Halobacterium, Haloferax, Halococcus*, Piezophiles (high pressure up 130 MPa) such as *Shewanella, Moritella, Pyrococcus*, Metallophiles (high metal concentration) such as *Ralstonia*,

Radiophiles (high radiation levels) such as *Deinococcus, Thermococcus and* Microaerophiles (growth in <21% O_2) such as *Campylobacter*.
- They have a wide range of applications such as food, detergent, chemical, paper industries and treatment of biomass (bioremediation).

References

Abe F, Horikoshi K (2001) The biotechnological potential of piezophiles. Trends Biotechnol 19:102–108. doi:10.1016/S0167-7799(00)01539-0

Al Khudary R, Venkatachalam R, Katzer M et al (2010) A cold-adapted esterase of a novel marine isolate, *Pseudoalteromonas arctica*: gene cloning, enzyme purification and characterization. Extremophiles 14:273–285. doi:10.1007/s00792-010-0306-7

Arpigny JL, Jaeger K-E (1999) Bacterial lipolytic enzymes: classification and properties. Biochem J 343:177–183. doi:10.1042/0264-6021:3430177

Atomi H, Imanaka T (2004) Thermostable carboxylesterases from hyperthermophiles. Tetrahedron: Asymmetr 15:2729–2735. doi:10.1016/j.tetasy.2004.07.054

Aurilia V, Parracino A, D'Auria S (2008) Microbial carbohydrate esterases in cold adapted environments. Gene 410:234–240

Biely P (2012) Microbial carbohydrate esterases deacetylating plant polysaccharides. Biotechnol Adv 30:1575–1588. doi:10.1016/j.biotechadv.2012.04.010

Blumer-Schuette SE, Brown SD, Sander KB et al (2014) Thermophilic lignocellulose deconstruction. FEMS Microbiol Rev 38:393–448. doi:10.1111/1574-6976.12044

Bornscheuer UT (2002) Microbial carboxyl esterases: classification, properties and application in biocatalysis. FEMS Microbiol Rev 26:73–81

Bornscheuer U, Buchholz K, Seibel J (2014) Enzymatic degradation of (ligno)cellulose. Angew Chemie—Int Ed 53:10876–10893. doi:10.1002/anie.201309953

Brault G, Shareck F, Hurtubise Y et al (2012) Isolation and characterization of EstC, a new cold-active esterase from *Streptomyces coelicolor* A3(2). PLoS One 7:e32041

Byun J, Rhee J, Kim ND et al (2007) Crystal structure of hyperthermophilic esterase EstE1 and the relationship between its dimerization and thermostability properties. BMC Struct Biol 11:1–11. doi:10.1186/1472-6807-7-47

Camacho R, Mateos J, González-Reynoso O et al (2009) Production and characterization of esterase and lipase from *Haloarcula marismortui*. J Ind Microbiol Biotechnol 36:901–909

Cantarel BI, Coutinho PM, Rancurel C et al (2009) The Carbohydrate-Active EnZymes database (CAZy): an expert resource for glycogenomics. Nucleic Acids Res. doi:10.1093/nar/gkn663

Capece MC, Clark E, Saleh JK et al (2013) Polyextremophiles and the constraints for terrestrial habitability BT—polyextremophiles. In: Polyextremophiles. Springer Netherlands, Dordrecht, pp 3–59

Cheng X, Wang X, Qiu T et al (2014) Molecular cloning and characterization of a novel cold-adapted family VIII esterase from a biogas slurry metagenomic library. J Microbiol Biotechnol 24:1484–1489

Cowan D a, Fernandez-Lafuente R (2011) Enhancing the functional properties of thermophilic enzymes by chemical modification and immobilization. Enzyme Microb Technol 49:326–346. doi:10.1016/j.enzmictec.2011.06.023

Cragg SM, Beckham GT, Bruce NC et al (2015) Lignocellulose degradation mechanisms across the Tree of Life. Curr Opin Chem Biol 29:108–119. doi:10.1016/j.cbpa.2015.10.018

Cuervo-Soto LI, Valdés-García G, Batista-García R et al (2015) Identification of a novel carbohydrate esterase from *Bjerkandera adusta*: Structural and function predictions through bioinformatics analysis and molecular modeling. Proteins Struct Funct Bioinf 83:533–546. doi:10.1002/prot.24760

Dalmaso G, Ferreira D, Vermelho A (2015) Marine extremophiles: a source of hydrolases for biotechnological applications. Mar Drugs 13:1925–1965. doi:10.3390/md13041925

Demirjian DC, Morís-Varas F, Cassidy CS (2001) Enzymes from extremophiles. Curr Opin Chem Biol 5:144–151

Eisenmenger MJ, Reyes-De-Corcuera JI (2009) High pressure enhancement of enzymes: a review. Enzyme Microb Technol 45:331–347. doi:10.1016/j.enzmictec.2009.08.001

Faulds CB (2010) What can feruloyl esterases do for us? Phytochem Rev 9:121–132. doi:10.1007/s11101-009-9156-2

Fazary AE, Ju YH (2007) Feruloyl esterases as biotechnological tools: Current and future perspectives. Acta Biochim Biophys Sin (Shanghai) 39:811–828. doi:10.1111/j.1745-7270.2007.00348.x

Fernandez-Lafuente R, Cowan DA, Wood ANP (1995) Hyperstabilization of a thermophilic esterase by multipoint covalent attachment. Enzyme Microb Technol 17:366–372. doi:10.1016/0141-0229(94)00089-1

Fuciños P, Pastrana L, Sanromán a et al (2011) An esterase from *Thermus thermophilus* HB27 with hyper-thermoalkalophilic properties: Purification, characterisation and structural modelling. J Mol Catal B: Enzym 70:127–137. doi:10.1016/j.molcatb.2011.02.017

Fuciños P, González R, Atanes E et al (2012) Lipases and esterases from extremophiles: overview and case example of the production and purification of an esterase from *Thermus thermophilus* HB27. Methods Mol Biol 861:239–266. doi:10.1007/978-1-61779-600-5_15

Gao R, Feng Y, Ishikawa K, Ishida H (2003) Cloning, purification and properties of a hyperthermophilic esterase from archaeon *Aeropyrum pernix* K1. J Mol Catal 25:1–8. doi:10.1016/S1381-1177(03)00064-X

Gerday C, Aittaleb M, Arpigny J et al (1997) Psychrophilic enzymes: a thermodynamic challenge. Biochim Biophys Acta—Protein Struct Mol Enzymol 1342:119–131

Ghanem EH, Al-Sayed HA, Saleh KM (2000) An alkalophilic thermostable lipase produced by a new isolate of *Bacillus alcalophilus*. World J Microbiol Biotechnol 16:459–464

Ghati A, Paul G (2015) Purification and characterization of a thermo-halophilic, alkali-stable and extremely benzene tolerant esterase from a thermo-halo tolerant *Bacillus cereus* strain AGP-03, isolated from "Bakreshwar" hot spring, India. Process Biochem 50:771–781. doi:10.1016/j.procbio.2015.01.026

Gomes J, Steiner W (2004) The biocatalytic potential of extremophiles and extremozymes. Food Technol Biotechnol 42:223–235

Handelsman J (2004) Metagenomics: application of genomics to uncultured microorganisms. Microbiol Mol Biol Rev 68:669–685. doi:10.1128/MMBR.68.4.669-685.2004

Hei DJ, Clark DS (1994) Pressure stabilization of proteins from extreme thermophiles. Appl Environ Microbiol 60:932–939

Hess M, Katzer M, Antranikian G (2008) Extremely thermostable esterases from the thermoacidophilic euryarchaeon *Picrophilus torridus*. Extremophiles 12:351–364. doi:10.1007/s00792-008-0139-9

Hotta Y, Ezaki S, Atomi H, Imanaka T (2002) Extremely stable and versatile carboxylesterase from a hyperthermophilic archaeon. Appl Environ Microbiol 68:3925–3931. doi:10.1128/AEM.68.8.3925-3931.2002

Ikeda M, Clark DS (1998) Molecular cloning of extremely thermostable esterase gene from hyperthermophilic archaeon *Pyrococcus furiosus* in *Escherichia coli*. Biotechnol Bioeng. doi:10.1002/(SICI)1097-0290(19980305)57:5<624::AID-BIT15>3.0.CO;2-B

Joseph B, Ramteke P, Thomas G (2008) Cold active microbial lipases: some hot issues and recent developments. Biotechnol Adv 26:457–470

Kim S, Lee SB (2004) Thermostable esterase from a thermoacidophilic archaeon: purification and characterization for enzymatic resolution of a chiral compound. Biosci Biotechnol Biochem 68:2289–2298. doi:10.1271/bbb.68.2289

Kim YO, Heo YL, Nam BH et al (2013) Molecular cloning, purification, and characterization of a cold-adapted esterase from *Photobacterium* sp. MA1-3. Fish Aquat Sci 16:311–318

Koseki T, Fushinobu S, Ardiansyah, et al (2009) Occurrence, properties, and applications of feruloyl esterases. Appl Microbiol Biotechnol 84:803–810. doi: 10.1007/s00253-009-2148-8

Kulakova L, Galkin A, Nakayama T et al (2004) Cold-active esterase from Psychrobacter sp. Ant300: Gene cloning, characterization, and the effects of Gly→Pro substitution near the active site on its catalytic activity and stability. Biochim Biophys Acta—Proteins Proteomics 1696:59–65. doi:10.1016/j.bbapap.2003.09.008

Lagarde D, Nguyen HK, Ravot G et al (2002) High-throughput screening of thermostable esterases for industrial bioconversions. Org Process Res Dev 6:441–445

Levisson M, van der Oost J, Kengen SW (2009a) Carboxylic ester hydrolases from hyperthermophiles. Extremophiles 13:567–581

Levisson M, Sun L, Hendriks S et al (2009b) Crystal structure and biochemical properties of a novel thermostable esterase containing an immunoglobulin-like domain. J Mol Biol 385:949–962. doi:10.1016/j.jmb.2008.10.075

Lombard V, Golaconda Ramulu H, Drula E et al (2014) The carbohydrate-active enzymes database (CAZy) in 2013. Nucleic Acids Res. doi:10.1093/nar/gkt1178

López-López O, Cerdán ME, Gonzalez-Siso MI (2014) New extremophilic lipases and esterases from metagenomics. Curr Protein Pept Sci:445–455. doi:10.2174/1389203715666140228153801

Lv XY, Guo LZ, Song L et al (2010) Purification and characterization of a novel extracellular carboxylesterase from the moderately halophilic bacterium *Thalassobacillus* sp. strain DF-E4. Ann Microbiol 61:281–290. doi:10.1007/s13213-010-0135-z

Maiangwa J, Ali MSM, Salleh AB et al (2015) Adaptational properties and applications of cold-active lipases from psychrophilic bacteria. Extremophiles 19:235–247. doi:10.1007/s00792-014-0710-5

Manco G, Giosue E, Auria SD et al (2000) Cloning, overexpression, and properties of a new thermophilic and thermostable esterase with sequence similarity to hormone-sensitive lipase subfamily from the archaeon *Archaeoglobus fulgidus*. Arch Biochem Biophys 373:182–192

Martínez-Martínez M, Alcaide M, Tchigvintsev A et al (2013) Biochemical diversity of carboxyl esterases and lipases from lake Arreo (Spain): a metagenomic approach. Appl Environ Microbiol 79:3553–3562. doi:10.1128/AEM.00240-13

Müller-Santos M, de Souza EM, Pedrosa F de O et al (2009) First evidence for the salt-dependent folding and activity of an esterase from the halophilic archaea *Haloarcula marismortui*. Biochim Biophys Acta—Mol Cell Biol Lipids 1791:719–729. doi:10.1016/j.bbalip.2009.03.006

Ollis DL, Cheah E, Cygler M et al (1992) The α/β hydrolase fold. Protein Eng 5:197–211

Pakchung A, Simpson P, Codd R (2006) Life on earth. Extremophiles continue to move the goal posts. Environ Chem 3:77–93

Park Y, Choi SY, Lee H et al (2006) A carboxylesterase from the thermoacidophilic archaeon *Sulfolobus solfataricus* P1; purification, characterization, and expression. Biochim Biophys Acta 1760:820–828. doi:10.1016/j.bbagen.2006.01.009

Pérez D, Martín S, Fernández-Lorente G et al (2011) A novel halophilic lipase, LipBL, showing high efficiency in the production of eicosapentaenoic acid (EPA). PLoS One 6:1–11. doi:10.1371/journal.pone.0023325

Pérez D, Kovačić F, Wilhelm S et al (2012) Identification of amino acids involved in the hydrolytic activity of lipase LipBL from *Marinobacter lipolyticus*. Microbiol (United Kingdom) 158:2192–2203. doi:10.1099/mic.0.058792-0

Rao L, Zhao X, Pan F et al (2009) Solution behavior and activity of a halophilic esterase under high salt concentration. PLoS One 4:e6980

Rhee JK, Ahn DG, Kim YG, Oh JW (2005) New thermophilic and thermostable esterase with sequence similarity to the hormone-sensitive lipase family, cloned from a metagenomic library. Appl Environ Microbiol 71:817–825. doi:10.1128/AEM.71.2.817-825.2005

Roh C, Villatte F (2008) Isolation of a low-temperature adapted lipolytic enzyme from uncultivated micro-organism. Proc Soc Agric Bacteriol 105:116–123. doi:10.1111/j.1365-2672.2007.03717.x

Schreck SD, Grunden AM (2014) Biotechnological applications of halophilic lipases and thioesterases. Appl Microbiol Biotechnol 98:1011–1021. doi:10.1007/s00253-013-5417-5

Shang YS, Zhang XE, Wang X De, et al (2010) Biochemical characterization and mutational improvement of a thermophilic esterase from *Sulfolobus solfataricus* P2. Biotechnol Lett 32:1151–1157. doi: 10.1007/s10529-010-0274-0

Sharma PK, Singh K, Singh R et al (2012) Characterization of a thermostable lipase showing loss of secondary structure at ambient temperature. Mol Biol Rep 39:2795–2804. doi:10.1007/s11033-011-1038-1

Tirawongsaroj P, Sriprang R, Harnpicharnchai P et al (2008) Novel thermophilic and thermostable lipolytic enzymes from a Thailand hot spring metagenomic library. J Biotechnol 133:42–49. doi:10.1016/j.jbiotec.2007.08.046

Tutino M, Parrilli E, De Santi C et al (2010) Cold-adapted esterases and lipases: a biodiversity still under-exploited. Curr Chem Biol 4:74–83

Udatha DB, Kouskoumvekaki I, Olsson L, Panagiotou G (2011) The interplay of descriptor-based computational analysis with pharmacophore modeling builds the basis for a novel classification scheme for feruloyl esterases. Biotechnol Adv 29:94–110. doi: 10.1016/j.biotechadv.2010.09.003

Van Den Brink J, De Vries RP (2011) Fungal enzyme sets for plant polysaccharide degradation. Appl Microbiol Biotechnol 91:1477–1492. doi:10.1007/s00253-011-3473-2

Wei X, Jiang X, Ye L et al (2013) Cloning, expression and characterization of a new enantioselective esterase from a marine bacterium *Pelagibacterium halotolerans* B2T. J Mol Catal B: Enzym 97:270–277. doi:10.1016/j.molcatb.2013.09.002

Wong DWS (2006) Feruloyl esterase: a key enzyme in biomass degradation. Appl Biochem Biotechnol 133:87–112. doi:10.1385/ABAB:133:2:87

Wu G, Wu G, Zhan T et al (2013) Characterization of a cold-adapted and salt-tolerant esterase from a psychrotrophic bacterium *Psychrobacter pacificensis*. Extremophiles 17:809–819

Yan S, Lin X, Chen X, Zhang S (2014) Purification and characterization of an esterase from *Halobacillus trueperi* whb27. J Pure Appl Microbiol 8:1–9

Yano JK, Poulos TL (2003) New understandings of thermostable and peizostable enzymes. Curr Opin Biotechnol 14:360–365. doi:10.1016/S0958-1669(03)00075-2

Zhang S, Wu G, Liu Z et al (2014) Characterization of EstB, a novel cold-active and organic solvent-tolerant esterase from marine microorganism *Alcanivorax dieselolei* B-5(T). Extremophiles 18:251–259

Zimmer C, Platz T, Cadez N et al (2006) A cold active (2R,3R)-(-)-di-O-benzoyl-tartrate hydrolyzing esterase from *Rhodotorula mucilaginosa*. Acta Crystallogr Sect F Struct Biol Cryst Commun 73:132–140. doi:10.1007/s00253-006-0463-x

Chapter 11
Extremophilic Esterases for Bioprocessing of Lignocellulosic Feedstocks

Juan-José Escuder-Rodríguez, Olalla López-López, Manuel Becerra, María-Esperanza Cerdán, and María-Isabel González-Siso

Abbreviations

AME	Acetyl mannan esterase
AXE	Acetylxylan esterase
CBM	Carbohydrate binding module
CE	Carbohydrate esterase
FA	Ferulic acid
FAE	Ferulic acid esterase
GE	Glucuronoyl esterase
GH	Glycoside hydrolase
MCA	Methyl 3,4-dihydroxycinnamate
MFA	Methyl 3-methoxy-4-hydroxycinnamate
MpCA	Methyl 4-Hydroxycinnamate
MSA	Methyl 3,5-dimethoxy-4-hydroxycinnamate
PMSF	Phenylmethylsulfonyl fluoride
SBP	Sugar beet pulp
WB	Wheat bran

What Will You Learn from This Chapter?

Esterase is the generic name given to any enzyme that catalyzes the hydrolysis (and formation) of an ester into its alcohol and acid. Esterases acting on lignocellulosic substrates remove the side chains of hemicelluloses, which are cross-linked with lignin, therefore favouring hemicellulose degradation, cellulose accessibility, and

J.-J. Escuder-Rodríguez • O. López-López • M. Becerra • M.-E. Cerdán • M.-I. González-Siso (✉)
Grupo EXPRELA, Centro de Investigacións Científicas Avanzadas (CICA), Departamento de Bioloxía Celular e Molecular, Facultade de Ciencias, Universidade da Coruña, Campus de A Coruña, 15071 A Coruña, Spain
e-mail: migs@udc.es

© Springer International Publishing AG 2017
R.K. Sani, R.N. Krishnaraj (eds.), *Extremophilic Enzymatic Processing of Lignocellulosic Feedstocks to Bioenergy*, DOI 10.1007/978-3-319-54684-1_11

biomass digestibility. Degradation of the complex lignocellulosic polymeric substrate into simple sugars is possible by synergic combination of these esterases with a series of other enzymes such as hemicellulases, cellulases, and oxidoreductases. This book chapter covers the enzymes acetyl mannan esterases that catalyze the deacetylation of 2,3-O-acetyl mannan, acetylxylan esterases that catalyze the deacetylation of 2,3-O-acetyl xylan and xylo-oligosaccharides, and ferulic acid esterases that catalyze the hydrolysis of the 4-hydroxy-3-methoxycinnamoyl (feruloyl) group from arabinose substitutions. Glucuronoyl esterases that degrade the ester bonds between glucuronic acids of xylan and alcohols of lignin are also covered. Several industries require the use of robust and stable enzymes for the degradation of hemicelluloses, with operative activity at high temperature and extreme pH. An insight focusing on extremophilic (mainly thermophilic) esterases in the field of lignocellulosic feedstocks bioprocessing is provided.

11.1 Introduction

Esterase is the generic name given to any enzyme (EC class 3.1) that catalyzes the hydrolysis (and formation in organic media) of an ester into its alcohol an acid. There is a wide variety of esterase types depending on the nature of their substrates. This book chapter focuses on members of the carbohydrate esterase (CE) families which acts on lignocellulosic substrates, i.e., acetyl xylan esterases (AXEs) that catalyze the deacetylation of 2,3-O-acetyl xylan and xylo-oligosaccharides (CE families 1–7, 12, 16), acetyl mannan esterases (AMEs) that catalyze the deacetylation of 2,3-O-acetyl mannan (CE family 1), and ferulic acid esterases (FAEs) that catalyze the hydrolysis of the 4-hydroxy-3-methoxycinnamoyl (feruloyl) group from arabinose substituents (CE family 1) (Ulaganathan et al. 2015). Inclusion into CE families follows the classification provided by the Carbohydrate-Active enZymes Server or CAZy (www.cazy.org) (Lombard et al. 2014).

These three types of esterases mentioned above remove the side chains of hemicelluloses (Fig. 11.1), which are cross-linked with lignin, therefore favouring cellulose accessibility and biomass digestibility. Degradation of the complex lignocellulosic polymeric substrate into simple sugars is possible by synergic combination of these esterases with other hydrolases (more than 20 different activities) mainly of the glycoside hydrolase (GH) families, and also oxidoreductases (Shallom and Shoham 2003; Ulaganathan et al. 2015).

The new CE family 15 of glucuronoyl esterases (GEs) has recently emerged (d'Errico et al. 2014). These enzymes degrade ester bonds between glucuronic acids of xylan and alcohols of lignin (Fig. 11.1). When used in conjunction with other hemicellulases, cellulases and oxidoreductases, GEs improve the delignification of lignocellulosic biomass.

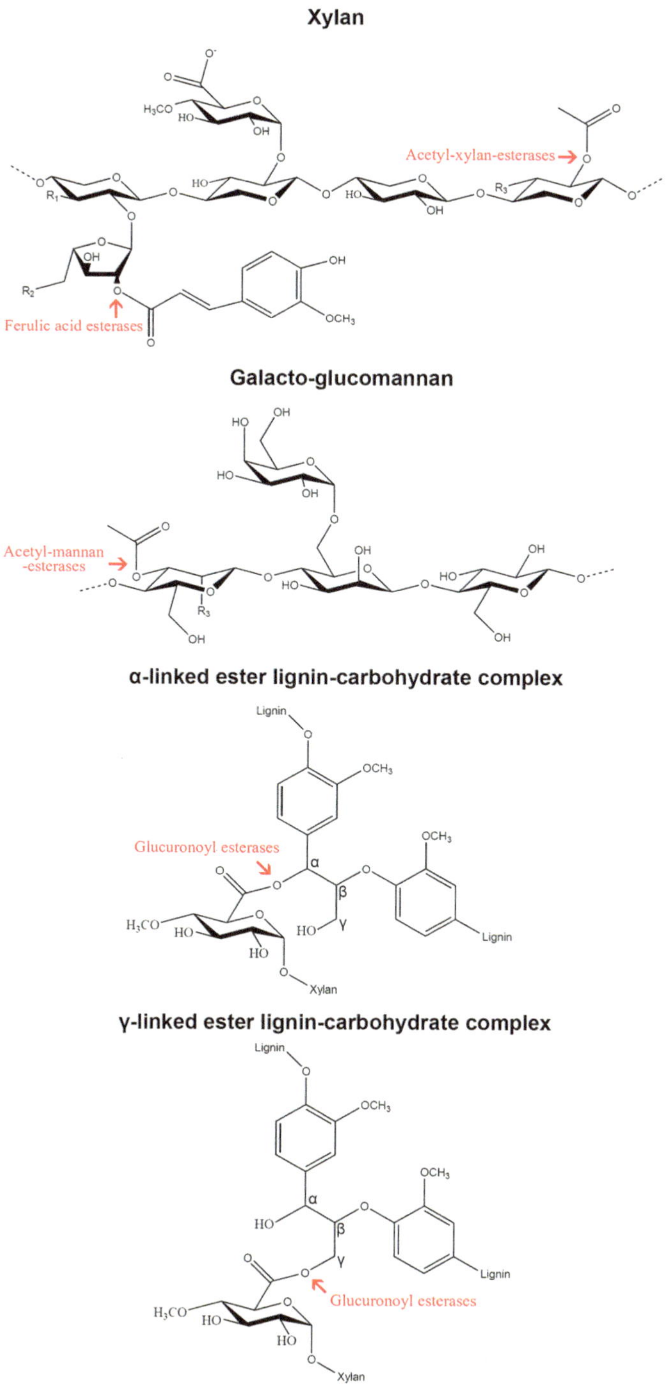

Fig. 11.1 Carbohydrate esterases acting on hemicelluloses and their complexes with lignin. R1: OH or arabinofuranosyl group; R2: OH or ferulic acid; R3: OH or acetyl group

Hemicelluloses rank second (after cellulose) as most abundant polysaccharides, constituting about 1/4th of the total biomass on Earth (Ulaganathan et al. 2015). The breakdown of these polysaccharides is carried out by microorganisms that exploit plants as nutrient sources, and that live either free in nature or in the digestive tract of higher animals. Such microorganisms have developed different strategies for plant cell wall degradation, as reported by Shallom and Shoham (2003), aerobic fungi such as *Trichoderma* and *Aspergillus* produce high concentrations of a variety of extracellular enzymes that act simultaneously and synergistically to degrade polysaccharides into monosaccharides or disaccharides. Aerobic bacteria such as *Bacillus subtilis* secrete a smaller number of extracellular degrading enzymes yielding oligosaccharides that continue their breakdown by cell-bound or intracellular enzymes. Anaerobic bacteria such as *Clostridium* have developed associations of cellulolytic and hemicellulolytic enzymes in multienzymatic complexes.

The above mentioned types of CE-hemicellulases (AXEs, AMEs, FAEs and GEs), in concerted and synergic action with other enzymes, can be useful not only for biofuels production by fermentation of hydrolyzed lignocellulosic feedstocks, but also for paper and pulp industry, animal food biotechnology, and drink and food industries (Ghatora et al. 2006). Some CE-hemicellulases commercially available hitherto together with their origin and mode of action are listed in Table 11.1.

There is a need for robust and stable enzymes capable of the degradation of hemicelluloses in an industrial context. Enzymes from extremophiles are of particular interest since these biocatalysts have evolved to be functional under the extreme conditions where these microorganisms survive. Specifically, it is known that enzymes from thermophilic microorganisms exhibit a greater degree of stability than those from their mesophilic counterparts. These enzymes usually are resistant to other denaturing agents such as solvents in addition to temperature (Vieille and Zeikus 2001). Herewith, an overview of the state of extremophilic (mainly thermophilic) esterases research in the field of lignocellulosic feedstocks bioprocessing is provided.

Before the advent of metagenomics, the sources of new enzymes were technically limited to the culturable microorganisms, which have been estimated to represent less than 1% of the total diversity in most environments (López-López et al. 2013). Metagenomics consists in the study of the metagenome, the pool of genomes in an environmental microbial community, including the genomes of unculturable organisms, thus allowing the discovery of unknown enzymes with potentially new and unique properties. Genes encoding new enzymes can be mined in the metagenome sequence or isolated by functional screening of metagenomic libraries. Metagenomics has been used in the isolation of new thermophilic esterases, mostly lipolytic (López-López et al. 2013, 2014). Esterases derived from metagenomic researches and acting on hemicelluloses are included in this review.

Table 11.1 Esterases commercially available in market

Carbohydrate esterase (CE) families	Source organism	Enzyme Commission (E.C) number	Company	Specific activity (U/mg)	Mode of action
Acetyl xylan esterases (AXEs)	*Cellvibrio japonicas NCIMB 10462*	E.C. 3.1.1.72	Prozomix	410	Catalyses the hydrolysis of acetyl groups from acetylated xylose, acetylated glucose, α-napthyl acetate, and p-nitrophenyl acetate.
	Clostridium thermocellum	E.C. 3.1.1.72	Prozomix	175	Deacetylation of xylans and xylo-oligosaccharides. Also catalyses the hydrolysis of acetyl groups from acetylated xylose, acetylated glucose, α-napthyl acetate, and p-nitrophenyl acetate.
	Opitutus terraes PB90-1	E.C. 3.1.1.72	Prozomix	NA	Deacetylation of xylans and xylo-oligosaccharides. Also catalyses the hydrolysis of acetyl groups from acetylated xylose, acetylated glucose, α-napthyl acetate, and p-nitrophenyl acetate.
	Streptomyces coelicolor A3(2)	E.C. 3.1.1.72	Prozomix	NA	Deacetylation of xylans and xylo-oligosaccharides. Also catalyses the hydrolysis of acetyl groups from acetylated xylose, acetylated glucose, α-napthyl acetate, and p-nitrophenyl acetate.
	Streptomyces avermitilis MA-4680	E.C. 3.1.1.72	Prozomix	NA	Deacetylation of xylans and xylo-oligosaccharides. Catalyses the hydrolysis of acetyl groups from polymeric xylan, acetylated xylose, acetylated glucose, alpha-napthyl acetate, p-nitrophenyl acetate but not from triacetylglycerol. Does not act on acetylated mannan or pectin.
	Microbial	E.C. 3.1.1.72	Creative Enzymes	50	Catalyses the hydrolysis of acetyl groups from acetylated xylan, ethyl acetate, cephalosporin C and derivatives.
	Thermotoga maritima	E.C. 3.1.1.72	Nzytech	NA	Participates in the deacetylation of xylans and xylo-oligosaccharides. Substrates: Variety of acetylated

(continued)

Table 11.1 (continued)

Carbohydrate esterase (CE) families	Source organism	Enzyme Commission (E.C) number	Company	Specific activity (U/mg)	Mode of action
					compounds, including cephalosporin; 4-nitrophenyl-β-D-xylopyranoside monoacetates.
	Cellvibrio japonicus	E.C. 3.1.1.72	Nzytech	NA	Participates in the deacetylation of xylans and xylo-oligosaccharides. Substrates: acetyl side-chains of acetylated xylans.
	Clostridium thermocellum	E.C. 3.1.1.72	Nzytech	NA	Bifunctional endo-1,4-β-xylanase and acetyl xylan/glucomannan esterase. Substrates: xylans, such as oat spelt xylan and arabinoxylan (GH11), and removes acetate from acetylated xylan (CE4).
	Bacillus subtilis	E.C. 3.1.1.72	Nzytech	NA	Participates in the deacetylation of xylans and xylo-oligosaccharides. Substrates: 7-aminocephalosporanic acid, cephalosporin C, p-nitrophenyl acetate, β-naphthyl acetate, glucose pentaacetate, and acetylated xylan.
	Ruminococcus flavefaciens	E.C. 3.1.1.72	Nzytech	NA	Participates in the deacetylation of xylans and xylo-oligosaccharides. Substrates: β-naphthyl acetate, lower activity against α-naphthyl acetate.
Ferulic acid ester-ases (FAEs)	*Clostridium thermocellum*	E.C. 3.1.1.73	Prozomix	0.1	Catalyses the hydrolysis of the 4-hydroxy-3-methoxycinnamoyl (feruloyl) group from an esterified sugar.
	Acetivibrio cellulolyticus	E.C. 3.1.1.73	Prozomix	1169	Biological ester hydrolysis.
	Clostridium thermocellum	E.C. 3.1.1.73	Creative Enzymes	0.5–28	Catalyzes the chemical reaction: feruloyl - polysaccharide + H_2O = ferulate + polysaccharide.
	Ruminococcus albus	E.C. 3.1.1.73	Creative Enzymes	NA	Catalyzes the chemical reaction: feruloyl - polysaccharide + H_2O = ferulate + polysaccharide.
	Acetivibrio cellulolyticus	E.C. 3.1.1.73	Creative Enzymes	1169	The two substrates of this enzyme are feruloyl-polysaccharide and H_2O, whereas its two products are ferulate and polysaccharide.

	Rumen microorganism	E.C. 3.1.1.73	Megazyme	40	Catalyses the hydrolysis of the 4-hydroxy-3-methoxycinnamoyl (feruloyl) group from an esterified sugar.
	Ruminococcus albus	E.C. 3.1.1.73	Nzytech	NA	Cleaves the ferulate groups involved in the crosslinking of hemicelluloses to lignin in plant cell walls. Substrates: ferulate crosslinks between xylans and lignin.
	Clostridium thermocellum	E.C. 3.1.1.73	Nzytech	NA	Attacks the ferulate groups involved in the crosslinking of hemicelluloses to lignin in plant cell walls.
Glucuronoyl esterases (GEs)	Opitutus terrae PB90-1 (4-O-Methyl-glucuronyl methyl esterase)	E.C.3.1.1.	Prozomix	NA	Removal of methyl esters from xylans.

NA Not available

11.2 Structure and Catalytic Mechanism of Esterases

The structure and catalytic mechanism are described as part of a more comprehensive review about extremophilic lipolytic esterases recently published by López-López et al. (2014).

Esterases and lipases belong to the alpha/beta hydrolase family. Both these enzymes mediate the hydrolytic cleavage of an ester bond between an alcohol group and a carboxylic acid. These reactions arouse great interest in diverse industrial sectors.

Esterases and lipases share little primary sequence similarity but their tertiary structure is highly conserved. They present a typical alpha/beta hydrolase fold with eight beta sheets, all parallel except the anti-parallel second, connected through six surrounding alpha helices. This fold is responsible for positioning the residues of the catalytic site, which are not contiguous in the primary sequence, in the three-dimensional structure. The active site contains a catalytic triad of amino acid residues always arranged in the same order along the sequence: serine (Ser), aspartate (Asp) or glutamate (Glu), and histidine (His), with the catalytic serine embedded in the consensus motif Gly-X-Ser-X-Gly (López-López et al. 2014).

The catalytic mechanism is common to these two groups of enzymes. In the first step, the hydroxyl group of the catalytic serine nucleophilically attacks the carbonyl carbon of the lipid ester bond. A tetrahedral intermediate is thus formed, stabilized by the catalytic residues His and Asp/Glu, and by the presence of an oxyanion hole (López-López et al. 2014). Then, the alcohol component of the ester bond is cleaved and esterification of the acid component to the catalytic serine –OH forms a covalent intermediate. In the next step, a water molecule hydrolyzes this covalent intermediate and forms a new tetrahedral intermediate, releasing the acyl product (Fig. 11.2). In (trans-) esterification reactions the water molecule is replaced by an alcohol (or an ester).

Esterases and lipases differ in their biochemical properties, and there are several criteria for distinguishing between them. One differential characteristic is substrate preference. Esterases hydrolyze only short-chain (<12 carbon atoms) water-soluble fatty acid esters, while lipases show preference for long-chain (≥ 12 carbon atoms) fatty acid esters, with low water solubility. Other differential trait is the phenomenon called interfacial activation that occurs in most lipases but not in esterases that show classical Michaelis–Menten kinetic behavior. Most lipases possess a lid or loop covering the active site and its opening leads to sudden activation of the enzyme. This change from closed to open conformation, making the active site accessible to the substrate so that the enzyme can transform it, is driven by the lipidic interface of a substrate emulsion (López-López et al. 2014).

Fig. 11.2 Molecular mechanism of the action of esterases and lipases (For a description see the text)

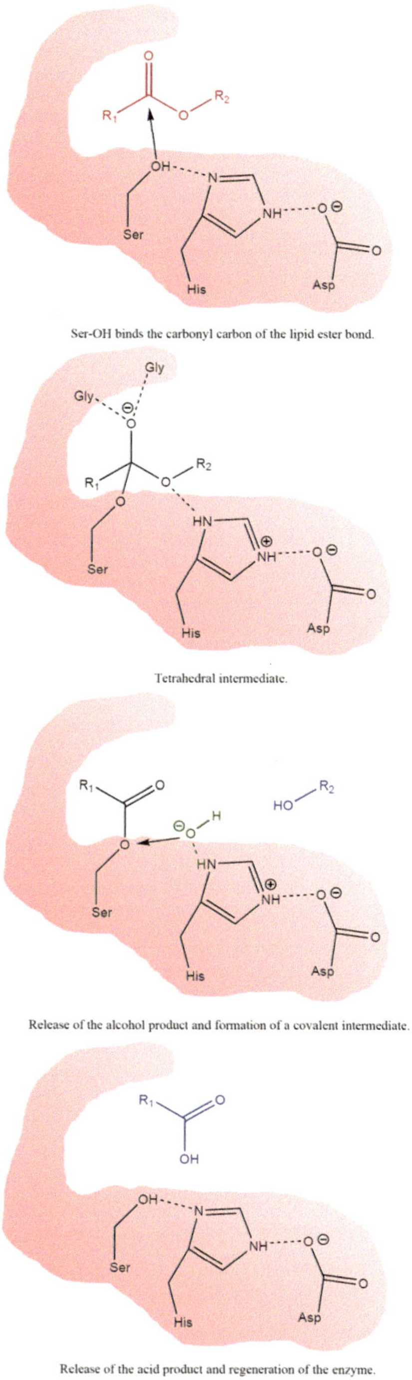

Ser-OH binds the carbonyl carbon of the lipid ester bond.

Tetrahedral intermediate.

Release of the alcohol product and formation of a covalent intermediate.

Release of the acid product and regeneration of the enzyme.

11.3   Acetyl Xylan Esterases (AXEs)

Xylan consists of beta (1→4)-linked xylose chains that carry acetyl, methyl-glucuronyl, and arabinosyl substituents (Fig. 11.1). Acetylated arabinoxylan is the major type of hemicellulose in hardwood and grasses (Schubot et al. 2001).

AXEs (EC 3.1.1.72) deacetylate xylan and deacetylation enhances the hydrolysis of xylan by xylanases, and xylan removal in its turn enhances cellulose hydrolysis by cellulases (Ulaganathan et al. 2015).

General AXEs 3D-structure comprises three conserved domains, an esterase domain, a carbohydrate binding module (CBM) and an unknown function domain. The esterase domain usually has, like in other esterases and lipases, an alpha/beta hydrolase fold and the catalytic triad Ser-His-Asp (Ulaganathan et al. 2015). Most of CBMs contain a beta-jelly-roll structure (Shallom and Shoham 2003).

A remarkable AXE, ascribed to CE family 7 and designated AxeA, was isolated from the anaerobe hyperthermophilic bacteria *Thermotoga maritima* and characterized by Drzewiecki et al. (2010). It is encoded within a chromosomal gene cluster for the breakdown and utilization of complex xylans (encoding 22 proteins), and it is the most thermoactive and thermoresistant AXE currently reported. The extreme stability of AxeA makes this enzyme particularly valuable for several biotechnological applications; for example for lignocellulose degradation processes, modification of cephalosporin antibiotics, and for the introduction of acetyl functional groups into polysaccharides The recombinant form, expressed in *Escherichia coli*, has an optimum activity at 90 °C with an inactivation half-life in the absence of substrate of near 67 h at 90 °C. Moreover, AxeA showed broad substrate specificity, being active on the synthetic substrate p-nitrophenyl-acetate, several acetylated sugars, and xylan of different origins, as well as cephalosporin C. The AxeA monomer is structurally similar to cephalosporin C deacetylases. Therefore AxeA can be useful not only in lignocellulose degradation processes, but also in the modification of cephalosporin antibiotics.

AxeA is an intracellular enzyme whose physiological function would be the deacetylation of xylooligosaccharides produced by extracellular xylanases from complex xylans and entering into the cells through transporters. All proteins involved in the process being encoded by the same chromosomal gene cluster. AxeA is active in homodimeric and homohexameric forms, while crystallographic data suggest an oligomeric state of 12 monomers arranged as two homohexamers in the asymmetric unit of the protein crystal. AxeA shows 42% identity and 58% similarity in amino acid sequence to the multifunctional xylo-oligosaccharide/cephalosporin C deacetylase from *Bacillus subtilis* (Drzewiecki et al. 2010). Crystallography of this enzyme also shows a hexameric quaternary structure, in this case formed by a trimer of dimers, with the active sites pointing towards the center of the hexameric ring (Vincent et al. 2003). The thermophilic intracellular acetyl-xylo-oligosaccharide esterase from *Geobacillus stearothermophilus* Axe2 belongs to the lipase GDSL family (characterized by a modified pentapeptide containing the nucleophilic Ser) and has a "doughnut-shaped" homo-octameric structure; the eight active sites are organized in four closely situated pairs, facing the internal cavity (Lansky et al. 2014).

There is a diversity of other thermophilic microorganisms producing AXEs with different characteristics that in addition to high temperature are active in acid or alkaline reaction media (Ghatora et al. 2006). Recently, the *Aspergillus niger* acetyl xylan esterase (AnAXE1) was expressed and targeted to the apoplast in *Arabidopsis thaliana* which caused reduced plant xylan acetylation (Pawar et al. 2015). Xylans of transgenic lines resulted to be easily enzymatically digested and extracted by hot water, acids or alkali. In fact, the fermentation by the mushroom *Trametes versicolor* of tissue hydrolysates from a transgenic plant with 30% less xylan acetylation resulted in about 70% more ethanol than from a wild type plant, while transgenic plants grew and developed normally. Therefore, endogenous xylan deacetylation in woody tissues of transgenic plants is an alternative way to facilitate degradation of lignocellulosic biomass and its fermentation to biofuels.

11.4 Acetyl Mannan Esterases (AMEs)

Mannans are made of beta-1,4-linked D-mannose backbone combined with glucose and galactose residues (Fig. 11.1). They constitute the main type of hemicellulose in softwood and plant structures like seeds and fruits (Chauhan et al. 2012). Those including 1,3 and 1,4 linked beta-D-glucans are mostly found in Poales (Ulaganathan et al. 2015). AMEs (EC 3.1.1.6) act on galacto-glucomannans and release the acetyl group. In the same way like xylan, the mannan structure demands the synergistic action of a variety of main- and side-chain-cleaving glycosyl hydrolases and esterases (Moreira and Filho 2008). Although some scientific literature is available about other enzymes degrading mannan, specific information about AMEs is very scarce, and they are much lesser known than the counterpart AXEs. Several AXEs and AMEs are unspecific and active on both xylan and mannan as substrates. Tenkanen (1998) reported two extracellular AMEs: an acetyl esterase from *T. reesei,* which is only active towards short oligomeric and monomeric acetates but both derived from xylan and glucomannan, and the acetyl glucomannan esterase of *A. oryzae,* mostly active towards polymeric glucomannan, but able to remove acetyl groups from xylan. High enzymatic AME activity is detected in the filtrate of a culture of the mushroom *Schizophyllum commune* (Tenkanen et al. 1993). The characterization of remarkable thermophilic AMEs has not been reported up to our knowledge.

11.5 Ferulic Acid Esterases (FAEs)

Alpha-L-arabinofuranosyl residues of xylan are esterified with ferulic acid (FA) at O-5 or O-2 positions (Fig. 11.1). Dimers of FA cross-link the xylan chains and also attach them to lignin in gramineaceous plants (Schubot et al. 2001; Rakotoarivonina et al. 2011).

FAEs (EC 3.1.1.73), also known as feruloyl esterases, cinnamoyl esterases or cinnamic acid hydrolases, cleave ester bonds between arabinose and FA, or hydroxycinnamic acids in general, sidegroups of xylan. FAEs also catalyze the release of the hydroxycinnamic acids sterified with pectin. They enhance the action of xylanases and cellulases by increasing the accessibility of these hydrolases to hemicelluloses and cellulose after catalyzing the removal of FA (Ulaganathan et al. 2015). And *vice versa*, FAEs alone usually produce only small amounts of free FA from natural substrates, being more effective in cooperation with xylanases (Debeire et al. 2012). A particular biotechnological application of FAEs is the release of FA from crop residues to use it as an effective antioxidant in food, cosmetics and pharmaceutical industries (Sang et al. 2011). The majority of FAEs shows also AXE activity, although rather low.

Crepin et al. (2004) classified FAEs into four types (A–D) based on their substrate specificity towards mono- and diferulates, substitutions on the phenolic ring, and on their amino acid sequence identity (Table 11.2). Topakas et al. (2005) compared FAEs of the four types, two mesophilic and two thermophilic from *Sporotrichum thermophile* and found that the substrate specificity was in accordance with their classification over a wide range of phenylalkanoate substrates. Moreover, thermophilic FAEs showed a lower catalytic efficiency than their mesophilic counterparts, but released more FA from plant cell walls within a shorter time interval at comparable temperatures.

Ghatora et al. (2006) analyzed the production of plant cell wall acting esterases by a series of 16 thermophilic and thermotolerant fungal strains isolated from composting soil and observed that, in general, the thermophilic strains were better producers of esterases than the thermotolerant strains, and also that most of the esterase isoforms produced by thermophilic fungi were FAEs with high affinity for p-nitrophenyl-ferulate. A few esterases with novel properties were identified in this work.

The active site triad and reaction mechanism of FAEs is thought to be similar to lipases and cutinases (Andersen et al. 2002). The crystal structures of the FAE modules from two xylanases (XynY, XynZ) of *Clostridium thermocellum* display a typical (beta/alpha)$_8$ fold, with a classical Ser–His–Asp catalytic triad (Shallom and Shoham 2003). Nonetheless, none of the fungal esterases analyzed by Ghatora et al. (2006) were active against p-nitrophenyl myristate, i.e., none showed lipolytic activity.

Thermobacillus xylanilyticus is a strict aerobic hemicellulolytic thermophilic spore-forming bacterium that produces a single domain FAE, characterized by Rakotoarivonina et al. (2011). The enzyme, named Tx-Est1, showed 52% amino acid sequence similarity to the previously characterized FAE domains of the bifunctional enzyme XynY from *C. thermocellum*. Tx-Est1 contained, in its C-terminal part, the conserved putative lipase catalytic triad residues Ser202, Asp287, and His322. The putative catalytic nucleophile Ser202 belongs to the Gly-X-Ser-X-Gly lipase pentapeptide consensus. The conserved putative residues forming the oxanion hole in the N-terminal part of the protein are probably constituted by the consensus Gly-Val-Gly-Gly-Asp at position 91–95. Optimum

Table 11.2 The classification of FAEs according to Crepin et al. (2004)

Parameter		Type A *Aspergillus niger* FaeA	Type B *Penicillium funiculosum* FaeB, *Neurospora crassa* Fae-1	Type C *A. niger* FaeB, *Talaromyces stipitatus* FaeC	Type D *Pseudomonas fluorescens* XylD
Preferential induction medium		WB (Wheat Bran)	SBP (Sugar Beet Pulp)	SBP-WB (Sugar Beet Pulp-Wheat Bran)	WB (Wheat Bran)
Hydrolysis of methyl esters	MCA (Methyl 3,4-dihydroxycinnamate)	NO	YES	YES	YES
	MFA (Methyl 3-methoxy-4-hydroxycinnamate)	YES	YES	YES	YES
	MpCA (Methyl 4-Hydroxycinnamate)	YES	YES	YES	YES
	MSA (Methyl 3,5-dimethoxy-4-hydroxycinnamate)	YES	NO	YES	YES
Release of diferulic acid		Yes (5–5′)	No	No	Yes (5–5′)
Sequence similarity		Lipase 26–40%	Carbohydrate esterase family 1 acetyl xylan esterase 45–46%	Chlorogenate esterase tannase 43–60%	Xylanase 91%

pH and temperature of Tx-Est1 expressed in *E. coli* were 8.5 and 65 °C respectively, maintaining an 80% of maximal activity at 80 °C and a thermostability above 24 h at 50 °C. The enzyme can therefore be considered thermoalkaliphilic. Tx-Est1 activity was strongly inhibited by phenylmethylsulfonyl fluoride (PMSF), a serine protease inhibitor, and by diethylpyrocarbonate, a histidine modifier, showing the importance of Ser and His in its activity. Tx-Est1 displayed a very high affinity for feruloylated arabino-xylotetraose, which could be correlated with the physiological role of this intracellular enzyme, whose preferential substrates might be oligosaccharides. The enzyme was also able to efficiently release phenolic acids from complex agricultural by-products of different compositions. The characteristics of Tx-Est1 are presumably advantageous for industrial or biotechnological applications based on lignocellulosic feedstocks bioconversion.

Metagenomics has been used to search for novel FAEs from unculturable microorganisms in several environments. EstF27 is a FAE found by metagenomic techniques, isolated from agricultural soil with treatment of mechanized straw returning; 4-methylumbelliferyl p-trimethylammonio cinnamate chloride was used as specific substrate for FAE activity detection (Sang et al. 2011). EstF27 showed one conserved active site with the pentapeptide motif G-X-S-X-G (amino acid positions from 149 to 153) and a putative catalytic triad comprising His73, Asp123 and Ser151, in a distribution similar to that present in lipases. However, it was inactive towards p-nitrophenyl laurate (C12), indicating that it exhibited no lipase activity. According to its substrate specificity, EstF27 was classified as a type A FAE. The enzyme showed highly soluble expression in *E. coli* at 37 °C, high activity and stability in alkaline conditions and remarkable stability in the presence of high salt concentrations, but was neither active nor stable at thermophilic (higher than 55 °C) temperatures.

Other than soils, hemicellulases with FAE activity have also been isolated from the rumen of several animals. Cheng et al. (2012) isolated a protease-insensitive FAE from cow rumen, the enzyme was named FAE-SH1, expressed in *E. coli* and purified. The recombinant enzyme showed broad specificity against the four methyl esters of hydroxycinnamic acids (Table 11.2) and optimum activity conditions at pH 8 and 40 °C. It retained 45% of activity after 3 h of incubation at 50 °C and over 70% of activity after 3 h of incubation at 4 °C and pH from 2 to 9. FAE-SH1 improved the cleavage of FA from wheat straw in combination with cellulase, β-1,4-endoxylanase, β-1,3-glucanase, and pectinase that converts pectin into pectic acid. Recently, a library of the termite enteric flora metagenome was screened for FAE activity (Rashamuse et al. 2014) and seven positive fosmid clones were found. Six of the seven FAE genes were expressed in *E. coli*, and the purified enzymes exhibited optimal conditions for activity from 40 to 70 °C and pH from 6.5 to 8.0. The analysis of substrate specificity corroborated the requirement for at least one methoxy group on the aromatic ring of the hydroxycinnamic acid ester substrate for optimal FAE activity. The six new FAEs contained the classical G-X-S-X-G pentapeptide sequence with the catalytic Ser and also the other classical lipase conserved regions; although they showed no lipase activity.

11.6 Glucuronoyl Esterases (GEs)

GEs (EC 3.1.1.-) belong to CE family 15 and break ester bonds between glucuronyl substituents in xylan and lignin alcohols, whereas other ester bonds including esters of galacturonic acid are not recognized by these enzymes (d'Errico et al. 2014). The 4-O-methyl substituent in the glucuronic acid residue is the key structural determinant for the specificity of GEs. Eight GEs have been characterized hitherto. They belong to serine type esterases requiring no metal ion co-factors for catalytic activity and their typical structure is modular with a catalytic core and a N-terminal CBM linked by a serine and threonine rich region prone to O-glycosylation. The active site of GEs is exposed to the surface of the enzyme, therefore potentially providing access to large substrates such as lignin ester carbohydrate complexes (d'Errico et al. 2014). Among the GEs characterized so far there are several from thermophilic (*Myceliophthora thermophila*, *Sporotrichum thermophile*) and thermotolerant (*Phanerochaete chrysosporium*) microorganisms.

11.7 Complexed Hemicellulases

A xylanosome is an extracellular multi-enzymatic complex that synergistically degrades xylan. Xylanosomes contain xylanases and side-chain acting enzymes like ferulic acid esterases or acetyl xylan esterases.

The organization of the multi-enzyme complexes cellulosome or xylanosome offers a number of advantages for the effective hydrolysis of polysaccharides. It allows optimum concerted activity and synergism of the enzymes, avoiding non-productive adsorption of the enzymes, limiting competition between the enzymes for the sites of adsorption on the substrates, and facilitating the processivity of the exo-enzymes along the polysaccharides.

The cellulosome and xylanosome of *Clostridium* sp. have been recently reviewed with a focus on strain improvement for bioethanol and biobutanol production from lignocellulosic agro-industrial residues (Thomas et al. 2014). Much of our understanding of these multiprotein complexes: catalytic components, architecture, and mechanisms of attachment to the bacterial cell and to substrate, has been derived from the study of *Clostridium thermocellum*. *Clostridia* contain multifunctional enzymes that consist of cellulases with mannanase and xylanase units. Cellulosome from *Clostridium* sp. contains a non-catalytic scaffolding protein complexed with a number of cellulosomal enzymes (multiple endo-glucanases, cellobiohydrolases, xylanases and other degradative enzymes). Man-designed cellulosomes have been constructed and, for example, enhanced synergistic activity was observed on wheat straw (a natural recalcitrant substrate) by engineering tetravalent cellulosome complexes containing two different types of cellulases and two distinct xylanases. The cellulosome complex is constructed extracellularly, probably at the cell surface and during the log phase of growth, being released to the medium and attached to cellulose in the stationary phase of growth. *Clostridium* sp. lacks true xylanosomes; instead xylanases are often found associated with the cellulosome.

Streptomyces olivaceoviridis xylanosomes have been characterized. Strikingly, a scaffoldin subunit has not been found and therefore the mechanism of xylanases aggregation is unknown (Jiang et al. 2006). A scaffoldin is a large non-catalytic glycoprotein, which integrates the various subunits into the cohesive complex (cellulosome or xylanosome), by combining its "cohesin" domains with a typical "dockerin" domain present on each of the subunit enzymes. The high-affinity of the cohesin-dockerin interactions determines the structure of the multienzymatic complex. Scaffoldins usually contain also a CBM and may contain a module for anchoring to the bacterial cell. *S. olivaceoviridis* xylanosomes include eight subunits, with two major xylanases (named FXYN and GXYN), two truncated forms that are products of proteolytic activity on FXYN and the xylan-binding domain of FXYN. FXYN contains both xylanase and endoglucanase activities but cellulase activity of the xylanosome is very low (Jiang et al. 2006).

Other characterized xylanosomes are those from *Butyrivibrio fibrisolvens*, *Bacillus subtilis,* and the fungus *Penicillium purpurogenum*. The subunit composition is different depending on the species and the carbon source used for their growth. Also, one fosmid clone was isolated from a metagenomic library of the termite gut microbial community, which carried three contiguous xylanase genes putatively comprising a xylanosome operon (Nimchua et al. 2012).

11.8   Conclusions and Future Perspectives

Pretreatment processes precede most bioprocesses used for biotechnological valuation of lignocellulosic feedstocks. Their aim is to make the very recalcitrant supramolecular structure of these materials more accessible to the hydrolytic enzymes. One added difficulty is the different composition of lignocelluloses depending on their source. The conversion of polysaccharides into simple fermentable sugars needs the synergistic concerted action of several GHs, hemicellulases, oxidoreductases and accessory enzymes that break the interactions among cellulose, hemicellulose and lignin; and in some cases also the interactions with pectin. These accessory enzymes are carbohydrate esterases with different substrate specificity (AXEs, AMEs, FAEs and GEs). Their 3D structure and mechanism of catalysis are similar to those described for lipolytic esterases, although they usually lack lipolytic activity. These enzymes are associated in extracellular complexes, such as xylanosomes and cellulosomes, in anaerobic bacteria.

New enzymes with improved characteristics for industrial applications are still needed to create a highly efficient enzymatic cocktail that degrades lignocellulosic feedstocks for further bioprocessing. These improved characteristics include, among others, stability, resistance to high temperature and to extreme pH. Such new enzymes can be constructed modifying the ones existing in nature by protein engineering, or can be searched in the microbial population present in extremophilic environments by metagenomics, allowing the discovery of completely new enzymes with unique features even from unculturable microorganisms. Metagenomics and protein engineering can also be combined in the search for a biocatalyst with the desired performance. Development of suitable substrates and high-throughput screening methods (like fluorescent substrates together with microfluidic devices that are promising systems) is a challenge to take the maximum profit from the possibilities offered by the metagenomics tools.

Take Home Message

- Esterases are those enzymes which hydrolyses the esters to acids and alcohols. There are different types of esterases acting on lignocellulosic substrates, namely acetyl xylan esterases, acetyl mannan esterases, feruloyl esterases and glucuronoyl esterases. The enzyme acetyl mannan esterases catalyzes the deacetylation of 2/3-O linked acetyl mannan. Acetyl xylan esterases mediates the catalysis of the deacetylation of 2/3-O linked acetyl xylan and xylo-oligosaccharides. The ferulic acid esterases catalyzes the hydrolysis of the 4-hydroxy-3-methoxycinnamoyl (feruloyl) group from arabinose substitutions. Glucuronoyl esterases mediated the degradation of the ester bonds between glucuronic acids of xylan and alcohols of lignin.
- Esterases and lipases belong to the alpha/beta hydrolase family. Both these enzymes mediate the hydrolytic cleavage of an ester bond between an alcohol group and a carboxylic acid.
- Esterases and lipases share little primary sequence similarity but their tertiary structure is highly conserved. Esterases and lipases differ in their biochemical properties and there are several criteria for distinguishing between them. Esterases hydrolyze only short-chain (<12 carbon atoms) water-soluble fatty acid esters, while lipases show preference for long-chain (≥12 carbon atoms) fatty acid esters, with low water solubility. Other differential trait is the phenomenon called interfacial activation that occurs in most lipases but not in esterases that show classical Michaelis–Menten kinetic behavior.
- A xylanosome is an extracellular multi-enzymatic complex that synergistically degrades xylan. Xylanosomes contain xylanases and side-chain acting enzymes like ferulic acid esterases or acetyl xylan esterases.
- Metagenomics consists in the study of the metagenome, the pool of genomes in an environmental microbial community, including the genomes of unculturable organisms, thus allowing the discovery of unknown enzymes with potentially new and unique properties. Metagenomics has been used in the isolation of new thermophilic esterases.

Acknowledgement Funding both from the European Union Seventh Framework Programme (FP7/2007-2013) under Grant Agreement n° 324439, and from Xunta de Galicia (Consolidación D.O.G. 10-10-2012. Contract Number: 2012/118) co-financed by FEDER.

References

Andersen A, Svendsen A, Vind J et al (2002) Studies on ferulic acid esterase activity in fungal lipases and cutinases. Colloids Surf B 26:47–55

Chauhan PS, Puri N, Sharma P, Gupta N (2012) Mannanases: microbial sources, production, properties and potential biotechnological applications. Appl Microbiol Biotechnol 93:1817–1830

Cheng F, Sheng J, Dong R et al (2012) Novel xylanase from a Holstein cattle rumen metagenomic library and its application in xylooligosaccharide and ferulic acid production from wheat straw. J Agric Food Chem 60:12516–12524

Crepin VF, Faulds CB, Connerton IF (2004) Functional classification of the microbial feruloyl esterases. Appl Microbiol Biotechnol 63:647–652

Debeire P, Khoune P, Jeltsch J-M, Phalip V (2012) Product patterns of a feruloyl esterase from *Aspergillus nidulans* on large feruloyl-arabino-xylo-oligosaccharides from wheat bran. Bioresour Technol:425–428

d'Errico C, Jørgensen JO, Krogh KBRM et al (2014) Enzymatic degradation of lignin-carbohydrate complexes (LCCs): model studies using a fungal glucuronoyl esterase from *Cerrena unicolor*. Biotechnol Bioeng 112: 914-922

Drzewiecki K, Angelov A, Ballschmiter M et al (2010) Hyperthermostable acetyl xylan esterase. Microb Biotechnol 3:84–92

Ghatora SK, Chadha BS, Saini HS et al (2006) Diversity of plant cell wall esterases in thermophilic and thermotolerant fungi. J Biotechnol 125:434–445

Jiang Z, Deng W, Yan Q, Zhai Q, Li L, Kusakabe I (2006) Subunit composition of a large xylanolytic complex (xylanosome) from *Streptomyces olivaceoviridis* E-86. J Biotechnol 126:304–312

Lansky S, Alalouf O, Vered H et al (2014) A unique octameric structure of Axe2, an intracellular acetyl-xylooligosaccharide esterase from *Geobacillus stearothermophilus*. Acta Crystallogr D 70:261–278

Lombard V, Ramulu HG, Drula E et al (2014) The carbohydrate-active enzymes database in 2013. Nucleic Acids Res 42:D490–D495

López-López O, Cerdán ME, González Siso MI (2013) Hot spring metagenomics. Life 2:308–320

López-López O, Cerdán ME, González Siso MI (2014) New extremophilic lipases and esterases from metagenomics. Curr Protein Pept Sci 15:445–455

Moreira LRS, Filho EXF (2008) An overview of mannan structure and mannan degrading enzyme systems. Appl Microbiol Biotechnol 79:165–178

Nimchua T, Thongaram T, Uengwetwanit T et al (2012) Metagenomic analysis of novel lignocellulose-degrading enzymes from higher termite guts inhabiting microbes. J Microbiol Biotechnol 22:462–469

Pawar PM-A, Derba-Maceluch M, Chong S-L et al (2015) Expression of fungal acetyl xylan esterase in *Arabidopsis thaliana* improves saccharification of stem lignocellulose. Plant Biotechnol J. doi:10.1111/pbi.12393

Rakotoarivonina H, Hermant B, Chabbert B et al (2011) A thermostable feruloyl-esterase from the hemicellulolytic bacterium *Thermobacillus xylanilyticus* releases phenolic acids from non-pretreated plant cell walls. Appl Microbiol Biotechnol 90:541–552

Rashamuse K, Ronneburg T, Sanyika W et al (2014) Metagenomic mining of feruloyl esterases from termite enteric flora. Appl Microbiol Biotechnol 98:727–737

Sang SL, Li G, Hu XP, Liu YH (2011) Molecular cloning, overexpression and characterization of a novel feruloyl esterase from a soil metagenomic library. J Mol Microbiol Biotechnol 20:196–203

Shallom D, Shoham Y (2003) Microbial hemicellulases. Curr Opin Microbiol 6:219–228

Schubot FD, Kataeva IA, Blum DL et al (2001) Structural basis for the substrate specificity of the feruloyl esterase domain of the cellulosomal xylanase Z from *Clostridium thermocellum*. Biochemistry 40:12524–12532

Tenkanen M (1998) Action of *Aspergillus oryzae* and *Trichoderma reesei* esterases in the deacetylation of hemicelluloses. Biotechnol Appl Biochem 27:19–24

Tenkanen M, Puls J, Rättö M, Viikari L (1993) Enzymatic deacetylation of galactoglucomannans. Appl Microbiol Biotechnol 39:159–165

Thomas L, Joseph A, Gottumukkala LD (2014) Xylanase and cellulase systems of *Clostridium* sp.: an insight on molecular approaches for strain improvement. Bioresour Technol 158:343–350

Topakas E, Christakopoulosa P, Fauldsb CB (2005) Comparison of mesophilic and thermophilic feruloyl esterases: characterization of their substrate specificity for methyl phenylalkanoates. J Biotechnol 115:355–366

Ulaganathan K, Goud BS, Reddy MM et al (2015) Proteins for breaking barriers in lignocellulosic bioethanol production. Curr Protein Pept Sci 16:100–134

Vieille C, Zeikus GJ (2001) Hyperthermophilic enzymes: sources, uses, and molecular mechanisms for thermostability. Microbiol Mol Biol Rev 65:1–43

Vincent F, Charnock SJ, Verschueren KHG et al (2003) Multifunctional xylooligosaccharide/cephalosporin C deacetylase revealed by the hexameric structure of the *Bacillus subtilis* enzyme at 1.9 A resolution. J Mol Biol 330:593–606

Chapter 12
An Overview on Extremophilic Chitinases

Mohit Bibra, R. Navanietha Krishnaraj, and Rajesh K. Sani

What Will You Learn from This Chapter?

Cellulose, hemicellulose and chitin are the three most abundant polysaccharides on the earth. Due to the increase in the energy demand chitinases have become relatively important in the past few decades. Chitinases, produced by bacteria, fungi, insects, and plants, have been used in several applications ranging from anti-phytopathogen to antitumor cancer agents. Evolution in molecular biology and genetic engineering has given new prospects in understanding the structure of chitinases' functionality and catalytic mechanisms at molecular level. Owing to their applicability in different fields, studies related to chitinases have become very important. Synergism between the cellulases and chitinases has supported in understanding the catalytic mechanism and developing a deep understanding of the both types of the enzymes. Literature suggest that there are few reports on extremophilic chitinases production. Molecular biology and genetic engineering have only helped in the expression and understanding of extremophilic chitinases so far. This chapter focuses on the various sources of chitinases, their production and catalytic mechanisms, extremophilic chitinases, molecular studies and their potential applications. This chapter also provides an insight on the various aspects of chitinases with emphasis on both research and industry.

M. Bibra • R. Navanietha Krishnaraj • R.K. Sani (✉)
Department of Chemical and Biological Engineering, South Dakota School of Mines and Technology, 501 East St. Joseph Street, Rapid City, SD 57701-3995, USA
e-mail: Rajesh.Sani@sdsmt.edu

© Springer International Publishing AG 2017   225
R.K. Sani, R.N. Krishnaraj (eds.), *Extremophilic Enzymatic Processing of Lignocellulosic Feedstocks to Bioenergy*, DOI 10.1007/978-3-319-54684-1_12

12.1 Introduction

In the past 150 years agriculture, pharmaceutical, transport, health sectors etc. have developed at a faster rate providing the mankind with greater life expectancy, better food sources, increased standard of living and other innumerable advantages. Most of these developments have involved chemical processes which over the course of time have polluted the natural resources drastically. The extensive use of chemicals for various processes and post effluent management not only requires huge capital input but also poses threat for the environment due to improper handling. Human activities have further resulted in a profound impact on the local, regional and global environment. A vision of greener world for coming generations has necessitated development of bioprocesses to replace chemicals being used for the production of pharmaceuticals, plastics, rubber, paper, oils etc. Microorganisms and their enzymes have emerged as a viable option for the change that is required. Several key industrial processes example production and development of biofuels , acetic acid, lactic acid, milk based products (Lemes et al. 2016), biological control (Herrera-Estrella and Chet 1999), baking (Bueno et al. 2016) etc. involve use of microorganisms and their enzymes. Extremophilic enzymes have further enhanced the opportunities of their use in several chemical processes for which biological option was an outlier for last few decades. Cellulose and chitin are most abundant materials on the earth, and are being extensively investigated for their use in different fields. Chitin, has made a huge impact in several industrial sectors. The chapter discusses about the sources for chitinases and their related functions in the respective sources. It also provides different aspects of production, catalysis, developments, and applications of various chitinases including extremophilic chitinases.

12.2 Chitin and Its Derivatives

Chitin is the most abundant polysaccharide in marine environment and second most abundant polysaccharide in the terrestrial environments after cellulose. It is a crystalline, water insoluble, and recalcitrant cellulose derived homopolymer where the 2-OH group is substituted by an acetamido group on the β-1, 4-linked N-acetylglucosamine units (Eijsink et al. 2008; Bhattacharya et al. 2007).

Naturally, chitin exists in two conformations: α chitin and β chitin. In α configuration, the individual polymeric chains are arranged in antiparallel fashion whereas in β configuration these are arranged in parallel fashion (Hamid et al. 2013). Chitosan, a water-soluble chitin derivative, is derived from chitins by removing the N-acetyl groups which render in less bulky amino groups on the polymer. The solubility of chitosan in water makes it a favorable substrate in many different applications e.g., gels, fibers, and films (Rinaudo 2006).

It is commonly found as a key component in the structural make up of insects, fungi, yeast, algae, and in the internal structures of vertebrates where it functions. As per estimates, the amount of chitin observed are of the order of 10^{10}–10^{11} tons on an annual basis (Tanaka et al. 1999).

Production, and processing of sea food, exoskeleton shedding, and production of other products from chitinaceous organisms generates huge quantities of chitin that pose a threat for the marine, and terrestrial environment as a potent source of pollution. Hence, the degradation of chitin materials is not only important for recycling of nutrients, but also to prevent any potential environmental hazards (Chakrabortty et al. 2012). Conventional treatment methods employed in industry involve pretreatment of chitin with HCl for demineralization, and NaOH for deproteinization (Jagadeeswari et al. 2011; Oku and Ishikawa 2006). However, due to the various hazards associated with the use of chemicals, biological control (e.g. chitinase producing organisms or direct application of enzymes) offers a potential sustainable, and green solution for the chitin waste disposal. Enzymes, such as lysozyme, some glucanases, and chitinase can hydrolyze this linear chitin polymer and among these chitinases can specifically degrade the chitin and chitin based materials.

12.3   Chitinases

Chitinases (EC 3.2.1.114) are extracellular, inducible, enzymes which hydrolyze the β-(1-4) glycosidic bond between the C1 and C4 of two consecutive N-acetylglucosamine groups producing chitooligomers (Liu et al. 2003). They are members of glycosyl hydrolase group, which is classified into 136 families (http://www.cazy.org/Glycoside-Hydrolases.html, accessed on June 16, 2016), based on amino acid sequence similarities. Most of the chitinases belong to glycosyl hydrolase family 18 and 19 (GH 18 and GH 19). GH 18 includes chitinases from bacteria, fungi, viruses, animals, and few plants. They are non-catalytic endo-acting and catalytic with exo and endo binding preferences producing chitobiose as the main product. However certain endoacting chitinases are not able to cleave trimers and tetramers yielding longer products. On the other hand, GH 19 includes almost exclusively plant chitinases (Tanaka et al. 1999).

Chitin-hydrolyzing enzymes are classified into three categories (endochitinases, exochitinases, and N-acetyl-β-glucosaminidases-GlcNAc) on the basis of cleavage mechanism (Neeraja et al. 2010; Dahiya et al. 2006) (Fig 12.1).

1. Endochitinases randomly cleave β-(1-4) glycosidic bonds of chitin
2. Exochitinases cleave the chain from the non-reducing end to form diacetyl-chitobiose ($GlcNAc_2$).
3. N-Acetyl-β-glucosaminidases hydrolyze $GlcNAc_2$ into GlcNAc or produce GlcNAc from the non-reducing end of N-acetyl-chitooligosaccharides.

228   M. Bibra et al.

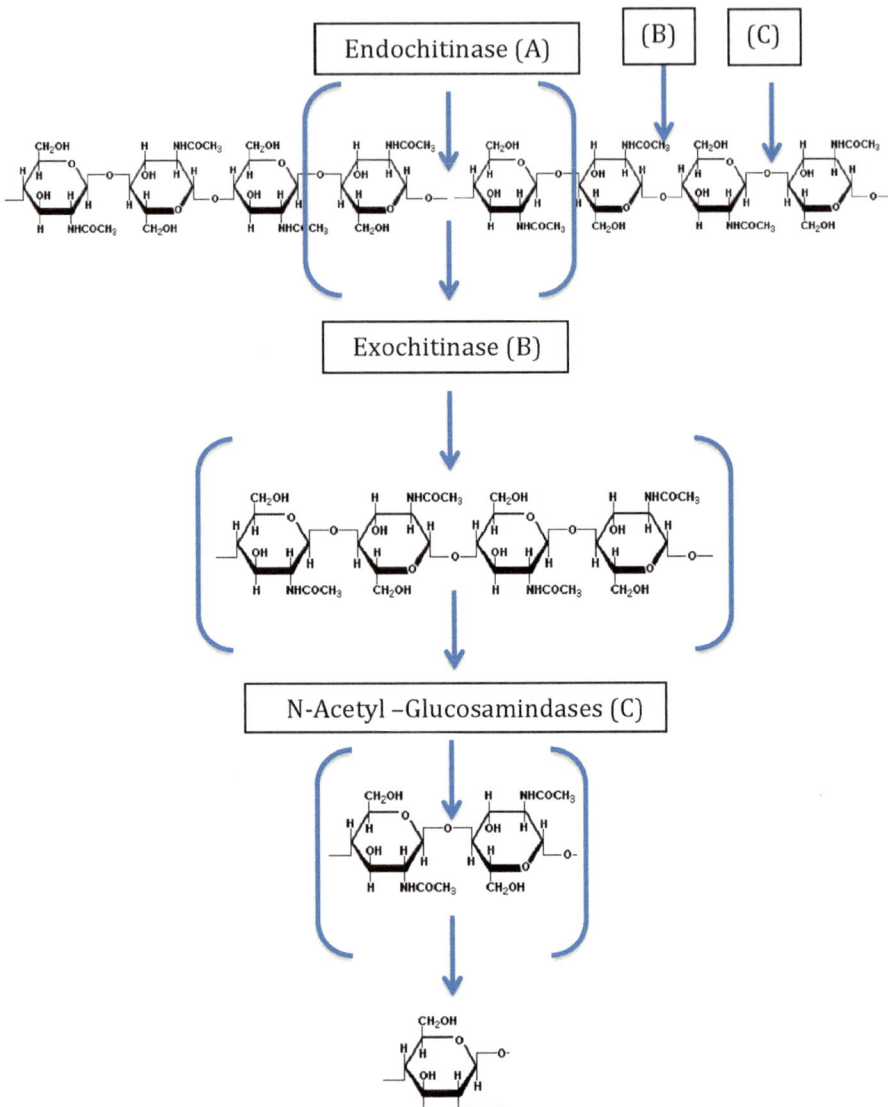

Fig. 12.1 Mechanistic action of chitinases on chitin (**A**) acts on non reducing ends of chitin giving oligochitomers; (**B**) acts on non reducing end to form chitobioside (GlcNAc)$_2$ and (**C**) acts in the (GlcNAc)$_2$ to produce GlcNAc

12.4   Sources of Chitinases

Chitinases can be obtained from a wide range of organisms including bacteria, fungi, insects, plants and animals. Depending on the source of their production, they perform different functions ranging from acting as an agent in anti-phytopathogenic to that in cell differentiation. The chitinase producing organisms can be isolated from the sites having sufficient amount of chitin that includes chitin waste sites, crab and shrimp food industries, trees infected with fungi and others. A chitinase producing gene Chi18H8 isolated from soil and was considered to be suppressive towards club root disease of cabbage (Hjort et al. 2014). A thermostable and alkaline chitinase producing strain *Bacillus thruingiensis* subsp. *kurstaki* HBK-51 was isolated from a mixture of crabs, campus soil, and compost (Kuzu et al. 2012). Chitinase BoCHI3-1 was obtained from the suspension cultured bamboo cells (Onaga et al. 2011). Chitinase producing *Paenibacillus* sp. D1 was obtained from a common effluent plant (Singh and Chhatpar 2011). A broad classification of the chitinases on the basis of their source is discussed in the following sections.

12.4.1   Bacterial Chitinases

Bacterial chitinases are produced both by all three domains—Bacteria, Archaea, and Eukarya. Chitinases from bacterial origin belong to GH (glycosyl hydrolase) family 18 with an exception of chitinase from *Streptomyces griseus* HUT 6037 which belongs to GH family 19 (Ohno et al. 1996). Bacterial chitinases have a molecular weight of about 20–60 kDa comparable to that of chitinases obtained from plant species (40–85 kDa). Bacterial chitinases have been reported from various genera including *Serratia, Thermococcus, Pyrococcus, Streptomyces, Bacillus,* and *Aeromonas* (Bhattacharya et al. 2007).

   The optimum temperature and pH for bacterial chitinases can vary with the bacterial species in the same genera. An endochitinase from *Streptomyces violaceusniger* has optimum temperature of 28 °C whereas the optimum temperature of 80 °C has been reported for a thermostable chitinase produced by *Streptomyces thermoviolaceus* OPC-520 (Shekhar et al. 2006; Tsujibo et al. 1993). Similarly chitinase from *Bacilus cereus* 6E1 had optimum temperature of 35 °C as compared to 50 °C of *Bacillus* sp. BG-11 (Bhushan and Hoondal 1998; Wang et al. 2001). Extremophilic chitinases, which can withstand extreme conditions for example extreme temperature and pH, high salt concentrations etc., are obtained mostly from archaea and actinomycetes. There are few reports of extremophilic chitinases from fungi as well as from plants. Table 12.1 shows the list of different extremophilic chitinases available.

   The chitinase from hyperthermophile archaeon, *Pyrococcus furiosus*, is active at 90 °C and pH 6.0–7.5. This chitinase gene has two catalytic domains and two substrate-binding domains, which is uncommon in bacterial chitinases. There is only one report of chitinase having more than one catalytic domain from a bacterium *Thermococcus kodakaraensis* (Oku and Ishikawa 2006). A thermostable chitinase found in *Bacillus*

Table 12.1 Extremophilic chitinases produced by various organisms

Organisms	Optimum temperature	Tmax	Temperature range	Residual activity	Optimum pH	pH range	References
Ananas comosus	70 °C	85 °C	20–85 °C	90% at 80 °C after 1 h	3.0	3.0–12.0	Onaga et al. (2011)
Bacillus sp BG-11	50 °C	90 °C	45–85 °C	50% at 80 °C after 20 min	7.5–9.0	6.0–9.0	Bhushan and Hoondal (1998)
Bacilus cereus 6E1	35 °C	70 °C	4–70 °C	70% at 65 °C after 1 h	5.8	2.5–8.0	Wang et al. (2001)
Bacillus pumilus SG2	55 °C	60 °C	30–70 °C	60% at 60 °C and 78% at 0.5 M Nacl after 1 h	7.0	4.0–10.0	Vahed et al. (2013)
Bacillus thuringiensis subsp. kurstaki HBK-51	110 °C	110 °C	30–120 °C	96% after 3 h	9.0	3.0–12.0	Kuzu et al. (2012)
Chaetomium thermophillum	60 °C	60 °C	30–80 °C	96.7% at 60 °C after 1 h	5.5	4.0–8.0	Li et al. (2010)
Paenibacillus sp D1	50 °C	60 °C	40–60 °C	86.07 ± 2.04% at 60 °C	8.0	4.0–8.0	Singh and Chhatpar (2011)
Rhizopus oryzae	60 °C	60 °C	40–60 °C	50% at 70 °C after	5.5–6.0	5.0–8.5	Chen et al. (2013)
Streptomyces thermoviolaceus OPC-520	80 °C	80 °C	NA	Active at 50 °C for 14 days	9.0	4.0–12.0	Tsujibo et al. (1993)
Talaromyces emersonii	65 °C	76 °C	20–80 °C	NA	5.5–6.5	3.0–7.0	McCourmack et al. (1991)
Thermoascus aurantiacus var. levisporus	50 °C	60 °C	30–80 °C	95.6% at 50 °C after 1 h	8.0	4.0–12.0	Li et al. (2010)
Thermococcus kodakaraensis KOD 1 ChiAΔ4[a]	90 °C	NA	NA	Active at 100 °C for >7 h	4.5	4.0–9.0	Tanaka et al. (2001)
Thermococcus kodakaraensis KOD 1 ChiAΔ5[a]	85 °C	100 °C	NA	Inactivated at 100 °C in 1 min	5.0	4.5–8.0	Tanaka et al. (2001)
Thermomyces lanuginosus	55 °C	70 °C	NA	24% at 70 °C after 20 min	4.5	NA	Guo et al. (2005)

[a]Two different catalytic domains of same enzyme having different operational conditions

thuringiensis subsp. *Kustaki* HBK-51 is isolated from chitin wastes (crabs, campus soil and compost) on chitinase detection agar. The chitinase produced has an optimal activity at 110 °C and at pH 9.0. The enzyme could retain about 75–98% of activity over a wide range of temperature 30–120 °C and pH 3–12 after 3 h incubation (Kuzu et al. 2012).

Another thermostable chitinase Tk-ChiA obtained from *Thermococcus kodakarensis* has optimal temperature 85 °C and pH 5.0 for colloidal chitin. This enzyme has two catalytic domains and three substrate binding domains. Mutation studies with deletion of certain sequences showed that the two catalytic sites in the enzyme function independent of each other. Their combined hydrolytic effect is additive instead of being synergistic (Tanaka et al. 1999). *Paenibacillus* sp., a ubiquitously found bacterial species, not only produce chitinases but also aid in the plant growth by nitrogen fixation, mineral solubilization and production of siderophores and phytohormones. *Paenibacillus* sp. D1 isolated from a common effluent treatment plant produced chitinase which could withstand the pH 8.0 at a temperature of 45–60 °C in the presence of fungicides (Singh and Chhatpar 2011).

12.4.2   Fungal Chitinases

Chitinases have been found in several fungi including *Trichoderma, Oenicillium, Penicillium, Lecanicillium, Neurospora, Mucor, Beauveria, Lycoperdon, Aspergillus, Myrothecium, Conidiobolus, Metharhizium, Stachybotrys, Agaricus* (Karthik et al. 2014; Hamid et al. 2013). The cell wall in certain fungi is composed of chitin, which is water insoluble. Thus it is highly essential to break down the chitin into its precursor components, which then can be used by fungi for growth. Chitinases can effectively break down the chitin into precursor components, which can then be utilized by fungi for hyphal growth, hyphal extension, hyphal fusion, autolysis etc. Fungal chitinases mostly belong to GH 18 family and show high amino acid homology with class III plant chitinases (Dahiya et al. 2006; Takaya et al. 1998). Chitinases have vital physiological and biological roles including morphogenetic, autolytic, nutritional, and parasitic roles. A chitinase gene (CTS1) is required for the cell separation after division and cell clumping in *Saccharomyces cerevisiae,* while functional expression of chitosanase and chitinase have been reported to influence morphogenesis in the yeast (*Schizosaccharomyces pombe*) (Shimono et al. 2002).

Optimum temperature for most fungal chitinases is 40–50 °C. Certain thermophilic fungal species are known to produce chitinases, which can function at high temperature and pH (Li et al. 2010; Karthik et al. 2014). Chitinase from *Rhizopus oryzae* is active at temperature of 60 °C (Chen et al. 2013). Two thermophilic fungal species, *Thermoascus aurantiacus var. levisporus*, and *Chaetomium thermophillum*, have been reported to produce two novel thermostable chitinases TaCHIT1 and CtCHIT1 which can withstand a temperature of 60°C (Li et al. 2010). Chitinase from *Talaromyces emersonii* has an optimum activity at a temperature of 65°C and pH 5.5–6.5 (Mccormack et al. 1991). A chitinase of *Thermomyces lanuginosus* exhibits optimum catalytic activity at 55 °C and the half-life time of the enzyme at 65 °C is 25 min (Guo et al. 2005).

12.4.3 Plant Chitinases

Chitinases are found in higher plants monocotyledons and dicotyledons. They play a imperative role in the defensive mechanism of plants and are also included in the class of pathogen related (PR) proteins. On the basis of the amino acid sequences, plant chitinases have been categorized into 5 or 6 classes. The key structure of the class I, II and IV enzymes includes a main structural unit consisting of two α-rich globular domains. While 8 α-helices and 8 β-strands form the class III and V plant chitinases. The former carries out the hydrolysis of the β-1, 4-glycosidic linkage by means of an inverting mechanism, and the latter through a retaining mechanism (Tamo et al. 2003). Chitinases from plant origin generally have molecular weight from 25 to 36 kDa and may be either acidic or basic depending on the amino acids present (Punja and Zhang 1993).

Many plant endochitinases, especially those with a high isoelectric point, exhibit an additional lysozyme or lysozyme like activity (Brunner et al. 1998). Plant chitinases have shown inhibitory activity towards the fungal spore germination and mycelial growth in disc plate diffusion against common fungal pathogens *Trichoderma, Fusarium, and Alternaria* (Hamid et al. 2013; Onaga et al. 2011; Dahiya et al. 2006). In addition to their roles with the pathogen management in the plants, these enzymes also play a significant role in certain physiological activities including embryogenesis and ethylene synthesis, legume nodulation (Fukamizo et al. 2003; Kasprzewska 2003; Punja and Zhang 1993). They degrade and deactivate the bacterial chitooligosaccharides when the bacterial-plant interaction is not compatible. In mycorrhizal fungi, the chitinases degrade the inducers in compatible interactions and bind to receptors during non-compatible interactions chitinases inducing an immune reaction, which in turn enhances the production of pathogen related proteins (Kasprzewska 2003; Collinge et al. 1993).

The constitutive production of chitinase is very low but it is triggered and enhanced along with other PR proteins in presence of various, abiotic agents (ethylene, salicylic acid, salt solutions, ozone, and UV light) and by biotic factors (fungi, bacteria, viruses, viroids, fungal cell wall components, and oligosaccharides) in their environment (Kasprzewska 2003). Two chitinase genes FaChi2-1 and FaChi2-2, from strawberry plants were effective against *Colletotrichum fragariae* or *Colletotrichum acutatum*, which are known to cause severe strawberry disease anthracnose crown rot in southeastern US and in many parts of the world (Khan and Shih 2004). Most of the chitinases that are obtained from plants are active in moderate temperature and pH range. Only a few reports of extremophilic chitinases from plant origin are available in literature. A thermostable chitinase PLChi A obtained from *Ananas comosus* has half-life of more than 5.2 days at 70 °C and pH 4.5. Even at 80 °C, the half-life is 65 min at pH 4.0 (Onaga et al. 2011). Two different isoforms of chitinases obtained from cultured cells of *Bambusa oldhamii* were stable at 70 °C and 80 °C at a pH of 3.0 and 4.0 respectively. The two isoforms were observed to effectively control the growth of *Scolecobasidium longiphorum* and stable after storing for 1 year at 4 °C (Kuo et al. 2008).

12.4.4   Insect Chitinases

Chitin forms a crucial component of insect endoskeleton and its appropriate level should be maintained for the upholding of insect growth. Insect chitinases have a molecular weight range of 40–85 kDa. These play important roles as degradative enzymes during ecdysis where endochitinases randomly break the cuticle to chitooligosaccharides which are subsequently hydrolyzed by exoenzymes to N-acetyl-glucosamine. The monomers are reused for new cuticle synthesis (Koga et al. 1997; Kramer et al. 1993). As the new cuticle is produced it is protected from the chitinolytic activity of chitinases by a cuticle organizing protein Chaudhari et al. (2011). Hormones regulate the enzyme production during the transformation of the larvae. Two main sources of insect chitinases are *Manduca sexta* and *Bombyx mori* (Koga et al. 1997; Kramer et al. 1993). The expression and levels of the chitinase enzyme in *Manduca sexta* is very tightly regulated during the morphogenesis of insect. The chitinase is active for very short time during the larval-larval, larval-pupal and pupal-adult molting (Kramer et al. 1993). Insect chitinases have been found to possess transglycosylation activity that can produce oligosaccharides of pharmaceutical significance (Lee et al. 2002; Shen et al. 2009). No extremophilic chitinases have been reported from insects so far.

12.5   Catalysis Mechanism and Molecular Insights of Chitinases

Chitinases from different sources including microbes, fungi, plants and insects belong to either family GH 18 or GH 19. Different isoforms of chitinases are found in different organisms e.g. *Serratia marcecens, Aeromona, Bacillus licheniformis* X-74, and *Streptomyces griseus* etc. These isoforms are believed to be the product of post-translational modification when glycosylation or proteolysis occurs (Dahiya et al. 2006). Chitin and cellulose share common similarities in terms of abundance, water insolubility, β-1, 4 glycosidic bonds, and crystalline structure. A similar correlation has been found in the chitinases studied so far showing the presence of catalytic and substrate binding domains in a similar fashion to the cellulases. Reese et al. (1950) first demonstrated that an accessory protein (now called as carbohydrate binding module -CBM) is involved in the cellulose hydrolysis by cellulases, which makes the cellulose more accessible for the hydrolytic enzymes. It was only in 2005 after the discovery of an accessory protein CBP 21 in chitinase from *Serratia marscens* that this hypothesis was accepted (Eijsink et al. 2008). However, the number of catalytic and substrate binding domains varies in different chitinases.

The chitinases have CBMs which can adhere and disrupt the surface of polysaccharide (Merzendorfer 2013). The substrate-binding domain accumulates catalytic sites on the surface of substrates and also disrupts the hydrogen bonds in the crystalline region of substrates and thereby facilitating subsequent hydrolysis by the catalytic domains (Tanaka et al. 2001). Chitin-binding domain of ChiB, a chitinase from *Serratia marcecens*, interacts with the reducing end of substrates that extend beyond subsites in the active-site cleft (Zakarlaseen et al. 2009). During hydrolysis, chitinases show processive action in which they remain attached to the substrate in subsequent hydrolysis. This processive action is energetically favorable for chitinases as the individual polymer chains do not attach to the insoluble material during hydrolysis.

The substrate-binding sites in processive chitinases have more aromatic (tryptophan) residues. These residues function as a flexible and hydrophobic sheath along which the polymer chain can slide during the processive mode of action (Zakarlassen et al. 2009). Point mutations in the tryptophan residues near the catalytic center resulted in loss of processivity of ChiB for water insoluble substances (e.g. chitin) but a 29-fold increase in activity towards the water soluble oligomeric and polymeric substances (e.g. chitosan) (Horn et al. 2006). Thus processivity is a requirement for hydrolyzing insoluble substrates but can reduce the efficiency of the enzyme toward more accessible substrates.

CBP 21, a binding protein, helps in binding to β-chitin while certain other CBP21 like proteins aid in binding to the α chitin. CBP has been observed as a member of CBM family 33 (Levasseur et al. 2013). It has been observed that CBP21 like proteins can enhance the infectivity of insect virus suggesting its possible role in substrate binding. Several crystal structure and site-directed mutagenesis studies confirm the above made suggestion regarding the chitinases activity towards insoluble substrates. It has been found that CBP21 contains conserved polar residues that are crucial for its synergistic effect. Catalytic efficiency of *Sp*ChiD, chitinase D from *Serratia proteamaculans,* in degradation of insoluble chitin substrates was improved by fusing polycystic kidney disease (PKD) domain and chitin binding protein 21 (CBP21) (Madhuprakash et al. 2015). Point mutation in an aromatic residue near to the catalytic center in chitobiohydrolase chitinase B (ChiB) from *Serratia marcescens* resulted in loss of activity against the water insoluble chitin but enhanced activity for polymeric chitin and chitin hexamer oligosaccharide. Thus it was established that enzyme speed for chitin processivity is compromised at the expense of processing speed for chitosan and other related chitin isomers (Horn et al. 2006; Katouno et al. 2004).

The chitinases carry out their function by acid catalyzed glucose hydrolysis, which could be achieved by two methods. Stereochemistry retention of the anomeric oxygen at C-1 relative to the initial configuration (i) and inversion of stereochemistry (ii). Stereochemistry retention involves the double displacement mechanism in which the β-1, 4 glycosidic oxygen is protonated to produce an oxocarbenium intermediate. This oxocarbenium intermediate is stabilized by a second carboxylate from nearby sugar residue by covalent or electrostatic interactions. The nucleophilic attack by water produces the hydrolysis products that retain the stereochemistry. This type of mechanisms is commonly found in family 19 chitinases. During stereochemistry inversion for family 18 chitinases a bound water molecule acts as a nucleophile. The reaction initiates with the protonation of glycosidic oxygen by a protonated residue. N-acetyl moiety of the -1 amino acid residue near the catalytic center carries a nucleophilic attack, which results in cleavage of the sugar chain and formation of an oxazolinium ion intermediate. Subsequent hydrolysis of this ion commences the reaction completion (Merzendorfer 2013; Dahiya et al. 2006; Horn et al. 2006; Brameld and Goddard 1998a, b; Tews et al. 1997).

Molecular insights have demonstrated that the stearic movement of residues Asp140, Asp142, and Glu144 is very critical in bringing the water molecule close to the reaction center and further nucleophilic attack. Ser 93 and Tyr 10 aid in the catalysis of the chitinases by stabilizing Asp 140 and Asp 142. Conformational changes are observed in Trp-97 and Trp-220 as soon as the chitin binds the protein creating a hydrophobic sandwich between sugar residues at +1 and +2. In double displacement reaction, the anomeric carbon is positively polarized due to the electron withdrawing effect of the oxazolinium ring. Glu-144 polarizes the water molecule and two other water-mediated hydrogen bonds to the protein and place the water molecule in vicinity of the catalytic site to favor the nucleophilic attack on the anomeric carbon C1, retaining the β-anomeric stereochemistry (Zakarlassen et al. 2009). The activity of the chitinases is also dependent on the amount of chitin present and the ease of accessibility of it to the enzymes. Studies on a rice chitinase cloned in *Pichia pastoris* showed that the chitinase activity on fungi *R. stolonifer, B. squamosa, A. niger, and P. aphanidermatum* was dependent on exposure of the chitin to the chitinase and amount of chitin present in the fungal cell wall (Yan et al. 2008).

12.6 Chitinase Production

Fermentation, e.g. solid state, submerged, is the main process involved in the chitinase production. Chitinases are extracellular enzymes and their production is governed by a cumulative effect of several physical and biochemical factors e.g. carbon and nitrogen sources, agitation rate, pH, dissolved oxygen (DO), temperature, media composition, and inoculum levels. These affect the success and efficiency of a fermentation process and hence enzyme production. Table 12.2 provides information regarding the various conditions maintained for chitinase production in different organisms. Maximum chitinase production has been observed when chitin is used as a sole carbon and nitrogen source. Easily metabolizable sugars e.g. glucose supported growth but reduced the chitinase production (Jholapara et al. 2013; Chakrabortty et al. 2012; Faramarzi et al. 2009; Patidar et al. 2005). No significant change in chitinase production by *Masilla tominae* was observed with different organic nitrogen sources (Faramarzi et al. 2009).

The chitinases production is also influenced by agitation rate. Instead of a low (75 RPM) and high (225 RPM) a moderate (150 RPM) agitation rate enhanced the chitinase production and activity in *Verticillium lecanii* F091 (Liu et al. 2003). High agitation rates can cause negative effects such as rupture of cells, change in morphological state, decrease in productivity, vacuolation, and autolysis whereas at lower agitation rates there is not enough mixing of the components. pH not only

Table 12.2 Various physical and biochemical parameters observed for chitinase production

Organism	Fermentation type	Reactor	pH	Temp	RPM	Volume: working/total	Activity	References
Bacillus alvei NRC-14	Submerged fermentation	Erlenmeyer flasks	6.0	28 °C	130	100 ml/250 ml	1.5 U/ml	Abdel-Aziz et al. (2012)
Bacillus cereus 6E1	Submerged fermentation	NA	5.8	35 °C	NA	600 ml	84 Units/g substrate	Wang et al. (2001)
Bacillus pumilus	Submerged fermentation	Erlenmeyer flasks	6.5	30 °C	150	100 ml/500 ml	97.67 U/100 ml	Tasharrofi et al. (2011)
Isaria Fumosorosea	Submerged fermentation	NA	5.7	25 °C	NA	NA	186.34 ± 3.81 mU/ml	Ali et al. (2010)
Massila timonae	Submerged fermentation	Erlenmeyer flasks	6.5	30 °C	150	100 ml/500 ml	10.1 U/ml	Faramarzi et al. (2009)
Penicillium chrysogenum PPCS1	Solid substrate fermentation	Erlenmeyer flasks	5.0	24 °C	150	5 g/150 ml	3809 units/g initial dry substrate)	Patidar et al. (2005)
Penicillium chrysogenum PPCS2	Solid substrate fermentation	Erlenmeyer flasks	4.0	24 °C	150	5 g/150 ml	2516 units/g initial dry substrate)	Patidar et al. (2005)
Serratia marscenes CBC5	Submerged fermentation	Erlenmeyer flasks	7.0	30 °C	200	50 ml/250 ml	10.1 U/ml	Chakrabortty et al. (2012)
Trichoderma harzianum TUBF966	Submerged fermentation	Erlenmeyer flasks	5.0	30 °C	180	40 ml/NA	14.7 U/ml	Sandhya et al. (2004)
Verticillium lecanii F091	Submerged fermentation	Erlenmeyer flasks	4.0	24 °C	200	200 ml/500 ml	9.95 mU/ml	Liu et al. (2003)
Verticillium lecanii F091	Submerged fermentation	Stirred-tank bioreactor	4.0	24 °C	150	3 l/5 l	18.2 mU/ml	Liu et al. (2003)
Verticillium lecanii F091	Submerged fermentation	Airlift bioreactor	4.0	24 °C	150	15 l/30 l	19.9 mU/ml	Liu et al. (2003)

decides the growth of the organism and chitinase production but also the secretion of chitinase from the organism into the medium. Optimum pH and temperature range varies with the different organisms. Usually chitinases from organisms functional at high temperatures have also shown wide range of favorable pH range (Table 12.2). Effect of inoculum size on chitinase production depends on the species type. Two different isolates of *Penicillium* PPCS1 and PPCS2 showed different effects of inoculum size on the chitinase production. In PPCS1 the chitinase production increased up to a point and then started decreasing while in PPCS2 the production continuously was increasing with increase in inoculum size (Patidar et al. 2005).

Optimum levels of $MgSO_4$ enhance the chitinase production in submerged and submerged substrate fermentation both. As observed the chitinase production increases to a level with increasing concentrations of $MgSO_4$ after which it becomes constant (Jholapara et al. 2013; Bhushan and Hoondal 1998). Addition of non-ionic detergents e.g. Tween 80, Tween 20, Triton X-100 also increase the chitinase production (Patidar et al. 2005). These non-ionic detergents disrupt the cell walls and aid in extracellular secretion of the chitinases that in turns results in higher chitinase production. Lab scale to pilot scale production of chitinases from *Verticillium lecanii* F091 has resulted in higher chitinase activity (Liu et al. 2003). Following production chitinases can be purified using different methods ammonium chloride precipitation, ammonium sulfate precipitation, ion exchange chromatography and ethanol-glycol etc. (Faramarzi et al. 2009).

12.7 Chitinase Assay

Enzyme assays help in finding the amount, type, and various factors affecting the activity. To determine the chitinase activity, different methods have been proposed using the colloidal chitin, carboxymethylchitin-Remazol Brilliant Violet 5R (CM-chitin- RBV; Loewe Biochemica GmbH, Otterfing, Germany), chitin-azure or p-nitrophenol based substrate (Chakrabortty et al. 2012; Dahiya et al. 2005; Tanaka et al. 1999). Insoluble colloidal chitin has been most commonly used as a substrate for determining the chitinase activity. Colloidal chitin is prepared by processing the chitin obtained from of shells crabs, oysters, shrimps etc. thorough different steps that involve the use of acids and precipitation. As per method developed by Rodriguez-Kabana et al. (1983), 50 gm chitin is finely crushed with pestle and mortar and followed by grinding in mixer. This chitin is partially hydrolyzed with 400 ml 10 N HCl for 2–3 h with continuous shaking at room temperature. After hydrolysis the chitin appears in colloidal form. The colloidal chitin is washed several times with large volumes of distilled water to adjust pH 7.0 (Khan and Khan 2014). It is very important during preparation of colloidal chitin to ground it to a fine powder in order to separate the chitin chunks from precipitated chitin.

A modified approach to prepare colloidal chitin given by Murthy and Bleakly (2012) is economical and less time consuming. Crab shell flakes, are grinded in a mortar and pestle for 5 min and sieved through 130 mm two piece polypropylene Buchner filter. Twenty grams of sieved flakes are treated with 150 ml of ~12 M HCl, which is added and stirred continuously with a glass pipette for 5 min and later is done at an interval of 5 min for 60 min at room temperature. The mixture is then passed through 8 layers of cheesecloth to remove large chitin chunks. The filtrate obtained is treated with 2 l of ice-cold water to allow colloidal chitin precipitation. The filtrate with ice water can be kept overnight at 4 °C to get better precipitation. The mixture is then filtered through two layers of coffee filter paper under vacuum and is washed with 3 l of tap water to raise the pH to 7.0. The colloidal chitin obtained is then pressed between the coffee filter papers to remove the moisture and placed in glass beaker and autoclaved at 121 °C, 15 psi (STP) for 20 min. The autoclaved colloidal chitin has cake like texture and can be stored at 4 °C until further use (Murthy and Bleakley 2012).

For chitinase assay the cultures (microbial or fungal) are grown to obtain an extract by filtration, separation, or centrifugation. The cultural filtrate serves as the source for the extracellular chitinase enzymes. This cultural filtrate is incubated with colloidal chitin at ambient temperature and pH depending upon the organism from which they are obtained. Chitinase hydrolyze the colloidal chitin into chitooligosaccharides, which can be measured by different methods used to estimate the reducing sugars. Common method for reducing sugar measurement is by using dinitrosalicyclic acid (DNS). For this method 1 ml of the supernatant obtained after centrifuging the cultures at 5000 rpm for 20 min is added to 1 ml of 1% (w/v) colloidal chitin in citrate phosphate buffer pH 5.5 and incubated at 50 °C for 30 min. Incubating the reaction mixtures in boiling water bath for 3–5 min stops the reaction. The solutions are then centrifuged at 5000 rpm for 10 min. To 1 ml of the supernatant obtained 1.5 ml of dinitrosalycilic acid (DNS) is added and kept in boiling water bath for 5 min. The change in color is observed depending on the amount of reducing present in the solution. Absorbance at 540 nm using a UV-VIS spectrophotometer is taken to compare the amount of reducing sugars in the solution against the standard. Using different concentrations of the reducing sugars with dinitrosalycilic acid (DNS) a standard curve can be plotted following the same protocol as for the enzyme assays. One chitinase enzyme unit is defined as the amount of enzyme which catalyzes the release of one µg of reducing sugar per ml per minute under the reaction conditions (Chakrabortty et al. 2012).

Estimation of chitinase activity can also be done using a chromogenic substance. A chromogenic substrate on being incubated with the enzyme solutions releases chromogenic products, which are measured spectrophotometrically at different wavelengths. Measuring the chitinolytic activity by using p-nitrophenyl-N-acetyl-β-D-glucosaminide (pNP-GlcNAc) is based on estimating the amount of released p-nitrophenol (pNP). The reaction mixture consists of 0.5 ml enzyme solution, 0.5 ml 10 mm pNP-GlcNAc solution, and 0.5 ml 0.1 M citrate-phosphate buffer pH 5.5. The mixture is incubated at 60 °C for 30 min and the reaction is stopped by adding 0.5 ml 1 M Na_2CO_3 to the mixture. The release of pNP is spectrophotometrically measured at 400 nm, and enzyme activity is calculated using a standard curve for known concentrations of *p*NP. One chitinase enzyme unit is defined as the amount of enzyme that can release 1 μmole *p*NP per hour under assay conditions (Tasharrofi et al. 2011).

The chitinase activity can also be measured using the SDS-PAGE, which gives an advantage of molecular weight determination along with enzyme activity. In gel chitinase activity measurement is carried in polyacrylamide gel electrophoresis (PAGE) having 5% stacking gel, and 12% resolving gel, which has 0.66 mg/ml carboxymethylchitin-Remazol Brilliant Violet 5R (CM-chitin-RBV). After electrophoresis the gel is incubated with 100 mM sodium phosphate buffer and 0.1% of Triton X-100 for 2 h, at room temperature, which helps in the renaturation of the proteins. Clear zones are developed after the incubation due to the in-gel degradation of CM-chitin- RBV by chitinases. The gel is stained by Coomassie Brilliant Blue R-250 or silver nitrate, which terminates the enzyme reaction and aids in the appearance of protein bands (Wang et al. 2001).

12.8 Genetic Engineering and Molecular Biology

Techniques of protein engineering and directed protein evolution have been continuously used to maximize the effectiveness and efficiency of hydrolytic enzymes. Multiple approaches allow the identification of mutant enzymes possessing desirable qualities such as increased activity, modified specificity, selectivity, or cofactor binding. Site directed mutagenesis or deletions give an insight about the functioning of the individual amino acids and their role in the catalysis or substrate binding in the chitinases.

Recombinant technology has been critical in modification of chitinases in-terms of stability and overexpression. *Pyrococcus furiosus* and *Thermococcus kodakarensis* share gene homology among the two chitinases genes. Although no chitinase activity was observed in culture supernatant of *Pyrococcus furiosus* grown in media with chitin as the sole carbon source. Genetic engineering of the sequence encoding chitinases protein into *E. coli* after single nucleotide deletion at position 1006 resulted in recombinant strains able to show chitinase activity in cell culture extract. The cell culture extract from the recombinant *E. coli* strains showed significant chitinase activity with optimum temperature and pH 90 °C and 6.0–7.5 respectively (Oku and Ishikawa 2006). cDNA for two chitinase genes *Tachit1* and *Ctchit1* from thermophilic fungi *Thermoascus aurantiacus var. levisporus and Chaetomium thermophilum*, respectively was prepared and genetically engineered in a yeast *Pichia pastoris*. The genes were added to a shuttle vector pPIC9K which has AOX1 promoter and the *Saccharomyces cerevisiae* a-factor secretion signal located immediately upstream of the multiple cloning site. Restriction endonuclease digestion and later ligation of the amplified products to above mentioned plasmids resulted in new plasmids having both the chitinase genes. The new recombinant plasmids were cut with restriction endonuclease SacI and incorporated into competent *Pichia pastoris* cells by electroporation (Li et al. 2010).

Mostly chitinase reported are extracellular enzymes but both extracellular and intracellular enzymes were obtained in two plant origin chitinase genes *LbCHI31* and *LbCHI32* from *Limonium bicolor* on transformation in *E. coli*. However, the extracellular counterparts exr*CHI31* and exr*CHI32* of these two chitinase genes showed more activity than the intracellular counterparts inr*CHI31* and inr*CHI32* (Liu et al. 2013). Sometimes the recombination of the chitinase gene sequences into other organism does not result in chitinase activity. Molecular biology studies can be very helpful in studying the machinery for the synthesis and secretion of chitinases. Tanaka et al. (2001) observed that incorporation of *Chi A* gene isolated did not result in the periplasmic secretion of enzyme in recombinant *E. coli*. It was only after the signal sequence of *ChiA* was replaced by a bacterial signal sequence, periplasmic secretion of enzymes could be achieved. Thus a signal peptide, which directs the secretion of enzymes, can affect the success of the recombination. The wild type organism might have the suitable signal peptide for chitinase secretion but it might not be suited for the recombinant species. The recombinant species might be able to produce the gene but not able to secrete it in absence of proper signal peptide.

Genetic engineering and molecular biology studies can also give insights for the catalysis and substrate binding efficiency of the enzymes. Site directed and point mutations in the amino acids provide information about their role during catalysis and substrate binding. Mutants obtained by point mutations in wild-type chitinase A DNA from *Vibrio carchariae* helped in studying the catalysis mechanism of chitinases. Mutations in Trp 275 and Trp 397 emphasized their role in the binding of soluble substrates. Mutations of Trp168, Tyr171, Trp570, and D392N resulted in loss of the hydrolyzing activity against colloidal chitin, and reduced the hydrolyzing activity against the pNP substrate. Mutations in Trp 168 and Trp 171 to glycine indicated their importance in bringing the chitin chain to the binding cleft (Suginta et al. 2007).

Site directed mutation in the exposed aromatic amino acids of chitinase B gene from *Serratia marcescens* 2170 also provided information about the significant role of these residues in the chitin hydrolyzing and binding activity. Replacement of tyrosine and tryptophan with alanine residues resulted in reduction in substrate binding and chitin hydrolysis establishing their significance in these activities. Although no change was observed when an exposed phenylalanine residue was replaced with alanine. Thus the difference in substrate binding and catalysis of two different chitinases *Chi A* and *Chi B* from *Serratia marcescens* were studied with the help of these mutational studies (Katouno et al. 2004).

Mutation studies also help us to understand the difference in activity of several isoforms and hence differentiate them form one another. Deletion mutants in isoforms *Thermococcus kodakarensis* showed that the two isoforms work independently of each other. Mutant studies confirmed that the activity of *Tk-ChiA* is due to the additive effect of activities in region A (*Tk-ChiAA3*) and region B (*Tk-ChiAA2*). Only mutants *Tk-ChiAA2* and *ChiAA4* with region B., exhibited high thermostability and retain more than 70% activity even after heat treatment at 100 °C for 3 h (Tanaka et al. 1999, 2001).

Interspecies genetic engineering of chitinase genes helps in the enhancement of desirable characteristics of species. Transformation of a chitinase gene pGL2 from rice enhanced the antifungal property of grape vine. Somatic embryos obtained from grape vine were transformed with *Agarobacterium tumifaciens* strain LBA4404 having a vector pGL2 with chitinase coding region from rice. Up to two folds higher chitinase activity was observed among the transformed plants. Reduced rate of lesion formation was observed in the transformed plants as compared to the non-transformed plants that correlated with the increase of chitinase activity in the transformed plants (Nirala et al. 2010). Recombinant and genetic engineering studies assist in modification of chitinase producers at intra and interspecies level. Biochemical and molecular biology studies have helped us to understand the catalytic, secretory, and binding processes of chitinases.

12.9   Applications

Research over the years have identified several commercial applications for the chitinase enzymes. Use of biological measures (e.g. microorganism or microbial products) to control the plant pathogens offers sustainable solution without posing any threats to the natural soil, water and air resources. As an antiphytopathogenic agents role of chitinases have been well researched and established. Chitinase Chi18H8 isolated from soil showed antiphytopathogenic activity against *Alternaria alternata, Colletotrichum gloeosporiodies, Fusarium graminarium* and *Fusarium oxysporum* (Hjort et al. 2014). Due to the absence of chitin in plant tissues the chitinases are better suited for phytopathogenic control as compared to other glucanases (Neeraja et al. 2010).

The chitinase produced by *Enterobacter* sp. NRG4 shows antifungal activity towards *Fusarium moniliforme, Aspergillus niger, Mucor rouxi, and Rhizopus nigricans* (Dahiya et al. 2005). Recombinant rice chitinase from *Pichia pastoris* exhibited antifungal property against *Rhizopus stolonifer* (Ehrenb. et Fr.) Vuill, *Botrytis squamosa* Walker, *Pythium aphanidermatum* (eds.) Fitzp, and *Aspergillus niger* van Tiegh. The antifungal activity of the chitinase was affected by the ease of chitin availability to enzyme and chitin amount in the fungal cell wall (Yan et al. 2008). Culture filtrate of *Streptomyces hygroscopicus* strain SRA14 possess antifungal properties because of extracellular chitinase enzyme (Prapagdee et al. 2008).

Chitinase from *Paenibacillus* sp. D1 has high tolerance towards commonly used fungicides (example Captan, Carbendazim, and Mancozeb) in the fields. In presence of Captan half-life of chitinase was 119.17 min at 80 °C and was able to withstand wide range of temperature (40–60 °C) and pH (pH 4.0–8.0). Thus it is a suitable candidate for application in field where huge variations can be found. The chitinase from *Paenibacillus* sp. D1 thus can be used in integrated pest management to control of soil-borne fungal phytopathogens (Singh and Chhatpar 2011).

Chitinases can be very influential in studying the growth patterns in fungi. They along with other hydrolyzing enzymes can hydrolyze the chitin in the fungal cell wall giving access to the protoplast. Several fungal studies related to cell wall synthesis, enzyme synthesis and secretion have been done with the help of chitinases. Chitinase from *Enterobacter* sp. NRG4 was used to obtain protoplasts from *Trichoderma reesei*, *Pleurotus florida*, *Agaricus bisporus*, and *Aspergillus niger* (Dahiya et al. 2005; Hamid et al. 2013).

Chitinases have also found application in controlling the morphogenesis in mosquito and hence controlling the diseases transmitted by them. Chitinases obtained from a saprophytic fungus *Myrothecium verrucaria* can control the spread of *Aedes aegypti*, a vector of yellow fever and dengue (Mendonsa et al. 1996). A numerous medicinal applications of chitinases have been found. Chitinases can augment the activity in antifungal ointments and drugs for against several fungal diseases. Solid waste ($CaCO_3$, chitin, and protein) from shellfish processing has been used to produce single cell proteins with the help of chitinases. Chitinase from *Serratia marcescens* is used in combination with yeast, *Pichia kudriavzevii*, to produce SCP where the chitinase hydrolyzes the chitinous material and yeast produces the single cell protein. In a similar fashion the chitinase from *Myrothecium verrucaria* and *Saccharomyces cerevisiae* has been used to produce SCP from chitinous waste. *Myrothecium verrucaria* chitinase preparation is used for chitin hydrolysis, and *Saccharomyces cerevisiae* for SCP (Dahiya et al. 2006; Wang and Hawang 2001; Hamid et al. 2013).

Studies on Acid mammalian chitinase (AMCase) revealed that the chitinases are important in mediating several inflammatory responses in human beings example asthma, allergic diseases, atopic dermatitis etc. AMCase has been found to be involved in T helper cells 2 mediated inflammatory response responsible for mediating the onset of asthma. This was confirmed by the fact that administration of anti-AMCase antibody leads to a decrease of T helper type 2 (Th2)-inflammation, tissue eosinophilia and lymphocyte accumulation (Zhu et al. 2004). Certain medical applications for chitin have also been developed. Chitin film and fiber can be used as materials for wound dressing and controlled drug release (Kanke et al. 1989; Kato et al. 2003). Chitin is also used as an excipient and drug carrier in film, gel or powder form for applications involving mucoadhesivity. Chitin derivative hydroxyapatite-chitin-chitosan (composite bone-filling material) forms a self-hardening paste for guided tissue regeneration in treatment of periodontal bony defects (Ito et al. 1999).

Products of chitin hydrolysis chitooligosaccharides, glucosamines, and GlcNAc are used in different pharmaceuticals. Chitopentose and Chitoheptose have shown antitumor activities. Hydrolysate produced by crude enzyme solution from *Bacillus amyloliquefaciens* V656 had $(GlcNAc)_6$ showed higher antitumor activity. Hydrolysates of water soluble chitosan inhibited the growth rate of CT26 cells and survival rate to 34% in 1 day (Liang et al. 2007).

12.10 Conclusions

Chitinases have been studied for more than 40 years now and research is still being carried on these enzymes because of their several applications in various areas. They have found their roles in plant pathogenesis, morphogenesis, growth related studies, single cell protein formations, pharmaceutical industries and biofuels. Very few of the chitinases being used are from extremophilic species which can tolerate high temperature or wide range of pH's. Finding new robust enzymes capable of withstanding extreme conditions will definitely enhance their applications in already proved commercial aspects. Studies related to the structure, catalysis and substrate binding can aid us in better understanding of certain unearthed concepts which might create new milestones in research. Use of protein engineering and molecular biology can confer certain desirable characteristics to the existing chitinases. Thus in the future, there is a great possibility and opportunity for generating chitinases with novel functions.

Take Home Message

- Chitin is a crystalline, water insoluble, and recalcitrant cellulose derived homo-polymer. Chitin exists in two conformations namely α chitin and β chitin. The individual polymeric chains are arranged in antiparallel fashion in α configuration. The individual polymeric chains are arranged in parallel fashion in β configuration.
- Chitosan, a water-soluble chitin derivative, is derived from chitins by removing the N-acetyl groups which render in less bulky amino groups on the polymer. The solubility of chitosan in water makes it a favorable substrate in many different applications e.g., gels, fibers, and films. It is commonly found as a key component in the structural make up of insects, fungi, yeast, algae, and in the internal structures of vertebrates.
- Enzymes, such as lysozyme, some glucanases, and chitinase can hydrolyze this linear chitin polymer and among these chitinases can specifically degrade the chitin and chitin based materials. Chitinases are classified into three categories namely endochitinases, exochitinases, and N-acetyl-β-glucosaminidases based on the cleavage mechanism. The enzyme endochitinase randomly mediates the cleave β-(1-4) glycosidic bonds of chitin, the enzyme exochitinases catalyzes cleave the chain from the non-reducing end to form diacetyl-chitobiose and the enzyme N-Acetyl-β-glucosaminidases hydrolyzes diacetyl-chitobiose into N-Acetyl-D-glucosamine or produce N-Acetyl-D-glucosamine from the non-reducing end of N-acetyl-chitooligosaccharides. The chitinolytic activity of the enzyme can be assessed by using p-nitrophenyl-N-acetyl-β-D-glucosaminide (pNP-GlcNAc) based on estimating the amount of released p-nitrophenol (pNP) using spectrophotometric technique.
- Thermostable chitinases can be obtained from bacterial sources such as *Thermococcus kodakarensis, Pyrococcus furiosus* and *Bacillus thuringiensis* subsp., *Paenibacillus* sp and fungal species such *Thermoascus aurantiacus*

var. levisporus, Chaetomium thermophillum, Talaromyces emersonii, and *Thermomyces lanuginosus.*

References

Abdel-Aziz SM, Moharam ME, Hamed HA, Mouafi FE (2012) Extracellular metabolites produced by a novel strain, *Bacillus alvei* NRC-14:1. Some properties of the chitinolytic system. New York Sci J 5:53–62

Ali S, Wu J, Huang Z, Ren SX (2010) Production and regulation of extracellular chitinase from the entomopathogenic fungus *Isaria fumosorosea.* Biocontrol Sci Tech 20(7):723–738

Bhattacharya D, Nagpure A, Gupta RK (2007) Bacterial chitinases: properties and potential. Crit Rev Biotechnol 27:21–28. doi:10.1080/07388550601168223

Bhushan B, Hoondal GS (1998) Isolation, purification and properties of a thermostable chitinase from an alkalophilic *Bacillus* sp. BG-11. Biotechnol Lett 20:157–159. doi:10.1023/A:1005328508227

Brameld KA, Goddard WA III (1998a) Substrate distrortion to a boat conformation at subsite-1 is critical in the mechanism of family 18 chitinases. J Am Chem Soc 120:3571–3580

Brameld KA, Goddard WA III (1998b) The role of enzyme distortion in the single displacement mechanism of family 19 chitinases. Proc Natl Acad Sci U S A 120:4276–4281

Brunner F, Stintzi A, Fritig B, Legrand M (1998) Substrate specificities of tobacco chitinases. Plant J 14:225–234

Chakrabortty S, Bhattacharya S, Das A (2012) Optimization of process parameters for chitinase production by a marine isolate of *Serratia marcescens.* Int J Pharm Biol Sci 2:8–20

Chaudhari SS, Arakane Y, Specht CA et al (2011) Knickkopf protein protects and organizes chitin in the newly synthesized insect exoskeleton. Proc Natl Acad Sci U S A 108:17028–17033

Chen W, Chen C, Jiang S (2013) Purification and characterization of an extracellular chitinase from the entomopathogen *Metarhizium anisopliae.* J Mar Sci Technol 21:361–366. doi:10.6119/JMST-012-0518-2

Collinge DB, Kragh KM, Mikkelsen JD et al (1993) Plant chitinases. Plant J 3:31–40

Dahiya N, Tewari R, Hoondal GS (2006) Biotechnological aspects of chitinolytic enzymes: a review. Appl Microbiol Biotechnol 71:773–782. doi:10.1007/s00253-005-0183-7

Dahiya N, Tewari R, Tiwari RP et al (2005) Production of an antifungal chitinase from *Enterobacter* sp. NRG4 and its application in protoplast production. World J Microbiol Biotechnol 21(8–9):1611–1616

Eijsink VGH, Vaaje-Kolstad G, Vårum KM, Horn SJ (2008) Towards new enzymes for biofuels: lessons from chitinase research. Trends Biotechnol 26:228–235. doi:10.1016/j.tibtech.2008.02.004

Faramarzi MA, Fazeli M, Yazdi MT et al (2009) Optimization of culture conditions for production of chitinase by a soil isolate of *Massilia Timonae.* Biotechnology 8(1):93–99

Fukamizo T, Sakai C, Tamoi M (2003) Plant chitinases: structure-function relationships and their physiology. Foods Food Ingredients J Jpn 208:631–632

Hamid R, Khan MA, Ahmad M et al (2013) Chitinases: an update. J Pharm Bioallied Sci 5 (1):21–29. doi:10.4103/0975-7406.106559

Herrera-Estrella A, Chet I (1999) Chitinases in biological control. EXS 87:171–184

Horn SJ, Sikorski P, Cederkvist JB et al (2006) Costs and benefits of processivity in enzymatic degradation of recalcitrant polysaccharides. Proc Natl Acad Sci U S A 103(48):18089–18094

Guo RF, Li DC, Wang R (2005) Purification and properties of a thermostable chitinase from thermophilic fungus *Thermomyces lanuginosus.* Acta Microbiol Sin 45:270–274

Hjort K, Prest I, Elväng A et al (2014) Bacterial chitinase with phytopathogen control capacity from suppressive soil revealed by functional metagenomics. Appl Microbiol Biotechnol 98:2819–2828. doi:10.1007/s00253-013-5287-x

Ito M, Hidaka Y, Nakajima M, Yagasaki H, Kafrawy AH (1999) Effect of hydroxyapatite content on physical properties and connective tissue reactions to a chitosan–hydroxyapatite composite membrane. J Biomed Mater Res 45:204–208

Jholapara RJ, Mehta RS, Bhagwat AM et al (2013) Exploring and optimizing the potential of chitinase production by isolated Bacillus spp. Int J Pharm Pharm Sci 5(4):412

Kanke M, Katayama H, Tsuzuki S, Kuramoto H (1989) Appilcation of chitin and chitosan to pharmaceutical preparations. I.: film preparation and in vitro evaluation. Chem Pharm Bull 37(2):523–525

Karthik N, Akanksha K, Binod P, Pandey A (2014) Production, purification and properties of fungal chitinases – a review. Indian J Exp Biol 52:1025–1035

Kasprzewska A (2003) Plant chitinases-regulation and function. Cell Mol Biol Lett 8:809–824

Kato A, Kano E, Adachi I, Molyneux RJ, Watson AA, Nash RJ et al (2003) Australine and related alkaloids: easy structural confirmation by 13 C NMR spectral data and biological activities. Tetrahedron Asymmetry 14(3):325–331

Katouno F, Taguchi M, Sakurai K et al (2004) Importance of exposed aromatic residues in Chitinase B from Serratia marcescens 2170 for crystalline chitin hydrolysis. J Biochem 136:163–168. doi:10.1093/jb/mvh105

Khan AA, Shih DS (2004) Molecular cloning, characterization, and expression analysis of two class II chitinase genes from the strawberry plant. Plant Sci 166:753–762. doi:10.1016/j.plantsci.2003.11.015

Khan RS, Khan ZH (2014) Studies on chitinase isolation and thermostability between Bacillus circulance strain L2 and Bacillus Licheniformis strain 2J-1. Sci Res Report 4:1–7

Koga D, Sasaki Y, Uchiumi Y et al (1997) Purification and characterization of Bombyx mori chitinases. Insect Biochem Mol Biol 27:757–767

Kramer KJ, Corpuz LM, Choi H, Muthukrishnan S (1993) Sequence of a c-DNA and expression of the gene encoding epidermal and gut chitinases of Manduca sexa. Insect Biochem Mol Biol 23:691–701

Kuo CJ, Liao YC, Yang JH et al (2008) Cloning and characterization of an antifungal class III chitinase from suspension-cultured bamboo (Bambusa oldhamii) cells. J Agric Food Chem 56:11507–11514. doi:10.1021/jf8017589

Kuzu SB, Güvenmez HK, Denizci AA (2012) Production of a thermostable and alkaline chitinase by Bacillus thuringiensis subsp. kurstaki strain HBK-51. Biotechnol Res Int 2012:135498. doi:10.1155/2012/135498

Lee HW, Park YS, Jung JS, Shin WS (2002) Chitosan oligosaccharides, dp 2–8, have prebiotic effect on the Bifidobacterium bifidium and Lactobacillus sp. Anaerobe 8:319–324

Levasseur A, Drula E, Lombard V et al (2013) Expansion of the enzymatic repertoire of the CAZy database to integrate auxiliary redox enzymes. Biotechnol Biofuels 6:41. doi:10.1186/1754-6834-6-41. PMID: 23514094

Li AN, Yu K, Liu HQ et al (2010) Two novel thermostable chitinase genes from thermophilic fungi: cloning, expression and characterization. Bioresour Technol 101:5546–5551. doi:10.1016/j.biortech.2010.02.058

Liang TW, Chen YJ, Yen YH, Wang SL (2007) The antitumor activity of the hydrolysates of chitinous materials hydrolyzed by crude enzyme from Bacillus amyloliquefaciens V656. Process Biochem 42(4):527–534

Liu BL, Kao PM, Tzeng YM, Feng KC (2003) Production of chitinase from Verticillium lecanii F091 using submerged fermentation. Enzyme Microb Technol 33:410–415. doi:10.1016/S0141-0229(03)00138-8

Liu Z, Huang Y, Zhang R et al (2013) Chitinase genes lbchi31 and lbchi32 from Limonium bicolor were successfully expressed in Escherichia coli and exhibit recombinant chitinase activities. Sci World J 2013:648382. doi:10.1155/2013/648382

Madhuprakash J, El Gueddari NE, Moerschbacher BM, Podile AR (2015) Catalytic efficiency of chitinase-D on insoluble chitinous substrates was improved by fusing auxiliary domains. PLoS One 10:e0116823. doi:10.1371/journal.pone.0116823

Mccormack J, Hackett TJ, Tuohy MG, Coughlan MP (1991) Chitinase production by *Talaromyces emersonii*. Biotechnol Lett 13:677–682

Mendonsa ES, Vartak PH, Rao JU, Deshpande MV (1996) An enzyme from *Myrothecium verrucaria* that degrades insect cuticles for biocontrol of *Aedes aegypti* mosquito. Biotechnol Lett 18(4):373–376

Merzendorfer H (2013) Insect-derived chitinases. Adv Biochem Eng Biotechnol 136:19–50. doi:10.1007/10_2013_207

Murthy N, Bleakley B (2012) Simplified method of preparing colloidal chitin used for screening of chitinase-producing microorganisms. Int J Microbiol 10:7

Neeraja C, Anil K, Purushotham P et al (2010) Biotechnological approaches to develop bacterial chitinases as a bioshield against fungal diseases of plants. Crit Rev Biotechnol 30:231–241. doi:10.3109/07388551.2010.487258

Nirala NK, Das DK, Srivastava PS et al (2010) Expression of a rice chitinase gene enhances antifungal potential in transgenic grapevine (*Vitis vinifera* L.) Vitis 49:181–187

Oku T, Ishikawa K (2006) Analysis of the hyperthermophilic chitinase from *Pyrococcus furiosus*: activity toward crystalline chitin. Biosci Biotechnol Biochem 70:1696–1701. doi:10.1271/bbb.60031

Ohno T, Armand S, Hata T, Nikaidou N et al (1996) A modular family 19 chitinase found in the prokaryotic organism *Streptomyces griseus* HUT 6037. J Bacteriol 178:5065–5070

Onaga S, Chinen K, Ito S, Taira T (2011) Highly thermostable chitinase from pineapple: cloning, expression, and enzymatic properties. Process Biochem 46:695–700. doi:10.1016/j.procbio.2010.11.015

Patidar P, Agrawal D, Banerjee T, Patil S (2005) Optimisation of process parameters for chitinase production by soil isolates of *Penicillium chrysogenum* under solid substrate fermentation. Process Biochem 40:2962–2967. doi:10.1016/j.procbio.2005.01.013

Prapagdee B, Kuekulvong C, Mongkolsuk S (2008) Antifungal potential of extracellular metabolites produced by *Streptomyces hygroscopicus* against phytopathogenic fungi. Int J Biol Sci 4:330–337

Punja ZK, Zhang YY (1993) Plant chitinases and their roles in resistance to fungal diseases. J Nematol 25:526–540

Reese ET et al (1950) The biological degradation of soluble cellulose derivatives and its relationship to the mechanism of cellulose hydrolysis. J Bacteriol 59:485–497

Rinaudo M (2006) Chitin and chitosan: properties and applications. Prog Polym Sci 31:603–632. doi:10.1016/j.progpolymsci.2006.06.001

Rodriguez-Kabana R, Godoy G, Morgan-Jones G, Shelby RA (1983) The determination of soil chitinase activity: conditions for assay and ecological studies. Plant and Soil 75(1):95–106

Sandhya C, Adapa LK, Nampoothiri KM, Binod P, Szakacs G, Pandey A (2004) Extracellular chitinase production by *Trichoderma harzianum* in submerged fermentation. J Basic Microbiol 44(1):49–58

Shekhar N, Bhattacharya D, Kumar D, Gupta RK (2006) Biocontrol of wood- rotting fungi using *Streptomyces violaceusniger* XL-2. Can J Microbiol 52:805–808

Shen KT, Chen MH, Chan HY et al (2009) Inhibitory effects of chitooligosaccharides on tumor growth and metastasis. Food Chem Toxicol 47:1864–1871

Shimono K, Matsuda H, Kawamukai M (2002) Functional expression of chitinase and chitosanase, and their effects on morphologies in the yeast *Schizosaccharomyces pombe*. Biosci Biotechnol Biochem 2002(66):1143–1147

Singh AK, Chhatpar HS (2011) Purification and characterization of chitinase from *Paenibacillus* sp. D1. Appl Biochem Biotechnol 164:77–88. doi:10.1007/s12010-010-9116-8

Suginta W, Songsiriritthigul C, Kobdaj A et al (2007) Mutations of Trp275 and Trp397 altered the binding selectivity of *Vibrio carchariae* chitinase A. Biochim Biophys Acta Gen Subj 1770:1151–1160. doi:10.1016/j.bbagen.2007.03.012

Takaya N, Yamazaki D, Horiuchi H et al (1998) Cloning and characterization of a chitinase-encoding gene (ChiA) from *Aspergillus nidulans*, disruption of that decreases germination frequency and hyphal growth. Biosci Biotechnol Biochem 62:60–65

Tamo F, Chiye S, Masahiro T (2003) Structure-plant chitinases: structure – function relationships and their physiology. Foods Food Ingredients J Jpn 208:631–632

Tanaka T, Fujiwara S, Nishikori S (1999) A unique chitinase with dual active sites and triple substrate binding sites from the hyperthermophilic archaeon *Pyrococcus kodakar*. Appl Environ Microbiol 65:5338–5344

Tanaka T, Fukui T, Imanaka T (2001) Different cleavage specificities of the dual catalytic domains in chitinase from the hyperthermophilic archaeon *Thermococcus kodakaraensis* KOD1. J Biol Chem 276:35629–35635. doi:10.1074/jbc.M105919200

Tasharrofi N, Adrangi S, Fazeli M et al (2011) Optimization of chitinase production by *Bacillus pumilus* using Plackett-Burman design and response surface methodology. Iran J Pharm Res 10:759–768

Tews I, Scheltinga AC, Terwissscha V et al (1997) Substrate assisted catalysis unifies two families of chitinolytic enzymes. J Am Chem Soc 119:7954–7959

Tsujibo H, Minoura K, Miyamoto K et al (1993) Purification and properties of a thermostable chitinase from *Streptomyces thermoviolaceus* OPC-520. Appl Environ Microbiol 59:620–622. doi:10.1271/bbb.64.96

Vahed M, Motalebi E, Rigi G et al (2013) Improving the chitinolytic activity of *Bacillus pumilus* SG2 by random mutagenesis. J Microbiol Biotechnol 23:1519–1528. doi:10.4014/jmb.1301. 01048

Wang SL, Hwang JR (2001) Microbial reclamation of shellfish wastes for the production of chitinases. Enzyme Microb Technol 28(4):376–382

Wang S-Y, Moyne A-L, Thottappilly G et al (2001) Purification and characterization of a *Bacillus cereus* exochitinase. Enzyme Microb Technol 28:492–498. doi:10.1016/S0141-0229(00) 00362-8

Yan R, Hou J, Ding D et al (2008) In vitro antifungal activity and mechanism of action of chitinase against four plant pathogenic fungi. J Basic Microbiol 48:293–301. doi:10.1002/ jobm.200700392

Zakarlassen H, Aam BB, Hom SJ et al (2009) Aromatic residues in the catalytic center of chitinase A from *Serratia marcescens* affect processivity, enzyme activity, and biomass converting efficiency. J Biol Chem 284:10610–10617. doi:10.1074/jbc.M900092200

Zhu Z, Zheng T, Homer RJ, Kim YK et al (2004) Acidic mammalian chitinase in asthmatic Th2 inflammation and IL-13 pathway activation. Science 304(5677):1678–1682

Chapter 13
Extremophilic Lipases

Marcelo Victor Holanda Moura, Rafael Alves de Andrade, Leticia Dobler, Karina de Godoy Daiha, Gabriela Coelho Brêda, Cristiane Dinis AnoBom, and Rodrigo Volcan Almeida

What Will You Learn from This Chapter?

Lipases, or triacylglycerol ester hydrolases (E.C. 3.1.1.3), are one of the most important classes of biocatalysts. Their versatility allows them to be used in various applications, such as organic and fine chemical synthesis, and the production of biofuels, food, beverages, and cleaning products. Several industrial processes are carried out under specific conditions that may be too hostile to allow biocatalysis with known mesophilic enzymes, such that new and more stable enzymes are still required to better fulfill the different industrial requirements. Extremophilic organisms—organisms that inhabit harsh environments—have certain adaptations in their enzymatic machinery that enable them to support extreme conditions. These organisms have yielded enzymes with attractive features, such as greater specific activity in

M.V.H. Moura • L. Dobler • K. de Godoy Daiha • G.C. Brêda • R.V. Almeida (✉)
Departamento de Bioquímica, Instituto de Química, Laboratório de Microbiologia Molecular e Proteínas, Programa de Pós-graduação em Bioquímica, Universidade Federal do Rio de Janeiro, Av. Athos da Silveira Ramos, 149, Centro de Tecnologia, Bloco A, sala 541, Cidade Universitária, 21941-909 Rio de Janeiro, RJ, Brazil
e-mail: volcan@iq.ufrj.br; rodrigovolcan.almeida@gmail.com

R.A. de Andrade
Departamento de Bioquímica, Instituto de Química, Laboratório de Microbiologia Molecular e Proteínas, Programa de Pós-graduação em Bioquímica, Universidade Federal do Rio de Janeiro, Av. Athos da Silveira Ramos, 149, Centro de Tecnologia, Bloco A, sala 541, Cidade Universitária, 21941-909 Rio de Janeiro, RJ, Brazil

Departamento de Bioquímica, Instituto de Química, Laboratório de Biologia Estrutural de Proteínas, Programa de Pós-graduação em Bioquímica, Universidade Federal do Rio de Janeiro, 21941-909 Rio de Janeiro, RJ, Brazil

C.D. AnoBom
Departamento de Bioquímica, Instituto de Química, Laboratório de Biologia Estrutural de Proteínas, Programa de Pós-graduação em Bioquímica, Universidade Federal do Rio de Janeiro, 21941-909 Rio de Janeiro, RJ, Brazil

© Springer International Publishing AG 2017
R.K. Sani, R.N. Krishnaraj (eds.), *Extremophilic Enzymatic Processing of Lignocellulosic Feedstocks to Bioenergy*, DOI 10.1007/978-3-319-54684-1_13

lower or higher temperatures, different optimal pH ranges, and a high tolerance to salt concentrations. This chapter provides a review of the evolution of scientific publications on extremophilic lipases (e.g. thermophilic, alkaliphilic, psychrophilic, halophilic, and acidophilic lipases), as well as a review of the structural characteristics of these biocatalysts, some molecular reasons to explain their stability in such diverse and extreme conditions, and a few examples of their industrial applications.

13.1   Introduction

Lipases are triacylglycerol ester hydrolases (E.C. 3.1.1.3). They are one of the most important classes of biocatalysts that act in a range of different substrates, catalyzing hydrolysis, esterification, alcoholysis, acidolysis, interesterification, aminolysis, and other reactions (Daiha et al. 2015). Some important markets for lipases include the food, beverage, and cleaning products industries and applications that involve organic synthesis. The world demand for these enzymes is forecast to grow 6.2% per year through 2017, approaching US$345 million (Freedonia Group 2015). Considering that some industrial processes require harsh conditions in order to increase the production rate and substrate solubility and/or minimize the formation of undesirable by-products, enzymes capable of acting under such unfavorable circumstances are of great interest. In this context, the present chapter provides a review of the literature concerning extremophilic lipases.

There is no strict definition in the literature for an extremophilic lipase. Enzymes may be classified as extremophilic or not depending on their microorganism of origin, or the optimal conditions for their activity (i.e. temperature, pH, salinity, etc.). Indeed, some lipases not originated from extremophilic organisms may still present extremophilic properties. In this review, the criteria used to define extremophilic enzymes are based mainly on the microorganism of origin and follow the definitions provided by Madigan et al. (2000) and Schreck and Grunden (2014), as shown in Table 13.1. However, some of the examples given do not fall into this specific classification, but have been included because of their great commercial importance and their capacity to withstand harsh reaction conditions.

13.2   Structure and Catalytic Mechanism of Lipases

The first structure of a lipase was determined in 1990. Since then, more lipase structures have been determined, and 147 structures classified as triacylglycerol lipases (E.C. 3.1.1.3) are currently deposited in the Protein Data Bank (PDB), all of them using crystallography and X-ray diffraction methodologies. Despite this apparently large number, there is still little structural information available at an atomic level on lipases from extremophilic organisms. Only 55 such structures are to be found in the PDB, corresponding to ten different lipases from extremophile organisms (Table 13.2).

Table 13.1 Definitions of extremophilic lipases used in this chapter

Types of extremophilic lipases	Optimal growth conditions[a]	
Alkaliphilic	pH of between 10 and 11	
Acidophilic	pH of 5 or less	
Thermophilic	Temperatures between 45 and 80 °C (organisms that grow at temperatures above 80 °C are considered hyperthermophilic)	
Psychrophilic	Temperature below 15 °C	
Halophilic	NaCl concentration	1–6%: low 6–15%: moderate 15–30%: extreme

[a]Criteria based on Madigan et al. (2000) and Schreck and Grunden (2014)

Table 13.2 Triacylglycerol lipase (EC 3.1.1.3) structures from extremophiles deposited in the Protein Data Bank (PDB)

Lipases from extremophilic eukaryotes	PDB code
Lipase from *Thermomyces lanuginosus*	4ZGB, 4S0X, 4N8S, 4KJX, 4GHW, 4GWL, 4GI1, 4GLB, 4GBG, 4FLF, 4DYH, 4EA6, 1GT6, 1DT3, 1DT5, 1DTE, 1DU4, 1EIN, 1TIB
Lipase A from *Pseudozyma antarctica*	3GUU, 2VEO
Lipase B from *Pseudozyma antarctica*	3W9B, 4K5Q, 4K6G, 4K6H, 4K6K, 3ICV, 3ICW, 1LBS, 1LBT, 1TCA, 1TCB, 1TCC, 4ZV7, 5A6V, 5A71
Lipases from extremophilic bacteria	**PDB code**
T1 Lipase from *Geobacillus zalihae*	3UMJ, 2Z5G, 2DSN
Lipase from *Geobacillus stearothermophillus*	4FMP, 1JI3, 1KU0, 4X6U, 4X71, 4X7B, 4X85
Lipase 42 from *Bacillus* sp.	4FKB
Lipase L2 from *Bacillus* sp.	4FDM
Lipase from *Geobacillus thermocatenulatus*	2W22, 5CE5
Lipase from *Geobacillus* sp. SBS-4S	3AUK
Lipases from extremophilic archaea	**PDB code**
Lipase from *Archaeglobus fulgidus*	2ZYH, 2ZYI, 2ZYR, 2ZYS

From these crystal structures, lipases were classified as belonging to the structural family of α/β hydrolases. In addition to lipases, the other members of this family are esterases, proteases, dehalogenases, peroxidases, and epoxide hydrolases (Anobom et al. 2014).

13.2.1   α/β-Hydrolase Fold

The α/β domains are composed of a parallel or mixed β-sheet surrounded by an α-helix and are widespread in nature. A subclass of this fold was identified in 1992 by comparing the structures of different hydrolytic enzymes. Although these enzymes do not have sequence similarity, and show specificity for different types of substrates, they have structural similarity and conserved catalytic residues, suggesting that they evolved from a common ancestor. The canonical α/β-hydrolase fold presents a central β-sheet composed of seven parallel β-strands (β1, β3, β4, β5, β6, β7) and one antiparallelβ-strand (β2). Strands β3 to β8 are connected by six β-helices that surrounding this central β-sheet. On one side are helices αA and αC, and on the other are helices αB, αD, αE, and αF (Fig. 13.1a) (Nardini and Dijkstra 1999; Anobom et al. 2014).

The active site is composed of a nucleophilic residue, an acid residue (Asp or Glu), and a histidine residue, which form the catalytic triad. In lipases, the nucleophilic residue has so far been characterized as a serine. This residue is located in a highly conserved pentapeptide called the nucleophilic elbow, which displays the sequence Sm-X-S-X-Sm (Sm is a small residue, usually a glycine; X is any residue; and S is serine) (Nardini and Dijkstra 1999; Anobom et al. 2014).

The pentapeptide forms a tight γ turn after the strand-β5, in a strand-loop-helix motif, and induces the nucleophilic residue to adopt an energetically unfavorable conformation of angles φ and ψ in the main chain, imposing steric constraints on

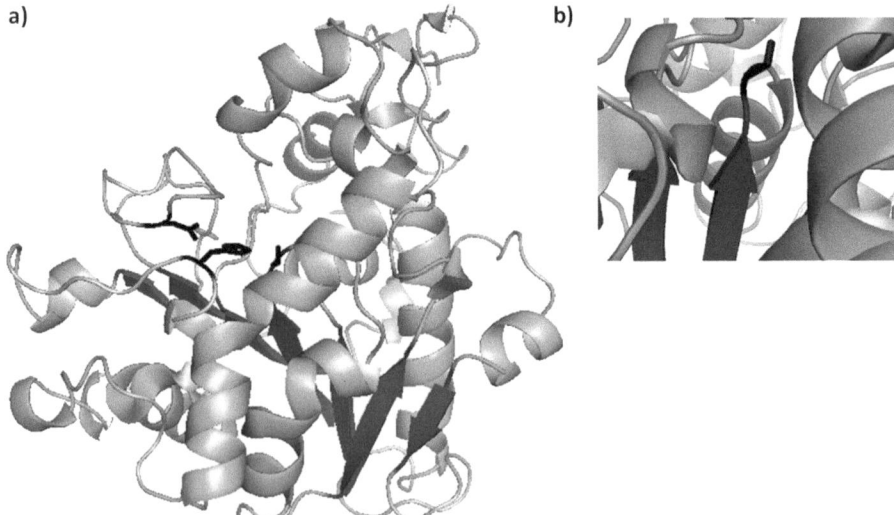

a)   b)

Fig. 13.1 (**a**) Crystallographic structure of lipase B from *Candida antarctica* (Uppenberg et al. 1994—PDB ID 1TCC) showing the characteristic α/β-hydrolase fold. The α-helices and unstructured loops are in *light gray*, the β-strands are *dark gray*, and the catalytic triad residues are *black* (Ser 105, Asp 187, His 224). (**b**) Detail showing the catalytic serine (represented as a *black stick*) in the nucleophilic elbow, in a strand-turn-helix motif

the residues located in its proximity (Fig. 13.1b). The acid residue may be an aspartic acid or glutamic acid residue, and is usually found after strand-$\beta 7$, but in some structures this residue has been found after strand-$\beta 6$. The histidine residue has been found in the loop region after the last β-strand (Nardini and Dijkstra 1999; Anobom et al. 2014).

The conformation of the nucleophilic elbow contributes to the formation of the oxyanion hole, required to stabilize the negatively charged transition state that occurs during hydrolysis. The oxyanion hole is normally formed by two nitrogen atoms located at the protein backbone. The first one is always located at the residue immediately after the nucleophile, while the second is typically found between strand $\beta 3$ and helix αA (Nardini and Dijkstra 1999).

The crystallographic structures of enzymes belonging to the α/β hydrolase family indicate that this fold may exhibit considerable variability. However, the presence of the catalytic triad, the nucleophilic elbow after the canonical strand $\beta 5$, and the presence of at least five parallel strands in the central β-sheet are characteristics common to all these structures (Nardini and Dijkstra 1999).

Several lipases can present a mobile amphipathic subdomain called a lid, which controls the access of substrate molecules to the active site. In the presence of a water-lipid interface, the lid opens and enzymatic activity is increased—a phenomenon called interfacial activation. The complexity and size of the lid of the different lipases whose structures have been determined varies greatly, and may be formed by a loop region, or by one or more α-helices (Anobom et al. 2014).

13.3 Literature Search: An Overview

With the aim of analyzing the evolution of the scientific work being published in relation to extremophilic lipases, a literature search was carried out of the Web of Science®, a research platform maintained by Thomson Reuters covering more than 12,000 journals in all subject areas.

The search covered publications filed between 1985 and July 2016. A specific strategy using keywords was set up for five major groups of extremophilic lipases: halophilic, psychrophilic, alkaliphilic, acidophilic, and thermophilic. The publications retrieved in the search had their title and abstract carefully examined to select only those that referred to one or more of the five groups (Table 13.3).

Extremophilic lipases are attracting increasing scholarly attention. More scientific documents were published in the last 8 years (from 2009 to 2016) than in the first 24 years of the analysis (Fig. 13.2). This increase may be associated with the acknowledged potential of using these enzymes in a variety of industrial processes.

The literature search revealed that thermophilic lipases seem to be the most abundant group of extremophiles lipases described in literature. Two other groups that are also gaining researchers' attention, mainly in recent years, are alkaliphilic and psychrophilic lipases. In comparison, fewer documents on halophilic and acidophilic lipases were retrieved. Table 13.3 depicts the search strategies used for each of the extremophilic lipase groups.

Table 13.3 Search strategies used for each of the extremophilic lipase groups

Extremophilic lipase groups	Search strategies	Number of documents (after analysis)
Halophilic lipases	(lipase* and (haloph* or halotolerant* or "salt tolerant*"))	50
Psychrophilic lipases	(lipase* and (psychroph* or "cold adapt*" or "cold activ*")) or ("low temperature" near/10 lipase*)	128
Alkaliphilic lipases	(lipase* and (alkaliphil* or "alkali* stab*")) or "alkaline lipase*"	213
Acidophilic lipases	(lipase* and (acidophilic or acidophile or "acid stable")) or "acidic lipase*" or ("acidic pH" near/10 lipase*)	50
Thermophilic lipases	(lipase* near/10 (*thermophil* or *thermostab* or "thermal stab*"))	366

"*" is a wildcard representing any group of characters, including no character; "near/n" is an operator indicating that two terms are within n words of each other

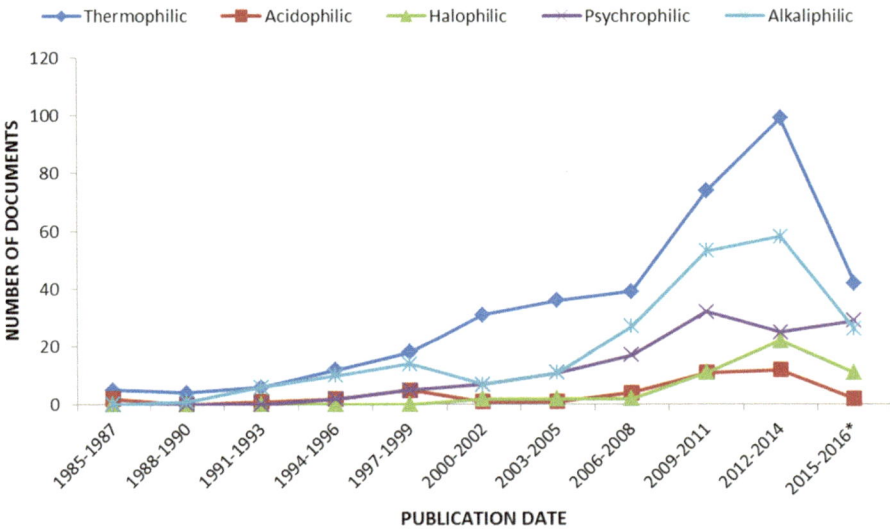

Fig. 13.2 Number of scientific documents found for each of the extremophilic lipase groups per publication year (grouped into triennia). The *asterisk* marks the current triennium, 2015–2017, which is not yet over

The countries of origin of the research groups were analyzed, as were any specificities related to the authorship of the publications. India and China are in the top five countries with most research in all types of extremophilic lipases. Researchers from Malaysia are third in the ranking of publications related to thermophilic lipases; South Korea is the third country in number of documents on psychrophilic and alkaliphilic lipases, and fourth in the thermophilic ranking; and Brazil ranks fifth in the number of publications on alkaliphilic lipases and second in the ranking of publications on acidophilic lipases (Table 13.4).

Table 13.4 Five top countries in publications per type of extremophilic lipase—from 1985 to 2016

Thermophilic lipases		Psychrophilic lipases		Alkaliphilic lipases		Acidophilic lipases		Halophilic lipases	
Country	Number	Country	Number	Country	Number	Country	Number	Country	Number
China	50	China	30	India	66	India	15	China	10
India	49	India	19	China	41	Brazil	4	India	8
Malaysia	35	S. Korea	12	S. Korea	15	China	3	Iran	7
S. Korea	25	Japan	12	Brazil	13	Japan	3	Spain	7
Spain	25	Italy	8	Taiwan	13	France	3	Brazil	3

Finally, it is important to point out that the searches conducted in this study recovered scientific publications containing the target keywords in the title and/or abstract. Therefore, they were not designed to identify all the documents published in the field of extremophilic lipases, but to map out this field of inquiry and compare the different groups.

13.4 Thermophilic Lipases

The literature search carried out (Fig. 13.1) reveals that the more sought-after extremozymes are the ones with the capacity to withstand and perform catalytic reactions at high temperatures. The interest in finding new thermophilic enzymes, particularly lipases, is due to the fact that industrial processes are generally conducted at temperatures above 45 °C, which rules out the use of mesophilic enzymes. There are a number of advantages in carrying out industrial bioconversions at higher temperatures, such as the thermodynamic increase in reaction rates, reduced contamination with foreign mesophilic microorganisms, reduced viscosity rates, and increased diffusion and solubility (López-López et al. 2014).

When thermophilic enzymes are used as biocatalysts, the temperature range at which the reactions can be performed is considerably higher, as their optimal reaction temperatures vary from 45 to 80 °C. This higher reaction temperature has some major impacts, affecting, for instance, the physico-chemical qualities of some lipid compounds, which need higher temperatures to be processed. However, this temperature range is still considerably lower than the temperatures that can be reached using inorganic catalysts—as high as 200 or 300 °C. Nonetheless, processes catalyzed by thermophilic enzymes have the further advantage of being cleaner, as they consume less energy and form fewer by-products (López-López et al. 2014).

Two of the most important genera of microorganisms utilized as sources of thermophilic lipases are *Bacillus* and *Geobacillus* (Haki and Rakshit 2003). Another important source of extremophilic enzymes are microorganisms from the Archaea domain, which are capable of surviving in very extreme conditions. These include *Pyrococcus furiosus*, whose optimal temperature of growth is 100 °C.

Handelsman and Shoham (1994) published an interesting study about the isolation of thermostable lipases. A bacterial, lipase-producing strain, H1, later identified as *Bacillus* sp., was identified, which had an optimal growth temperature of 65 °C. This isolated lipase's maximum activity was at 70 °C, pH 7.0, and it maintained at least 50% of its activity for 50 h at 60 °C.

A recent study by Mahadevan and Neelagund (2014) reported structural insights and characterization of a thermostable lipase from *Geobacillus* sp. Iso5, a typical thermophilic bacterium with optimal growth at 50 °C. The lipase was produced and purified was found to have an optimal temperature of 70 °C and optimal pH of 8.0. The biocatalyst was found to be thermostable (the ability to retain enzymatic activity) when 90% of its activity was maintained after 2 h incubation at 70 °C.

When studying hyperthermophilic microorganisms as sources of thermostable lipases, it is not easy to use the microorganisms themselves to produce the enzymes, since they often require extreme and costly cultivation conditions. With the development of molecular biology techniques, these hyperthermophilic enzymes are now being cloned and expressed in heterologous hosts, such as *Escherichia coli* and *Pichia pastoris* (Alquéres et al. 2011).

This technology was used by Almeida et al. (2006) to clone and express lipase PF2001Δ60 from *Pyrococcus furiosus* in *Escherichia coli*. Hyperthermophilic archaea *P. furiosus* was isolated from geothermally heated marine sediment of the coast of Italy, and presented an anaerobic metabolism with optimal growth occurring at 100 °C. When expressed in *E. coli* without its 60 initial base pairs, which hypothetically coded for a signal peptide, the lipase showed an optimal temperature of 80 °C and pH of 7.0. Furthermore, this lipase presented 100% thermostability after 6 h of incubation at 70 °C (Alquéres et al. 2011).

Lipase PF2001Δ60 was further covalently immobilized by Branco et al. (2015). The lipase, immobilized on glyoxyl-agarose, was found to have an optimal temperature of up to 90 °C, and around 80% residual activity after 48 h at 70 °C.

Thermophilic fungi are another important source of thermostable lipases. One remarkable example is *Thermomyces lanuginosus*, which produces a lipase widely known as TLL. Ávila-Cisneros et al. (2014) obtained TLL from *T. lanuginosus* using solid-state fermentation. This study showed that maximum lipase activity was achieved between 60 and 85 °C at pH 10. In thermostability studies, lipase activity rose from 30 to 50 °C, and stayed constant between 50 and 60 °C after 4 hours incubation. These characteristics show that TLL is a promising extremozyme for industrial applications. Indeed, it is already being marketed by Novozymes® under the trade name Lipozyme® TL. This product is widely used in studies related to the application of biocatalysis in the pharmaceutical field, such as the production of the anticonvulsant and antiepileptic drug pregabalin, and for producing biodiesel by transesterification reactions (Ávila-Cisneros et al. 2014).

The paper industry is one of the sectors that benefits from thermostable lipases. In the papermaking process, some impurities from the processing of lignocelullosic material tend to aggregate and form a substance called pitch. This substance is responsible for clogging machines and causing problems in subsequent process steps, such as bleaching. Many esters are present in the composition of pitch, and since the whole papermaking process is conducted at high temperatures, thermophilic lipases are used to control pitch formation (Gutiérrez et al. 2009). The first strategies for controlling pitch enzymatically were developed in the 1980s and continue to be used to this day (Gutiérrez et al. 2009). In the 1990s, Novozymes (Bagsvaerd, Denmark) launched a product called Resinase® A2X, a lipase from the fungus *Aspergillus oryzae*. This product's optimal temperature was 70 °C, and it was successfully used in pitch control in the paper industry in Asia. Since then, Novozymes has developed a variant of Resinase® that is more thermophilic than its predecessor, with an optimal temperature of 85 °C.

The studies mentioned above are just a few examples of the potential industrial applications of thermophilic lipases. There are many other opportunities that are yet to be explored, and the continuous search for new enzymes and technologies is likely to increase the use of these extremozymes as biocatalysts, offering excellent prospects for the implementation of cleaner and environmentally friendlier processes.

13.5 Psychrophilic Lipases

Psychrophiles are organisms that inhabit the cold environments of the Earth. They are able to live and thrive in temperatures between -20 °C and ≤ 0 °C. Since they are adapted to cold, they cannot normally grow at temperatures above 30 °C (Siddiqui and Cavicchioli 2006). Psychrophilic microorganisms are relatively abundant in nature, due to the large quantity living in the oceans. Early estimates indicated there were about 3.5×10^{30} cells in the subseafloor environment, which corresponded to 27–33% of the Earth's living biomass. Later, this number was estimated to constitute 0.6% of the Earth's total biomass (Kallmeyer et al. 2012). Other environments suitable for cold biospheres are the high altitudes of mountains, underground caves, and the polar areas of the globe. All three domains of life— Bacteria, Archaea, and Eukarya—have psychrophilic representatives. Many psychrophiles live in biotopes with more than one stress factor, such as low temperature and high pressure in deep underwater environments (piezo-psychrophiles), or high salt concentration and low temperature in sea ice (halo-psychrophiles). Cell-specific adaptation strategies related to their ability to with-stand such extreme conditions have been identified (Gomes and Steiner 2004).

   Psychrophilic organisms present several features that compensate for the slow metabolic rates that would occur at low temperatures. The majority of cold-adapted enzymes are characterized by having lower optimal activity temperatures, some-times with a concomitant decrease in stability at higher temperatures. Moreover, they tend to exhibit a high reaction rate (up to ten times higher k_{cat} than heat-stable homologs) by decreasing the activation barrier (measured in Gibbs free energy, ΔG) between the substrate and the transition state (Siddiqui and Cavicchioli 2006).

   Psychrophilic lipases have become consolidated for their excellent potential use in biotechnological processes. This is mainly because of their high catalytic activity at low temperatures and high enantioselectivity. These features render the use of such biocatalysts possible in a variety of industries, including detergent production (cold washing), the food industry (e.g., fermentation, cheese manufacture, bakery, meat tenderizing), environmental bioremediations (digesters, composting, oil or xenobiotic biology applications), and fine chemicals synthesis (e.g., organic syn-thesis of chiral intermediates) (Gomes and Steiner 2004; Joseph et al. 2008).

   Lipases A and B from the psychrophile *Candida antarctica* (CalA and CalB, respectively) are amongst the most widely used catalysts in the literature. Although isolated in the cold environment of Antarctica, they are remarkably stable and even have optimal activity at 40–45 °C. Indeed, most of the uses reported for these enzymes are at non-cold temperatures.

CalB, when used in organic solvents, has proved to be an excellent catalyst for the preparation of chiral drugs and fine chemicals precursors, due to its high stereoselectivity and yields in kinetic resolutions or desymmetrizations. One of the many uses of this catalyst is in the resolution of myo-inositol derivatives. An experimental design to optimize the enzymatic transformation of myo-inositol derivatives in hexane made with Novozyme 435 with vinyl acetate led to 15-fold increased productivity over the original protocol, with a high conversion rate (50% ± 1) and high selectivity (enantiomeric excess = 99%, enantioselectivity > 100) after about 20 h, with the catalyst being reused for seven cycles (Manoel et al. 2012).

In another example of psychrophilic lipases being used in kinetic resolution, Xu et al. (2010) cloned and expressed a *Proteus* sp. lipase (LipK107) in *E. coli*. They performed an *in silico* analysis of the 3D structure of the protein based on two homologous models of LipK107 made from X-ray structures of *Burkholderia glumae* lipase (BGL) and *Pseudomonas aeruginosa* lipase (PAL). By comparing the data from the *in silico* analysis with the experimental characterization of the recombinant protein, they determined the existence of a lid domain in LipK107, which was further proved by the interfacial activation assay of the recombinant enzyme. The structural model also predicted the enantioselectivity of LipK107 when the enzyme was used to catalyze the resolution of racemic 1-phenylethanol. The lid-open model of LipK107 identified the *R*-enantiomer as the preferred enantiomer, while the lid-closed model showed that the *S*-enantiomer was the most abundant.

The most commercially important field of application for lipases is their addition to detergents. Enzymes can reduce the environmental load of detergent products, since they save energy by enabling a lower wash temperature to be used, are biodegradable, leaving no harmful residues, and have no negative impact on sewage treatment processes (Joseph et al. 2008). Cold active lipases are expected to represent a larger share of industrially applicable enzymes in the coming years. These enzymes offer great industrial and biotechnological potential due to their capability to catalyze reactions at low temperatures. This would reduce the energy consumption and wear and tear of textile fibers, while minimizing the addition of toxic compounds used for the same purposes (Gomes and Steiner 2004).

13.6 Alkaliphilic Lipases

Madigan et al. (2000) define alkaliphiles as microorganisms that can successfully grow at pHs from 10 to 11 (Table 13.1). Nevertheless, the term "alkaline lipase" is widely used in the literature to refer to lipases whose optimal activity is above of pH 8. The literature search performed for this chapter therefore included pH 8- and pH 9-optimal enzymes in the group called alkaliphilic lipases, which were the subject of most of the publications on this group of lipases retrieved in the search. Most lipases prefer to catalyze reactions in alkaline pHs (Ohara et al. 2014).

Alkaline enzymes are widely used, especially in the cleaning industry, and are included in approximately 50% of detergent formulations produced in developed countries. Their industrial use started in the early 1900s, with a patent filed by Otto Rohm. Many products, such as soap powder and products for degreasing surfaces and cleaning glass, include lipases in their formulations. Some examples of the commercial detergents that contain lipases are: Ariel, Sunlight, Tide, Dixan, Nadhif, Surf, Wheel, Nirma, and Henko (Niyonzima and More 2015). All lipases used in detergent formulation have an optimal pH in the alkaline region, between pH 8.0 and 12.0. Since the main aim of adding lipases to these products is oil and fat stain removal, the main targets of the lipases tend to be carboxylic acids, which are more soluble in alkaline pHs. Moreover, the final products are generally formulations containing surfactants and salts (Niyonzima and More 2015).

In 2012, the cleaning industry accounted for around 27% of global lipase demand (Freedonia Group 2015). Novozymes offers some of the most important enzymes used in detergent formulations. Lipolase®, Lipoclean®, and Lipex® are genetically engineered variants of lipase from the fungus *Thermomyces lanuginosus* (Jurado-Alameda et al. 2012). These catalysts are regarded as significant enzymatic products.

A recent study called attention to the production, in solid-state fermentation, of a lipase from *Thermomyces lanuginosus* (TLL), which has an optimal pH of 10. What makes it remarkable is the fact that besides being active in alkaline pHs, the lipase also has other extremophilic characteristic, such as thermostability and high optimal temperature (85 °C) (Ávila-Cisneros et al. 2014).

Although the laundry industry accounts for a significant share of the industrial applications of lipases, there are many other applications for those enzymes that are already a reality in industry, such as the preparation of enantiopure compounds from racemates, the synthesis of esters, polyunsaturated fatty acid (PUFA) enrichment, and biodiesel synthesis (Liu et al. 2014).

Due to the great diversity of alkaline lipases, it is also possible to find research that investigates lipases that are both alkaline and psychrotrophic. Ji et al. (2015) produced and characterized an extracellular cold-adapted alkaline lipase from the psychrotrophic bacterium *Yersinia enterocolitica*. They found it had excellent activity at pH 9.0 and between 0 and 60 °C.

In view of the high number of known alkaline lipases, they are often found in combination with other extremophile characteristics, making them suitable for a variety of different bioprocesses.

13.7 Acidophilic Lipases

Acidophiles are organisms that thrive under highly acidic conditions, usually pH 2.0 or below (Madigan et al. 2000). They are found in all the domains of life—Archaea, Bacteria, and Eukaryotes—but their capacity to flourish in such environments is due to their capacity to pump protons out of the internal cellular space, thus keeping the internal pH close to neutral. Therefore, "acid lipases"—meaning enzymes that catalyze optimally at low pHs—are not easy to find even in the large group of hydrolases/lipases (EC 3.1.1.3). Consequently, it is more common to find research investigating neutral enzymes with high stability at low pHs.

The main problem affecting the viability of acid enzymes has to do with their structure. The most common catalytic triad that is seen in the (E.C. 3.1.1.3) enzyme family is the Ser-Asp-His triad, where the histidine has the role of proton acceptor from the catalytic serine, making the enzymes naturally alkaline (Ohara et al. 2014). Several studies have been conducted to better understand the structure of such enzymes. The aim is to produce information to enable the application of techniques such as site-directed mutation and then to achieve more successful catalysis, as in Ohara et al. (2014).

Ohara et al. made four changes to obtain SshEstI (an alkaline carboxylesterase from *Sulfolobus shibatae* DSM5389) with optimal activity at lower pHs. The strategy was based on literature that discovered one enzyme in the sedolisin family that has a Ser-Glu-Asp triad instead of the common catalytic triad (Ser-His-Asp). Since sedolisins are known to be acidic enzymes, and the His residue is related to the protonation of the catalytic triad, it was supposed that site-directed mutagenesis at the His might reduce the optimal pH of serine hydrolases (Ohara et al. 2014). Although the group has managed to decreased the optimal pH of enzyme activity from 8 to 6 (with maintenance of 80% activity at a pH of around 5), it has not been possible to effectively classify the enzyme as an acid enzyme. This work stands out for the novelty of using genetic engineering in the construction of acidic enzymes.

Other researchers have endeavored to transform naturally alkaline enzymes into acidic enzymes, but we are not aware of any that have yet achieved this goal. Other strategies used to decrease optimal pH or increase stability at low pH are directed evolution and immobilization on supports with acid micro-environments.

According to the findings of our literature search, the most intensively studied acidophilic lipase is the castor bean acid lipase, which was first discovered more than a decade ago. The enzyme has interesting features, such as the potential to purify an enzyme from mature castor bean seeds (prior to germination). Its optimal pH is 4.2, and when expressed in *E. coli* it conserved its acidophilic characteristics and hydrolyzed triolein at an optimal pH of 4.5 (Eastmond 2004).

Another successful case of the obtainment of an acid lipase from solid-state fermentation was the production of a lipase with an optimal pH of 2.5 that maintained 75% activity under extreme acidity (pH 1.5) (Mahadik et al. 2002).

It is clear, then, that although acidic lipases have many possible applications, they have not yet been fully explored. Their main uses are in medical studies for the treatment of lysosomal acid lipase deficiencies, like Wolman disease (first described by Patrick and Lake 1969) and cholesterol ester storage disease.

The former pathology, named after the first physician to describe it, is modernly called lysosomal acid lipase deficiency (LAL-D). Individuals with this disease cannot produce enough or any active lysosomal acid lipase, which is responsible for breaking down fatty material like cholesteryl esters and triglycerides (Patrick and Lake 1969) in the human body.

The enzyme replacement for these diseases was approved by the FDA in 2015 under the trade name Kanuma™. This drug, administrated intravenously, is composed of a hydrolytic lysosomal cholesteryl ester and triacylglycerol-specific enzyme and is produced from the egg white of transgenic hens (*Gallus gallus*). The production platform was chosen in order to achieve a more suitable glycosylation pattern for the enzyme (Sheridan 2016).

13.8   Halophilic Lipases

Halophilic organisms are defined by their ability to thrive in high concentrations of salt. They can be found in each of the three domains of life—Archaea, Bacteria, and Eukarya—and can be separated into three categories: slightly halophilic (0.2–0.5 M salt), moderately halophilic (0.5–2.5 M salt), and extremely halophilic (above 2.5 M salt) (Schreck and Grunden 2014). To withstand high salt conditions, some organisms use active transporters to pump salt out of the cell, while importing compatible solutes to the intracellular space, such as glycine, betaine, sugars, polyols, amino acids, and ectoines, which help them to maintain a more isotonic environment (Gomes and Steiner 2004; Schreck and Grunden 2014). Other organisms, in particular of the Archaea domain, do not have these mechanisms, but their proteins have some characteristic traits that explain their stability in such environments, such as a higher concentration of acidic amino acid residues and fewer aliphatic and hydrophobic short-chain amino acids.

Halophilic lipases started to be explored more widely in the last decade. Since the early 2000s, several metagenomic works have been published using isolates from highly saline environments, such as the Great Salt Lake in North America, and salt lakes in South American deserts and Asia (Babavalian et al. 2013).

Halophilic lipases are of industrial interest due to their high stability. Their stability is observed not only in high salt concentrations, but in various other respects, such as stability in high and low temperatures and towards organic solvents (Schreck and Grunden 2014). Such is the case of a crude preparation obtained by Boutaiba et al. (2006). In a metagenomic approach, they isolated 54 strains of an archaea from the Algerian desert. One of the strains, characterized as *Natronococcus* sp., was cultivated in Gibbons medium, generating a crude preparation that presented activity towards *p*-nitrophenyl palmitate (pNPP), indicating the presence of a lipase. This preparation showed optimal activity towards pNPP in the presence of 4 M NaCl, and showed no activity at all in the absence of salt. The preparation had maximum activity at 50 °C, and was highly thermostable, with more than 90% of original activity being retained when incubated for 60 min at 50 °C. The preparation was also stable at 4 °C, with no loss of function after 6 months of storage at 4 °C.

Kumar et al. (2012) isolated 108 halophilic strains with hydrolytic activity from various saline habitats in India, such as coastal regions of Gujarat, Goa, Kerala and Sambhar Salt Lake, Rajashthan. Twenty-three of them presented lipase activity;

nine showed protease activity; and one had protease, amylase, and lipase activity. Seven of the lipases were further characterized. Six of the isolates presented optimal temperatures above 50 °C, and five of them showed an optimal temperature of 65 °C. Four of the isolates also showed thermostability, being stable for 1 h when incubated at 50 °C or higher. Another common feature was stability in organic solvents, such as hexane, decane, and toluene. Of the seven lipases characterized, five were stable in 25% organic reaction conditions. These combined features showed that halophilic enzymes are extremely resilient and capable of withstanding several types of detrimental habitats.

Pérez et al. (2012) used a strain of *Marinobacter lipolyticus,* a moderate halophile isolated from Cádiz, Spain, to isolate and express lipase LipBL in *Escherichia coli.* This lipase, although not very stable in saline media, showed some interesting features when characterized. It was found to have an optimal temperature of 80 °C, the ability to hydrolyze olive oil and fish oil, and high stability in various organic solvents, such as DMSO (30%), N,N-dimethylformamide (30%), ethanol (30%), 2-propanol (30%), diethylether (30%), toluene (5%), and hexane (5%), when incubated for 30 min at room temperature. Due to its high industrial potential, the lipase was purified and immobilized on different inorganic supports, including octyl agarose (hydrophobic support), dextran-sulfate (anionic support), and CNBr-activated support, in which the enzyme was covalently linked. Immobilization was designed to enhance the enzyme's stability and facilitate its reuse. In this immobilized form, LipBL was also active towards different chiral and prochiral esters, such as butyroyl mandelic acid, methyl mandelate, dimethyl phenyl glutarate, and 4-phenyl-2-hydroxy ethyl butyrate (Pérez et al. 2011).

Although not widely explored, halophilic lipases are potentially good candidates for a wide range of industrial processes, constituting versatile, resilient catalysts.

13.9 Structural Characteristics

13.9.1 Structural Features that Contribute to Stability

The structural mechanisms responsible for the high stability of extremozymes is a matter of great biotechnological interest, as it allows the design of more stable proteins for industrial uses. The structures of enzymes from extremophilic organisms do not differ significantly from those of enzymes from other organisms (Vieille and Zeikus 2001). Several studies have been performed, mainly by comparing the structures of stable proteins with their non-stable counterparts, and by increasing the stability of some proteins by performing mutations.

Proteins from halophilic organisms, in particular those classified in the Archaea domain, have some characteristic traits that explain their stability in high salt conditions. In general, they have a higher concentration of acidic amino acid residues, which increases the negative charges on the protein. They also have fewer aliphatic and hydrophobic short-chain amino acids, such as Val, Gly, and Ala, thus diminishing their hydrophobicity. The reduced hydrophobicity decreases the tendency to aggregate at high salt concentrations and also minimizes detrimental electrostatic interactions between proteins (Madern et al. 2000; Gomes and Steiner 2004).

Some of these features have been demonstrated by Müller-Santos et al. (2009). In order to investigate the structure of halophilic lipases and esterases, they expressed gene *lipC* from the halophilic archaea *Haloarcula marismortui* in *E. coli*, naming the enzyme Hm EST. The enzymatic activity of Hm EST depended greatly on the salt concentration in the reaction conditions, with its optimal activity occurring at 2–3 M KCl. After purification, the protein was submitted to circular dichroism, showing that in environments with a lower salt concentration, it lost its structure, being unfolded. A 3D model was also constructed based on the surface properties of this enzyme. The model pointed to a structure enriched in acidic amino acids and depleted in basic residues. This result further indicates a "salting" adaptation, which granting the protein improved stability in high-salt environments.

Psychrophilic lipases are able to effectively function at cold temperatures and have higher rates of catalysis than lipases from mesophilic or thermophilic organisms, which show little activity at low temperatures. Psychrophilic lipases show decreased ionic interactions and hydrogen bonds, and longer surface loops, causing the increased flexibility of the polypeptide chain and enabling the easier accommodation of the substrates at low temperature (Gomes and Steiner 2004; Joseph et al. 2008). They also have fewer hydrophobic groups and more charged groups on their surface. Their primary structures include fewer arginine than lysine residues and a few proline residues, thereby decreasing the number of salt bridges, especially the Arg-mediated ones. The number of aromatic–aromatic interactions is also somewhat lower in this type of enzyme, which reduces the number of interactions and enables greater flexibility of the structure. These features are the explanations currently available for the adaptation to cold shown by these lipases (Siddiqui and Cavicchioli 2006; Joseph et al. 2008).

A strain of *Pseudomonas fragi*, the main spoiling agent of refrigerated meat and raw milk, has been isolated and one of its lipases has been cloned and expressed in *E. coli* (Alquati et al. 2002). The structures obtained by molecular modeling of this lipase were compared to lipases from *Pseudomonas aeruginosa* and *Burkholderia cepacia* in order to elucidate differences in their behavior. A large quantity of charged residues exposed at the protein surface was detected in the cold-active lipase, as well as fewer disulphide bridges and proline residues in loop structures. Arginine residues were distributed differently than in mesophilic enzymes, with only a few residues involved in stabilizing intramolecular salt bridges and a large proportion of them exposed at the protein surface that may contribute to the increased conformational flexibility of the cold-active lipase.

Most of the knowledge gathered so far was acquired in the study of thermophilic proteins. Nevertheless, many of the structural features responsible for high temperature stability can be extrapolated to explain stability in other extreme conditions.

In general, thermostable proteins have a relative lack of flexibility. Several strategies are used to increase protein rigidity, each modestly contributing to the stabilization energy. Stabilizing factors include: increased surface charge networks, hydrogen bonds, disulfide bonds, and secondary structure content, stabilization of helix dipoles, more proline residues, fewer glycine residues, fewer labile residues (e.g., Asn, Gln, Met, Cys), reduced surface area and volume, and stabilization by ligands such as metal. Most thermostable proteins exhibit some but not all of these properties (Tyndall et al. 2002).

In the following topic, the structure of lipase P1 from *Bacillus stearothermophilus* is shown as an example of an extremophilic bacterial lipase. The structure of the lipase from *Archaeoglobus fulgidus*, the only archael lipase determined so far, is also discussed.

13.9.2   *Lipase P1 from* Bacillus stearothermophilus

Lipase P1 from *Bacillus stearothermophilus* was the first structure determined for a thermophilic lipase. The structure consists of seven parallel β strands surrounded by α-helices 1 (αA) and 13 (αF) on one side, and α-helices 2 (αB), 4 (αC) 5, 9, 10 (αD) 11 and 12 (αE) on the other side (Tyndall et al. 2002).

As seen in the structures of various bacterial lipases, lipase P1 from *Bacillus stearothermophilus* lacks strands β1 and β2 of the canonical α/β-hydrolase fold. The catalytic triad consists of Ser113, Asp317, and His358. The catalytic serine in the nucleophilic elbow is located between the β5-strand and deep α4 helix inside the core of the structure. In the *Bacillus* lipase, the serine residue is incorporated into the consensus sequence Ala-Xaa-Ser-Xaa-Gly (where Xaa represents His and Gln, respectively). The α6-helix and the adjacent loop region constitute the flexible lid, which, in its closed conformation, isolates the substrate-binding cleft from solvent. The active site is also bounded by α11 and α12 (αE) helix, which are separated only by Cys295 residue. Lipase P1 was also the first structure determined for a lipase containing a Zn^{2+} binding site, besides the Ca^{2+} binding site normally found in bacterial lipases. The Zn^{2+} coordination provides a stabilizer mechanism to keep the structural elements together at the catalytic domain. The site consists of two histidine residues and two aspartic acid residues. His81 and His87 are located in a large insertion (approximately 25 residues) after the canonical helix αB, composed on the helix α3 and an antiparallel β-sheet (strands β1 and β2). Asp 61 and Asp 238 are located in helix αB and in the loop after helix α8, respectively. Another important deviation from the canonical α/β-hydrolase fold is the inclusion of a helix (α5) between canonical helix αC (α4) and strand β6. Helix α5 is near the zinc-binding region and makes hydrogen bonds with it. After helix α6 there is a significant deviation from the canonical fold: the insertion of about 50 residues composed of a large loop, helixes α7 and α8 (separated by two residues, Arg and Ser), and helix α9. This insertion allows the lid domain great mobility and is also potentially associated with the specificity of the enzyme (Tyndall et al. 2002).

The calcium binding site is formed by Gly286, located in the loop region after strand β7, and by Gly 360, Asp 365, and Pro366 in the loop situated after strand β9 (Tyndall et al. 2002).

The lipase P1 from *Bacillus stearothermophilus* is an example of an extremophilic lipase stabilized by ligands (Zn^{2+}). Furthermore, this lipase shows other stabilizing mechanisms, such as a higher number of salt bridges, and helix-forming, hydrophobic, and proline residues, than its mesophilic homolog (Tyndall et al. 2002).

13.9.3   Lipase from Archeoglobus fulgidus

Archaeoglobus fulgidus lipase shows optimal activity at 90 °C and pH 10–11, putting it in the most alkaline pH range detected for hydrolases. The overall structure presents a bipartite architecture consisting of an N-terminal α/β hydrolase domain and a C-terminal domain with β-barrel structure, which distinguishes it from all other lipases with determined structures (Chen et al. 2009).

The N-terminal domain (residues 1–237) is formed by a central β-sheet containing six parallel β-strands, surrounded by seven α-helixes: four on one side and three on the other. The lid is also found in the N-terminal domain, which comprises three α-helixes—α3, α4, and α5—and two hinge residues (Lys 61 and Asp 101) (Chen et al. 2009).

The C-terminal domain (residues 238–474) is of the β-sandwich type, consisting of two layers of seven β-strands. The antiparallel β-sheet front consists of strands G to M and the back β-sheet consists of strands N to T. A substrate-covering motif on top of the β-sandwich consists of α12, α13, α14, and α15, and forms part of the hydrophobic substrate binding tunnel. This motif was not recognized as a second lid, since it does not cover the active site of the enzyme and hinges have not been identified in this region (Chen et al. 2009).

The active site is at the bottom of a hydrophobic cleft covered by the lid. The catalytic triad comprises Ser136, Asp163 and His210, located in their canonical positions of the α/β-hydrolase fold. The oxyanion hole is formed by the backbone nitrogen atoms of Leu31 (βA-strand and α1 helix) and Met137 (after the nucleophile). Ser136 in the active site is at the entrance of a deep hydrophobic tunnel, 20 Å long and 7 Å wide, formed between the N-terminal catalytic domain and C-terminal domain. There is space for about 18 hydrocarbon units to be accommodated in this tunnel (Chen et al. 2009).

The hydrophobic substrate binding tunnel is covered by the lid domain and by the substrate-covering motif, which is directly above the hydrophobic tunnel. Access to the substrate-binding pocket is only possible through the lid opening in the upper part of the N-terminal domain (Chen et al. 2009).

The two hinge residues (Asp61 and Lys101) are located at the two ends of the lid. The opening of the lid is induced by the hinge and increases the hydrophobic surface exposed to the solvent. Lys101 and Asp61 interact with Ser64 via hydrogen bond under acidic conditions (Chen et al. 2009). However, in more basic conditions (pH above 8.5) Lys101 undergoes a 90° rotation to form a hydrogen bond with Glu109. This conformational change causes the lid to have a greater range of motions and makes it more easily opened for substrate binding to the active site. This is a possible explanation for the alkalophilic nature of the lipase from *Archaeoglobus fulgidus* (Chen et al. 2009).

The N- and C-terminal domains interact through a highly hydrophobic region and complementary charges. Four ion pairs are formed (Glu39-Arg317, Arg44-Asp472, Glu59-Arg328, and Lys184-Asp370) in this interaction and play an important role in the domain-domain interactions, forming a tight bipartite structure that results in a highly thermostable structure (Chen et al. 2009).

The C-terminal domain is essential for the substrate specificity and catalytic efficiency of the lipase from *Archaeoglobus fulgidus*. When it is deleted, the enzyme loses its function to hydrolyze long-chain esters (C16) and presents very low activity for medium-chain esters (C6). The C-terminal domain also contributes to the thermophilicity, alkalophilicity, thermostability, and pH stability of the enzyme (Chen et al. 2009).

13.10 Conclusions and Perspectives

There are many advantages and disadvantages of using enzymes in traditional chemical processes. The cost associated with the use of enzymes is still the principal obstacle to the widespread application of biocatalysis, as enzymes are more unstable than traditional catalysts. Consequently, extremophilic enzymes constitute an alternative to circumvent this bottleneck and increase the range of processes capable of being catalyzed by enzymes. Many extremophilic lipases— enzymes with extreme characteristics—are already on the market (e.g. CalB, CalA, TLL) and the number of publications on the subject is growing exponentially. Given the diversity of extremophilic organisms and their habitats, and the spread of molecular biology techniques and reduction of their costs, this technology has great as yet untapped potential. New structural research is needed to understand the molecular reasons behind the great stability of these organisms in such harsh conditions and develop ways of manipulating them by enzyme engineering.

Take Home Message

- Lipases are triacylglycerol ester hydrolases that act on different substrates and mediates the catalysis of wide range of reactions such as hydrolysis, esterification, alcoholysis, acidolysis, interesterification, aminolysis, and other reactions. Lipases has several applications such as in food, beverage, and cleaning products industries and applications that involve organic synthesis.
- The extremophilic lipases have five major groups namely halophilic, psychrophilic, alkaliphilic, acidophilic, and thermophilic. Thermophilic lipases have

been produced from bacterial sources such as *Geobacillus* sp. Iso5, *Pyrococcus furiosus*, *Bacillus*, *Geobacillus*, *Escherichia coli* and *fungal sources such as Thermomyces lanuginosus and Aspergillus oryzae*. Psychrophilic lipases can be produced from microbial sources such as *Candida antarctica, E.coli, Burkholderia glumae* lipase (BGL) and *Pseudomonas aeruginosa*. Alkaliphilic lipases *can be produced from Thermomyces lanuginosus and Yersinia enterocolitica*. Acidophilic lipases have been produced from *Sulfolobus shibatae* DSM5389 and *E. coli*. Halophilic lipases have been produced from *Marinobacter lipolyticus, Escherichia coli* and *Natronococcus* sp.

- Thermophilic lipases have applications in the paper industry, pharmaceutical field, such as the production of the anticonvulsant and antiepileptic drug pregabalin, and for producing biodiesel by transesterification reactions. Psychrophilic lipases have applications in a variety of industries, including detergent production (cold washing), the food industry (e.g., fermentation, cheese manufacture, bakery, meat tenderizing), environmental bioremediations (digesters, composting, oil or xenobiotic biology applications), and fine chemicals synthesis. Alkaliphilic lipases have applications in detergent industry, the synthesis of esters, polyunsaturated fatty acid (PUFA) enrichment, and biodiesel synthesis.

References

Almeida RV, Alquéres SMC, Larentis AL, Rössle SC, Cardoso AM, Almeida WI, Bisch PM, Alves TLM, Martins OB (2006) Cloning, expression, partial characterization and structural modeling of a novel esterase from *Pyrococcus furiosus*. Enzym Microb Technol 39 (5):1128–1136

Alquati C, De Gioia L, Santarossa G, Alberghina L, Fantucci P, Lotti M (2002) The cold-active lipase of *Pseudomonas fragi*. Eur J Biochem 269(13):3321–3328

Alquéres SMC, Branco RV, Freire DMG, Alves TLM, Martins OB, Almeida RV (2011) Characterization of the recombinant thermostable lipase (Pf2001) *from Pyrococcus furiosus*: effects of thioredoxin fusion Tag and triton X-100. Enzym Res 2011:1–7

Anobom CD, Pinheiro AS, De-Andrade RA, Aguieiras EC, Andrade GC, Moura MV, Almeida RV, Freire DM (2014) From structure to catalysis: recent developments in the biotechnological applications of lipases. Biomed Res Int 2014:1–11

Ávila-Cisneros N, Velasco-Lozano S, Huerta-Ochoa S, Córdova-López J, Gimeno M, Favela-Torres E (2014) Production of thermostable lipase by *Thermomyces lanuginosus* on solid-state fermentation: selective hydrolysis of sardine oil. Appl Biochem Biotechnol 174(5):1859–1872

Babavalian H, Amoozegar MA, Pourbabaee AA, Moghaddam MM, Shakeri F (2013) Isolation and identification of moderately halophilic bacteria producing hydrolytic enzymes from the largest hypersaline playa in Iran. Microbiology 82(4):466–474

Boutaiba S, Bhatnagar T, Hacene H, Mitchell DA, Baratti JC (2006) Preliminary characterization of a lipolytic activity from an extremely halophilic archaeon, *Natronococcus* sp. J Mol Catal B Enzym 41:21–26

Branco RV, Gutarra ML, Guisan JM, Freire DM, Almeida RV, Palomo JM (2015) Improving the thermostability and optimal temperature of a lipase from the hyperthermophilic *Archaeon Pyrococcus furiosus* by covalent immobilization. Biomed Res Int 2015:1–8

Chen CKM, Lee GC, Ko TP, Guo RT, Huang LM, Liu HJ, Ho YF, Shaw JF, Wang AHJ (2009) Structure of the alkalohyperthermophilic *Archaeoglobus fulgidus* lipase contains a unique C-terminal domain essential for long-chain substrate binding. J Mol Biol 390(4):672–685

Daiha KG, Angeli R, Oliveira SD, Almeida RV (2015) Are lipases still important biocatalysts? A study of scientific publications and patents for technological forecasting. PLoS One 10: e0131624

Eastmond PJ (2004) Cloning and characterization of the acid lipase from castor beans. J Biol Chem 279(44):45540–45545

Freedonia Group (2015) World enzymes report

Gomes J, Steiner W (2004) The biocatalytic potential of extremophiles and extremozymes. Food Technol Biotechnol 42(4):223–235

Gutiérrez A, José C, Martínez AT (2009) Microbial and enzymatic control of pitch in the pulp and paper industry. Appl Microbiol Biotechnol 82(6):1005–1018

Haki GD, Rakshit SK (2003) Developments in industrially important thermostable enzymes: a review. Bioresour Technol 89(1):17–34

Handelsman T, Shoham Y (1994) Production and characterization of an extracellular thermostable lipase from a thermophilic *Bacillus* sp. J Gen Appl Microbiol 40(5):435–443

Ji X, Chen G, Zhang Q, Lin L, Wei Y (2015) Purification and characterization of an extracellular cold-adapted alkaline lipase produced by psychrotrophic bacterium *Yersinia enterocolitica* strain KM1. J Basic Microbiol 65(9):3969–3975

Joseph B, Ramteke PW, Thomas G (2008) Cold active microbial lipases: some hot issues and recent developments. Biotechnol Adv 26(5):457–470

Jurado-Alameda E, Román MG, Vaz DA, Pérez JLL (2012) Fatty soil cleaning with ozone and lipases. Household Pers Care Today 7:49–52

Kallmeyer J, Pockalny R, Adhikari RR, Smith DC, D'Hondt S (2012) Global distribution of microbial abundance and biomass in subseafloor sediment. PNAS 109(40):16213–16216

Kumar S, Karan R, Kapoor S, Singh SP, Khare SK (2012) Screening and isolation of halophilic bacteria producing industrially important enzymes. Braz J Microbiol 43(4):1595–1603

Liu Y, Zhang R, Lian Z, Wang S, Wright AT (2014) Yeast cell surface display for lipase whole cell catalyst and its applications. J Mol Catal B Enzym 106:17–25

López-López O, Cerdán ME, Siso MIG (2014) New extremophilic lipases and esterases from metagenomics. Curr Protein Pept Sci 15(5):445

Madern D, Ebel C, Zaccai G (2000) Halophilic adaptation of enzymes. Extremophiles 4(2):91–98

Madigan MT, Martinko JM, Dunlap PV, Clark DP (2000) Bock biology of microorganisms, 9th edn. Prentice Hall, Upper Saddle River, NJ

Mahadevan GD, Neelagund SE (2014) Thermostable lipase from *Geobacillus* sp. Iso5: bioseparation, characterization and native structural studies. J Basic Microbiol 54(5):386–396

Mahadik ND, Puntambekar US, Bastawde KB, Khire JM, Gokhale DV (2002) Production of acidic lipase by *Aspergillus niger* in solid state fermentation. Process Biochem 38(5):715–721

Manoel EA, Pais KC, Cunha AG, Simas AB, Coelho MAZ, Freire DM (2012) Kinetic resolution of 1, 3, 6-tri-O-benzyl-myo-inositol by Novozym 435: optimization and enzyme reuse. Org Process Res Dev 16(8):1378–1384

Müller-Santos M, de Souza EM, Pedrosa FDO, Mitchell DA, Longhi S, Carrière F, Canaan S, Krieger N (2009) First evidence for the salt-dependent folding and activity of an esterase from the halophilic archaea Haloarcula marismortui. Biochim Biophys Acta Mol Cell Biol Lipids 1791(8):719–729

Nardini M, Dijkstra BW (1999) α/β hydrolase fold enzymes: the family keeps growing. Curr Opin Struct Biol 9(6):732–737

Niyonzima FN, More SS (2015) Microbial detergent compatible lipases. J Sci Ind Res 74:105–113

Ohara K, Unno H, Oshima Y, Hosoya M, Fujino N, Hirooka K, Takahashi S, Yamashita S, Kusunoki M, Nakayama T (2014) Structural insights into the low pH adaptation of a unique carboxylesterase from Ferroplasma altering the pH optima of two carboxylesterases. J Biol Chem 289(35):24499–24510

Patrick AD, Lake BD (1969) Deficiency of an acid lipase in wolman's disease. Nature 222:1067–1068. doi:10.1038/2221067a0

Pérez D, Martín S, Fernández-Lorente G, Fillice M, Guisán JC, Ventosa A, García MT, Mellado E (2011) A novel halophilic lipase, LipBL, showing high efficiency in the production of eicosapentaenoic acid (EPA). PLoS One 6(8):e23325

Pérez D, Kovačić F, Wilhelm S, Jaeger KE, García MT, Ventosa A, Mellado E (2012) Identification of amino acids involved in the hydrolytic activity of lipase LipBL from *Marinobacter lipolyticus*. Microbiology 158(8):2192–2203

Schreck SD, Grunden AM (2014) Biotechnological applications of halophilic lipases and thioesterases. Appl Microbiol Biotechnol 98(3):1011–1021

Sheridan C (2016) FDA approves farmaceutical drug from transgenic chickens. Nat Biotechnol 34 (2):117–119

Siddiqui KS, Cavicchioli R (2006) Cold-adapted enzymes. Annu Rev Biochem 75:403–433

Tyndall JD, Sinchaikul S, Fothergill-Gilmore LA, Taylor P, Walkinshaw MD (2002) Crystal structure of a thermostable lipase from *Bacillus stearothermophilus* P1. J Mol Biol 323 (5):859–869

Uppenberg J, Hansen MT, Patkar S, Jones TA (1994) The sequence, crystal structure determination and refinement of two crystal forms of lipase B from *Candida antarctica*. Structure 2 (4):293–308

Vieille C, Zeikus GJ (2001) Hyperthermophilic enzymes: sources, uses and molecular mechanisms for thermostability. Microbiol Mol Biol Rev 65:1–43

Xu T, Gao B, Zhang L, Lin J, Wang X, Wei D (2010) Template-based modeling of a psychrophilic lipase: conformational changes, novel structural features and its application in predicting the enantioselectivity of lipase catalyzed transesterification of secondary alcohols. Biochim Biophys Acta 1804:2183–2190

Chapter 14
Bioprospection of Extremozymes for Conversion of Lignocellulosic Feedstocks to Bioethanol and Other Biochemicals

Felipe Sarmiento, Giannina Espina, Freddy Boehmwald, Rocío Peralta, and Jenny M. Blamey

What Will You Learn from This Chapter?

Microbial enzymes are playing a preponderant role for diverse industrial applications, including second generation biofuels production. Because of the intrinsic properties of microbial enzymes such as consistency and versatility, they represent interesting environmentally-friendly additions or alternatives to current chemical biofuel production processes. In this context, extremozymes, or enzymes derived from extremophiles, play an even bigger role, because they are stable and able to catalyze reactions optimally under the harsh conditions of industrial processing. In the case of bioethanol production, extremophilic enzymes could be applied under the acidic or basic conditions of pretreatment, and also under the high temperatures of complete cellulose/hemicellulose hydrolysis.

So far, most industrial enzymes used in bioethanol production correspond to recombinant versions of mesophilic bacteria or fungi enzymes, thus, there is a real need for novel extremozymes to improve current chemical processes, or to develop new cost-efficient and sustainable production processes. To unlock the microbial diversity and discover novel extremophilic biocatalysts, classic enzymatic bioprospection over culturable microorganisms and novel molecular techniques such as metagenomics and genomics are being applied. In addition, these

F. Sarmiento
Swissaustral USA, 111 Riverbend Rd., Office #271, Athens, GA 30602, USA

G. Espina • F. Boehmwald • R. Peralta
Fundación Científica y Cultural Biociencia, Jose Domingo Cañas, 2280 Ñuñoa, Santiago, Chile

J.M. Blamey (✉)
Swissaustral USA, 111 Riverbend Rd., Office #271, Athens, GA 30602, USA

Fundación Científica y Cultural Biociencia, Jose Domingo Cañas, 2280 Ñuñoa, Santiago, Chile

Faculty of Chemistry and Biology, University of Santiago, Santiago, Chile
e-mail: jblamey@swissaustral.com

© Springer International Publishing AG 2017
R.K. Sani, R.N. Krishnaraj (eds.), *Extremophilic Enzymatic Processing of Lignocellulosic Feedstocks to Bioenergy*, DOI 10.1007/978-3-319-54684-1_14

271

biocatalysts are being further improved by using different enzyme engineering techniques. A review of these different approaches with a focus on extremozymes is presented in this chapter.

14.1 Introduction

It is widely known that there is a limited reservoir of fossil fuels in our planet. After reaching its extraction peak, its production will enter terminal decline, threatening the international energy security system and geopolitical stability. In addition, the excessive use of fossil fuels and their combustion have caused substantial environmental damage, pollution, acid rain and global warming.

These issues have started to seriously affect our society, and the spiraling environmental impact as well as the uncertain long term financial cost of energy has raised serious concerns in the public and governments across the world. Consequently, in response to the energy crisis alert and immediate environmental deterioration associated with conventional non-renewable fuel usage, there is a strong drive to find viable alternatives to the use of fossil fuels. Novel renewable sources of energy are required to decrease climate change, reduce environmental pollution and to sustainably satisfy the increasing demand for energy. In this context, lignocellulosic (plant) biomass is gaining increased industrial interest and, in the form of biofuels, is considered to be one of the most auspicious energy alternatives available in the short-term.

Lignocellulose is the most abundant renewable natural resource on earth and is mainly composed of three types of polymers: cellulose, hemicellulose and lignin, which are connected in a complex matrix. Cellulose represents 35–50% of the plant biomass by weight and is a linear polysaccharide composed of β-1,4 linked glucose units aggregated into microfibrils. These are composed of approximately 30–36 glucan chains aggregated laterally by means of non-covalent interactions to finally produce a stable crystalline lattice structure, insoluble in water and most organic solvents (Arantes and Saddler 2010). Hemicelluloses, such as xylan, xyloglucan, mannan and glucomannan, represents 20–35% of plant biomass and are highly-branched heteropolysaccharides composed of pentoses, hexoses and/or uronic acids. Their exact composition and structure varies strongly among different plant species, tissues and cell types (Scheller and Ulvskov 2010). On the other hand, lignin is a complex organic aromatic heteropolymer, which represents 10–25% of plant biomass. It consists of three methoxylated monolignols incorporated into lignin in the form of guaiacyl, syringyl and p-hydroxyphenyl present in diverse amounts, depending on the source of lignin (Chen and Dixon 2007). In addition to the main polymers of the lignocellulosic matrix, other minor components are present such as pectin, proteins, lipids, soluble sugars and minerals.

Plant cell walls are organized in a highly-ordered and tightly-packed intricate structure, where the crystalline cellulose is coated with hemicelluloses via hydrogen bonds (maintaining cell wall flexibility by preventing microfibrils adhering to each other), and both polysaccharides are embedded in lignin (adding strength and rigidity to the cell walls), which is cross-linked to hemicellulose via ferulic acid ester linkages (Viikari et al. 2012).

Plant biomass degradation is a fundamental process for the life of a myriad of microorganisms. During the course of evolution, fungi and bacteria have developed different enzymatic mechanisms to depolymerize plant cell walls in order to harness energy from lignocellulosic material.

Since lignocellulose is a very complex and recalcitrant material there are a great variety of cooperatively acting enzymes involved in its degradation. Among these enzymes the most notable are: Lignin-modifying enzymes (LMEs) [*e.g.* Lignin Peroxidases (EC 1.11.1.14), Manganese peroxidase (EC 1.11.1.13), Laccases (EC 1.10.3.2), Versatile peroxidase (EC 1.11.1.16), Glucose oxidase (EC 1.1.3.4), Glyoxal oxidase (EC 1.1.3.-), Aryl alcohol oxidase (EC 1.1.3.7)], Cellulases [*e.g.* endo-1,4-β-D-glucanases (EC 3.2.1.4), exo-1,4-β-D-glucanases (EC 3.2.1.74), exo-1,4-β-D-glucan cellobiohydrolase (EC 3.2.1.91) and β-D-glucosidases (EC 3.2.1.21)] and Hemicellulases (*e.g.* Endo-1,4-β-D-xylanases (EC 3.2.1.8), β-D-xylosidases (EC 3.2.1.37), α-L-arabinofuranosidases (EC 3.2.1.55), α-D-glucuronidases (EC 3.2.1.139), Acetyl xylan esterases (EC 3.1.1.72), Feruloyl esterases (EC 3.1.1.73), p-Coumaroyl esterases (EC 3.1.1.), Mannan endo-1,4-β-mannosidase (EC 3.2.1.78), exo-β-D-mannanase (EC 3.2.1.25), exo-1,4-β-mannobiohydrolase (EC 3.2.1.100), Acetyl esterase (EC 3.1.1.6), Xyloglucan-specific endo-β-1,4-Glucanase (EC 3.2.1.151), Glucuronoarabinoxylan endo-1,4-β-Xylanase (EC 3.2.1.136).

In the case of lignocellulosic feedstocks with a high content of pectin (*e.g.* sugar beet pulp) pectinases [Endo-pectin lyases (EC 4.2.2.10), Exo-pectin lyases (EC 4.2.2.9), Endo-pectate lyases (EC 4.2.2.2), Rhamnogalacturonan lyase (EC 4.2.2.24), Endo-polygalacturonase (EC 3.2.1.15), Exo-polygalacturonase (EC 3.2.1.67), endo-1,4-β-galactanase (EC 3.2.1.89), β-galactosidase (EC 3.2.1.23), Arabinan Endo-1,5-α-L-arabinanase (EC 3.2.1.99), α-L-arabinofuranosidases (EC 3.2.1.55), pectin esterase (EC 3.1.1.11), acetyl esterase (EC 3.1.1.6), rhamnogalacturonan acetyl esterase (EC 3.1.1.86)] enzymes are also required.

In the present chapter selected and relevant technological aspects of the discovery and development of novel enzymes for application in the production of biofuels from lignocellulosic feedstocks will be discussed with a focus on enzymes derived from extremophiles or extremozymes.

14.2 Extremozymes for the Production of Bioethanol from Lignocellulosic Feedstocks

To date, bioethanol is the most widely used liquid biofuel and lignocellulosic bioethanol represents a much better and environmentally acceptable alternative, in terms of long term sustainability than first generation biofuels made from agricultural crops. The global production and use of bioethanol have increased dramatically in the past few years. The main drivers of this market the increasing investments in research and technology development in the field and the government support of various countries that have set ambitious goals to substitute a part of the fossil fuels used for transportation (which represent about 34% of the total world energy consumption) with liquid biofuels. This includes those countries of the European Union that aim to reach a quota of 10% biofuels in the transport sector by 2020 and the US Department of Energy Office with a scenario for supplying 30% of the 2004 petrol demand with biofuels by the year 2030.

Nonetheless, the production of bioethanol from lignocellulosic biomass is a complex task which requires a multi-step process that includes: (i) pretreatment for the breakdown of lignin and subsequent recovery of cellulose and hemicellulose, (ii) enzymatic hydrolysis of cellulose and hemicellulose into fermentable mono sugars, and (iii) fermentation of the resulting monosaccharides by ethanologenic microorganisms to produce high yields of bioethanol.

The first step, pretreatment of the plant biomass, is strictly required in order to reduce its size and open its structure to facilitate rapid and efficient hydrolysis of carbohydrates to fermentable sugars. Several pretreatment methods are used, including physical pretreatments (*e.g.* microwave irradiation, pyrolysis, extrusion and freezing), chemical pretreatments (*e.g.* using alkali, concentrated or diluted acids, organic solvents, ethylene diamine, hydrogen peroxide, ionic liquids or ozonolysis) and physico-chemical pretreatments (*e.g.* steam explosion, ammonia fiber explosion (AFEX), wet oxidation, liquid hot water and super-critical carbon dioxide explosion) (Viikari et al. 2012). Most of these methods employ harsh conditions (*e.g.* high temperature, high pressure, extreme pHs, protein denaturing solvents) while mild biological pretreatments utilize microorganisms (mainly white-, brown- and soft-rot fungi, actinomycetes, and bacteria) that produce several of the enzymes involved in the lignin degradation mentioned above. To date it is still not very common to pretreat biomass only through the action of lignin-degrading enzymes; however, a combination of traditional physico-chemical pretreatments followed by enzymes adapted to harsh conditions may become an interesting alternative.

A second step of enzymatic hydrolysis is necessary for the complete deconstruction of the plant cell wall polysaccharides to their constituent fermentable hexose and pentose sugars. To date, the efficient saccharification of lignocelluloses into fermentable sugars requires a whole set of biocatalysts, which brings substantial economic implications due to the cost of these commercial enzymes. The worldwide bioethanol enzymes market in 2013 was worth $360.5 million and is expected to grow up to $548.6 million by the end of 2018 with a CAGR of 8.8% (Dewan 2014).

The third step of this process, fermentation of the C6 and C5 sugars, is out of the scope of this chapter because it depends on the specific fermentative microorganism selected as well as its specific metabolic pathways and constituent enzymes.

For the first two steps, it has been suggested that the utilization of a mixture of enzymes derived from extremophilic microorganisms can potentially increase efficiency and reduce process costs during the bioconversion of plant cell wall to biofuels. Therefore, there has been growing interest in the use of extremozymes (*e.g.* thermophilic, acidophilic or alkaliphilic enzymes) as they are tolerant to otherwise harsh industrial conditions for biological systems. Extremozymes and their unique characteristics allow them to be superior in their hydrolytic activity to their mesophilic counterparts currently used in most bioethanol production processes, having the potential to enhance plant biomass degradation and its use and conversion into bioethanol. See Table 14.1 in order to find a list of the main enzymes involved during the pretreatment and enzymatic hydrolysis of lignocellulose that could benefit from having an extremophilic version.

14.3 Classic Microbial/Enzymatic Search of Extremozymes

Classic enzymatic screening over cultured microorganisms is still an efficient methodology to find novel biocatalysts that are active under the extreme conditions of temperature, pH and other physico-chemical parameters of many industrial processes (Table 14.2). Typically, this methodology involves the following steps: first, environmental samples are collected from particular sites with conditions that match the characteristics of the target enzyme to be developed. For example, if the goal is to develop an enzyme with optimal activity at high temperatures, the best place to collect samples are hot environments such as hot springs or hydrothermal vents, where there are higher chances to isolate thermophilic and hyperthermophilic microorganisms that may harbor thermoresistant enzymes. In addition to natural environments, man-made systems such as industrial waste streams or mining effluents are of high interest to isolate particular extremophiles and extremozymes. Second, microorganisms are isolated from the environmental samples by applying specific selective pressures such as pH, temperature, substrate composition, and presence/absence of metals. These selective pressures are chosen to match the target characteristics of the biocatalyst to be developed. Third, a functional screening where enzymatic assays are performed over the supernatant and/or the crude extract of each selected microorganism is conducted to find the enzymatic activity of interest (Fig. 14.1). Through this "functional approach", the presence and the functionality of the target enzyme are confirmed under a particular set of selected conditions that can match industrial requirements. Also, enzymatic properties such as specific activity and thermal stability of the target protein can be determined early on a project without the need of first going through a cloning and expression phase of the candidate gene. Therefore, early confirmation of these enzymatic parameters under the target industrial conditions is a powerful asset to discover optimal biocatalysts for industrial applications, and it represents a strong advantage over molecular and metagenomic approaches for discovering novel enzymes.

Table 14.1 Main extremozymes of interest for lignocellulosic bioethanol production

Processing steps	Catalytic requirements[a]	Enzyme type		EC number	Brief description	CAZy family (Lombard et al. 2014)
Pretreatment	Acidophilic Thermophilic Solvent-tolerant enzymes	Ligninases	Lacasses	1.10.3.2	Copper-containing oxidase enzymes, act on both o- and p-quinols, and often acting also on aminophenols and phenylenediamine. The semiquinone may react further either enzymically or non-enzymically	AA1, NC
			Lignin Peroxidases	1.11.1.14	Hemoprotein, involved in the oxidative degradation of lignin	AA2
			Manganese peroxidase	1.11.1.13	Hemoprotein, involved in the oxidative degradation of lignin	AA2
			Versatile peroxidase	1.11.1.16	Hemoprotein that combines the substrate-specificity characteristics of the two other ligninolytic peroxidases, EC 1.11.1.13, manganese peroxidase and EC 1.11.1.14, lignin peroxidase. It is also able to oxidize phenols, hydroquinones and both low- and high-redox-potential dyes, due to a hybrid molecular architecture that involves multiple binding sites for substrates	AA2

Enzymatic hydrolysis	Acidophilic Thermophilic	Cellulase	Endo-1,4-β-D-glucanases	3.2.1.4	Randomly cleave internal β-1,4-glucosidic bonds at amorphous sites in the cellulose polysaccharide chain, generating oligosaccharides of various lengths and consequently new chain ends. Endohydrolysis of (1 → 4)-β-D-glucosidic linkages in cellulose, lichenin and cereal β-D-glucans	GH5, GH6, GH7, GH8, GH9, GH10, GH12, GH26, GH44, GH45, GH48, GH51, GH74, GH124, NC
			Exo-1,4-β-D-glucanases	3.2.1.74	Attack the non-reducing end of cellulose to yield cellobiose as the primary product, and liberate D-glucose from β-glucan and cellodextrins	GH1, GH3, GH5, GH9
			Exo-1,4-β-D-glucan cellobiohydrolase	3.2.1.91	Release cellobiose either from the reducing or non-reducing end of cellulose and liberate D-cellobiose from β-glucan in a processive manner	GH5, GH6, GH9
			β-D-glucosidases	3.2.1.21	Release D-glucose units from soluble cello-oligosaccharides, cellodextrins and a variety of glycosides	GH1, GH2, GH3, GH5, GH9, GH30, GH116, NC
		Hemicellulases	Xylan Endo-1,4-β-D-xylanases	3.2.1.8	Randomly hydrolyse the β-1,4 bond in the xylan backbone, yielding short xylo-oligomers	GH5, GH8, GH9, GH10, GH11, GH12, GH16, GH26, GH30, GH43, GH44, GH51, GH62, GH98
			Exo-1,4-β-D-xylosidases	3.2.1.37	Exo-type glycosidases that catalyse the successive removal of xylosyl residues from the non-reducing termini of xylobiose and linear xylo-oligosaccharides	GH1, GH3, GH5, GH30, GH39, GH43, GH51, GH52, GH54, GH116, GH120

(continued)

Table 14.1 (continued)

Processing steps	Catalytic requirements[a]	Enzyme type	EC number	Brief description	CAZy family (Lombard et al. 2014)
		α-L-arabinofuranosidases	3.2.1.55	Exo-type glycosidases that catalyse the successive removal of arabinose residue from the non-reducing termini of α-1,2-, α-1,3- and α-4,6-linked arabinofuranosyl residues	GH2, GH3, GH10, GH43, GH51, GH54, GH62
		α-D-glucuronidases	3.2.1.139	Catalyze the hydrolysis of α-1,2 glycosidic bonds between 4-O-methyl-D-glucuronic acid and xylan	GH4, GH67
		Acetyl xylan esterases	3.1.1.72	Hydrolyze the ester linkages between xylose units of xylan and acetic acid	CE1, CE2, CE3, CE4, CE5, CE6, CE7, CE12, CE15
		Feruloyl esterases	3.1.1.73	Hydrolyze the ester linkages between arabinofuranosyl side-chain residues and ferulic acid	CE1, NC
		p-Coumaroyl esterases	3.1.1.-	Hydrolyze the ester linkages between arabinofuranosyl side-chain residues and p-coumaric acid	–

[a]Depending on the reaction condition

Table 14.2 Summary of the two different approaches for the search and discovery of novel extremozymes

Search approach	Definition	Type of screening	Methodology	Challenges	Examples
Classic microbial/enzymatic screening	Efficient technique to find novel biocatalysts over cultured microorganisms	Functional	Environmental samples are collected from particular sites. Then, microorganisms are isolated from the environmental samples by applying specific selective pressures chosen to match the target characteristics of the biocatalyst to be developed. Finally, functional screening where enzymatic assays are performed over the supernatant and/or the crude extract of each selected microorganism is conducted to find the enzymatic activity of interest (Fig. 14.1)	– Its application is limited only to microorganisms that can be cultivated under laboratory conditions (1%) – Robust enzymatic assays are needed to screen for activities over microorganisms, and the assays must be adapted to work under extreme conditions – High-throughput systems for screening multiple samples simultaneously must be further developed and it is a requirement that the equipment and tools can cope with the extreme conditions required by extremophiles	– Three hyperthermophilic archaea growing as a consortium on crystalline cellulose at 90 °C (Graham et al. 2011) – A thermophilic xylanase found in *Geobacillus* sp. strain WSUCF1 when grown on xylan, (Bhalla et al. 2015)
Metagenomics	Cultivation-independent technique that consists in the extraction of all microbial DNAs in a certain environmental sample, constructing metagenomics libraries, and screening to seek novel functional genes	Sequence-based	Sequence-based screening is performed by using the polymerase chain reaction (PCR), hybridization and/or by using high-throughput sequencing which does not require cloning or PCR amplification, and can produce huge numbers of DNA reads at an affordable cost. Meaningful information is	– Its effectiveness depends on the concentration, quality, purity and fragment length of the starting DNA material – Novel sequences may not be (properly) annotated especially in the case of extremophilic genes – Databases does not always provide information about	– Functional expression of psychrophilic β-glucosidases and endoxylanases from metagenomes derived from enrichments from subseafloor sediments (Klippel et al. 2014) – Hydrolysis carried out at 50 °C on cellulose, xylan and corncob by four

(continued)

280   F. Sarmiento et al.

Table 14.2 (continued)

Search approach	Definition	Type of screening	Methodology	Challenges	Examples
			annotated by means of bioinformatics analysis based on sequence homology derived from known and characterized sequences	important catalytic and enzymatic properties	lignocellulose found in a metagenomics library of a long-term dry thermophilic methanogenic digester community (Wang et al. 2015)
		Functional	Function-based screening consists in the functional expression of metagenomics libraries in order to identify gene(s) or gene clusters of interest that display the desired activities in a specific industrial context. The most common screening approaches are direct phenotypical detection, heterologous complementation, induced gene expression and direct enzymatic activity detection	– Depends on the concentration, quality, purity and fragment length of the starting DNA material – A very large number of samples need to be screened – Robust enzymatic screening assays should be developed	

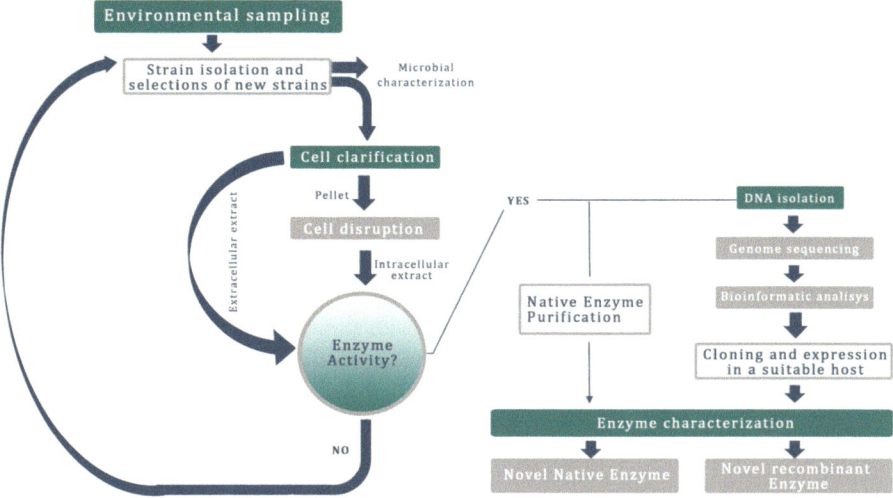

Fig. 14.1 Functional screening diagram

When searching for enzymes with potential applications for bioethanol production, the main targets correspond to microorganisms, mainly fungi and bacteria, living in decaying plant biomass where enzymes for the degradation of lignocellulosic material are probably active. There are many well-known bacterial genera that present cellulolytic activity such as *Clostridium, Butyrivibrium* and *Cellulomonas*. However, when the plant biomass is rich in lignin, fungus such as white-rot basidiomycetes are predominant because of their excellent production of ligninases. Also, some of the most studied microorganisms for the production of cellulolytic enzymes at industrial level (cellulases and hemicellulases) correspond to *Trichoderma reesei* and *Aspergillus niger*, two fungi which belong to the phylum *Ascomycota* (Stricker et al. 2008).

Nonetheless, the current trend in bioethanol production is to perform hydrolysis of lignocellulose under temperature close to 50 °C ensuring high yields of sugars (complete hydrolysis) and avoiding microbial contamination (Bhalla et al. 2013). In this context, the exploration of extreme environments for the search of enzymes for bioethanol production have yielded some interesting microbial resources that have helped to redefine the range of natural conditions under which cellulolytic organisms exist. For example, the work of Graham and collaborators described three hyperthermophilic archaea growing as a consortium on crystalline cellulose at 90 °C, which correspond to the first reported archaea able to optimally deconstruct lignocellulose above 90 °C (Graham et al. 2011). On the other hand, thermostable enzymes derived from thermophiles and hyperthermophiles are playing an important role, not only because of their robustness and their thermal stability, but also because of their high specific activity which allows reduced hydrolysis times and a reduction on the required amount of enzyme. Few examples of thermophiles and hyperthermophiles of interest for the production of thermostable cellulases, hemicellulases and ligninases belong to the genera *Bacillus, Geobacillus, Caldibacillus, Acidobacillus, Thermotoga, Anaerocellum* and *Rhodothermus*.

Even though classical bioprospection and functional screenings of enzymes in microorganisms is a straightforward methodology and it has been used for decades to develop novel biocatalysts with proven results, its application is limited only to microorganisms that can be cultivated under laboratory conditions, which corresponds to just 1% of the total number of prokaryotic species present in any given sample (Vartoukian et al. 2010). Classic laboratory techniques to grow microorganisms, such as liquid culture enrichment and plating, represent controlled systems that favor the microorganisms best suited for the particular tested conditions. This may result in biased sampling, where even the most dominant species in an environmental sample might not be represented. Many physicochemical variables affect the cultivation of microorganisms, such as temperature, pH, nutrient availability and oxygen levels. Biological factors, like the diverse interactions between different microorganisms and with other biological species, also play an important role. To take into account all of these different variables using classic culturing techniques is a complex task, so only a limited number of them are controlled when isolating novel microorganisms. Under these selected conditions, the majority of the microbes in an environmental sample cannot grow. This translates into the low diversity of cultured microorganisms. As a result, the majority of bacterial phyla identified so far have no cultured representatives.

Novel culturing approaches have been developed in recent years to advance the study of the unculturable microorganisms, often called "Microbial Dark Matter", in an effort to increase our knowledge of microbial diversity. These new approaches are largely based on growing the microorganisms *in situ* in the environment or by bringing the natural environment into the laboratory. In the latter approach, microorganisms grow with access to nutrients and chemicals from their natural environment but without the exposure to other external environmental forces. Few examples of these novel techniques are: diffusion chambers, gel microdroplets, dilution to extinction, and hollow-fiber membrane chambers (Vester et al. 2015). Another successful example corresponds to the iChip, a high-throughput *in situ* cultivation system based on miniature diffusion chambers which are inoculated with just one environmental cell (Nichols et al. 2010). The further development of novel culturing techniques like the iChip towards application for the cultivation of extremophilic microorganisms is needed to speed-up the discovery of novel extremozymes.

Additional technical challenges need to be addressed to successfully apply classic bioprospection towards developing novel extreme biocatalysts for biofuel applications. For example, robust enzymatic assays are needed to screen for activities over microorganisms, and the assays must be adapted to work under extreme conditions. Also, to expedite the discovery of novel enzymes, high-throughput systems for screening multiple samples simultaneously must be further developed. It is a requirement that the equipment and tools used in these systems can cope with the extreme conditions required by extremophiles. Hence, developing automated systems for working with extremophiles and extremozymes is a scientific and an engineering challenge.

Using extremophiles as producing strains for novel extremozymes is an ideal situation, however, normally they are not suitable for fermentation at large-scale in bioreactors, which translates in low cell mass yields. Indeed, certain hyperthermophiles do not grow optimally in bioreactors, potentially because of the accumulation of toxic compounds as result of Maillard reactions (Kim and Lee 2003). In addition, operating bioreactors under extreme conditions of temperature, pH and salt concentration shortens the life time of sensors and seals, which again exposes the need for novel tools to work with extremophiles. To avoid these issues, the current strategy is to clone and express the genes derived from extremophiles into well-known mesophilic hosts. Indeed, 90% of the enzymes currently used on industry correspond to recombinant versions (Adrio and Demain 2014). However, recombinant expression of genes that code for extremophilic enzymes in *E. coli* and other common bacterial and fungi hosts is also problematic. For example, a common issue during heterologous expression of genes in strains of *E. coli* is the incorrect folding of the expressed polypeptide and subsequent protein aggregation which translates into the formation of insoluble inclusion bodies (Rosano and Ceccarelli 2014). Other common issues correspond to protein toxicity, the presence or absence of chaperones, codon bias and poor secretion ability. There is a clear need for understanding and developing novel easy-to-use hosts which are able to express extremophilic enzymes. Recent efforts are reported for the use of *Talaromyces cellulolyticus*, a high cellulolytic enzyme-producing fungus, as a host for the expression of hyperthermophilic cellulases from the archaea *Pyrococcus* sp. (Kishishita et al. 2015). By using a glucoamylase promoter the recombinant expression yields in *T. cellulolyticus* were over 100 mg/L for two different hyperthermophilic cellulases. Other interesting hosts for industrial enzymes recombinant expression, including extremophilic enzymes for biofuels, correspond to *Bacillus subtilis, Saccharomyces cerevisiae, Pichia pastoris, Aspergillus niger* and *Trichoderma ressei* (Adrio and Demain 2014).

In spite of the above mentioned technical limitations, bioprospection of extremozymes from culturable microorganisms through the application of activity based enzymatic screenings is still an effective methodology to find novel biocatalysts for bioethanol production from plant biomass and other industrial applications. Recent research successfully reports the discovery and testing of novel extremophilic enzymes from isolated extremophiles for application in the production of bioethanol such as thermophilic xylanase cocktail found in *Geobacillus* sp. strain WSUCF1. When this thermophilic bacterium was grown on xylan or various inexpensive untreated and pretreated lignocellulosic biomasses, the enzyme cocktail with xylanase activity played an important role in hydrolyzing the hemicellulose component of lignocellulose to xylooligosaccharides and xylose at pH 6.5 and 70 °C, showing a half-life of 18 days. It was also reported that rates of hydrolysis on lignocellulosic material were better with the WSUCF1 secretome than those with commercial enzymes (Bhalla et al. 2015).

Other successful examples of thermostable enzymes for bioethanol production are the extracellular and intracellular hemicellulases and cellulases found in *Caldicellulosiruptor owensensis*. These enzymes deconstruct hemicellulose increasing the surface area and porosity of lignocellulose structure allowing the access for its cellulases to degrade cellulose. Through this enzymatic system *C. owensensis* was able to perform two-step hydrolysis to bioethanol production. Furthermore, the hydrolysis carried out by *C. owensensis* enzymes had a notable performance where commercial cellulases such as Cellic CTec2 (Novozymes) were utilized for lignocellulosic biomass hydrolysis. In this process, hyperthermal enzymolysis was performed at 70 or 80 °C by enzymes of *C. owensensis* followed by mesothermal enzymolysis (50–55 °C) with commercial cellulases. The advantages of this process are no sugar loss and few inhibitors generation (Peng et al. 2015).

Thermophilic laccases are also sought in thermophilic microorganisms because of their capacity in removing lignin from biomass, which enhances the efficiency of cellulose and hemicellulose hydrolysis and facilitates the utilization of carbohydrates in the production of lignocellulosic ethanol and other biofuels. *Myceliophthora thermophila* laccase–methyl syringate is an example of laccase-mediator system (LMS) applied to de-lignify unbleached eucalyptus kraft pulp. The potential of LMS to remove lignin from plant biomass could be exploited as an enzymatic pretreatment method in lignocellulosic ethanol production because the typical generation of inhibitory compounds such as furfural and phenols during thermochemical pretreatment is prevented (Christopher et al. 2014).

14.4 Metagenomics for the Discovery of Extreme Biocatalysts

In order to accelerate the process of enzyme discovery two factors should be taken in consideration: (1) efficiency and sensitivity of the screening strategy, and (2) the genetic diversity of candidate genes (Xing et al. 2012). Despite the fact that isolation and culture methods for the discovery of new microorganisms, as discussed above, are used as successful strategies for novel enzyme discovery, this approach imposes several constrains on the development of new enzymes, since nearly 99% of microorganisms from natural environments cannot be efficiently cultivated using current available methods (Vartoukian et al. 2010).

Metagenomics is a cultivation-independent technique that consists in the extraction of all microbial DNAs existing in a certain environmental sample, constructing metagenomic libraries, and screening to seek novel functional genes (Ferrer et al. 2005). This approach takes advantage of the rich diversity of genes and biochemical reactions of the millions of non-cultivated and uncharacterized microorganisms and offers an alternative to culturable-dependent approaches by greatly broadening the range of microbial resources that can be of use in the process of enzyme development. Metagenomics as a resource of great amount of genetic information emerged in the mid-late 1990s but it was back in 1995 when the first applications were reported, searching for cellulases encoding genes in microbial consortia from anaerobic digesters (Healy et al. 1995).

Metagenomic is a challenging technology with a great potential not only for industrial applications but also for the understanting of microbial adaptation and evolution. Following this approach several enzymes for biofuel production have been characterized from diverse extreme environments such as deep sea sediment, cold environments, alkaline lakes and volcanoes, among others.

In this section, research strategies and tools for the development of new biocatalysts from metagenomes will be analyzed; special emphasis will be on approaches for accessing novel biocatalysts from complex extremophilic environments and communities.

14.4.1 Sequence and Functional Screening Approaches

Screening using a metagenomic methodology can be performed on a sequence-based or a function-based approach (Table 14.2). Sequence-based screening is performed by using the polymerase chain reaction (PCR), hybridization and/or by using high-throughput sequencing (454 pyrosequencing, Ilumina, AB solid) which does not require cloning or PCR amplification, and can produce huge numbers of DNA reads at an affordable cost. Meaningful functional information is annotated by means of bioinformatic analysis based on sequence homology derived from known and characterized sequences. It is important to emphasize that using this approach novel sequences are possibly not to be annotated properly or simply not annotated at all (Vester et al. 2015). This is especially critical regarding the study of unexplored extremophilic environments, mainly due to the relative low amount of available extremophilic genes in annotated databases and the possible low phylogenetic affiliation of those extremophilic genes with their annotated mesophilic counterparts. This may create a bias and hinders the discovery of truly new enzymes since databases do not always give accurate information about important catalytic and enzymatic properties, such as substrate affinity and/or efficiency, optimal temperature, thermostability among others.

Function based screening consist in the functional expression of metagenomic libraries in order to identify gene(s) or gene clusters of interest that display the desired activities in a specific industrial context. Most common screening approaches are direct phenotypical detection, heterologous complementation, induced gene expression and direct enzymatic activity detection (Li et al. 2012). Functional-based screening holds the key to discover new biocatalysts and novel versions of known enzymes with potential biofuel applications. With this approach it is possible to solve problems such as substrate/product inhibition, stability, narrow substrate specificity or enantioselectivity by searching the right enzyme for a specific industrial setting. However, it is not that simple; since in order to find the right enzyme able to catalyze a specific transformation of industrial interest, commonly, a large number of samples need to be screened and robust enzymatic screening assays should be developed.

Recent examples for this activity-based approach are the functional expression of psychrophilic β-glucosidases and endoxylanases from metagenomes derived from enrichments from subseafloor sediments (Klippel et al. 2014). Functional analysis of the obtained enzymes revealed discrepancies and additional variability for the recombinant enzymes as compared to the sequence-based predictions.

Another example is the hydrolysis carried out at 50 °C on cellulose, xylan and corncob by four lignocellulose hydrolases (a cellulose, a xylanase, a β-xylosidase and a β-glucosidase) found in a metagenomic library of a long-term dry thermophilic methanogenic digester community (Wang et al. 2015). Optimal temperatures of these enzymes were found to be between 60° and 75 °C with more than 80% residual activities after 2 h at 50 °C. This work showed that screening of thermostable enzymes from microorganisms belonging to the same ecosystem could be a convenient strategy to degrade lignocellulose biomass.

Another effective functional approach is the use of heterologous complementation by foreign genes that are required for the growth of the host under specific selective conditions. Using this approach, it is possible to select the recombinant clones containing and producing the gene product in an active form.

14.4.2   DNA Extraction and Strategies for Sample Enrichment

The effectiveness of the metagenomic approaches is highly dependent of the concentration, quality, purity and fragment length of the starting DNA material. However, the yield and quality of DNA from environmental samples can be significantly affected by their chemical composition and degradation state, the type and abundance of the microbial community and the presence of interferents (*e.g.* humic compounds). In this context, direct DNA extraction from extremophilic environments is usually not a trivial task due to their inherent harshness. For example, high concentration of salts, metals, sulfur, and/or humic acids, and extreme pHs are common features of extremophilic environments. These extreme conditions may be difficult for DNA recovery and may affect several molecular methods such as downstream steps of PCR amplification, restriction, digestion and transformation. In some cases, extensive dilution of the crude DNA extract will allow direct PCR amplification, but this cannot fundamentally solve the problem. Therefore, further purification of DNA extracted from extreme environments is frequently mandatory for downstream processing. On the other hand, because of the difficulties to lyse some types of extremophilic microorganisms (*e.g.*

hyperthermophiles), biases may be introduced in the representation of individual genomes into the final obtained metagenome.

Despite the fact that many specific methods for the isolation of metagenomic DNA have been described, none of the methods reported hitherto are universally applicable and every type of sample requires optimization of DNA extraction methods. If the resultant concentration of metagenomic DNA is too low for downstream processing, as it is often the case when dealing with extremophilic samples, the metagenomic DNA can be greatly amplified by multiple displacement amplification (MDA) (Taupp et al. 2011). Even though this technique is biased and may yield short DNA fragments producing artificial sequences, this method can be considered as being able to deal with very dilute DNA samples for consequent PCR typing.

As metagenomes are large collections of genetic material, usually the target gene (s) may be problematic to identify, because they represent a very small portion of the total sequences contained in the metagenomic sample. To overcome this issue, several culture enrichment techniques have been developed, where microbial communities are exposed to physical, chemical and nutritional selective pressures, which increment the representation of desired phenotypes and dramatically enhance the gene hit rate. For example, a study from Grant and collaborators demonstrated that by using DNA isolated from enriched cultures grown on cellulose as their major carbon source, the cellulase activity found on the metagenomic samples increased three to four times when compared with metagenomic libraries isolated and prepared directly from total environmental DNA (Grant et al. 2009). However, culture enrichment techniques will promote the selection of fast-growing species, which is translated in the loss of a large proportion of the microbial diversity. This issue can be partially minimized by reducing the selection pressure to a mild level after a short period of stringent treatment.

14.4.3 Vectors and Host Selection

Appropriate vector selection plays an important role in metagenomic technology to successfully clone and express functional genes. The selection of vector systems depends on several factors such as the quality of the extracted DNA sample, the size of insert fragments, the needed number of copies of the vector, the host used, and the screening method. Plasmid, bacterial artificial chromosomes (BAC), cosmid, and fosmid are examples of the most frequently used vectors.

As important as choosing an appropriate expression vector, the selection of the host plays a major role on the successful functional expression of genes obtained by a metagenomic approach. Recombinant expression of genes can be biased, because of the large differences that can be found in the gene expression machineries among different taxonomic groups. Many times, the expression host does not recognize the sequence information and the enzyme expression is truncated or fails completely. As an example, *E. coli* is the most commonly used host due to the substantial genetic toolbox available. It has been suggested that only about 40% of the enzymatic activities found in metagenomic samples may be obtained by random cloning in *E. coli* (Gabor et al. 2004).

Alternative hosts for library construction and screening with different expression properties are under development and include *Bacillus subtilis*, *Pseudomonas putida*, *Streptomyces lividans* or *Rhizobium leguminosarum* (Wexler et al. 2005) Nonetheless, many improvements are still required for the expression of phylogenetically distant groups (*e.g.* Archaea).

Additionally, is important to consider that due to the phylogenetic/evolutionary distance between extremophiles and the available heterologous hosts from mesophilic origin, recombinant expression of extremophilic genes faces several additional challenges (*e.g.* toxicity of the gene product, breakdown of the gene product, protein misfolding leading to improper secretion and/or formation of inclusion bodies) based on the differences in their codon usage, recognition of promoters, missing initiation factors and/or cofactors, among others, which might affect the recombinant expression of extremophilic genes dramatically (Ekkers et al. 2012). Furthermore, it seems very difficult to predict any of these beforehand as the complete composition of metagenomics DNA is unknown and each challenge may vary from gene to gene and depends on the expression host used.

The efficiency of active enzyme expression derived from extremophilic metagenomes will be greatly improved as the technology continues to mature and new host bacteria are developed specifically for the expression of extremophilic genes (*e.g. Thermus thermophilus* or *Sulfolobus sulfataricus*). Nonetheless, in order to obtain a comprehensive solution, this should be complemented by developing optimal and broad cloning and expression vectors along with an improvement of fast functional screening methods for the detections of active extremozymes.

Recently a new thermostable screening system has been developed by Biométhodes (Evry, France; http://www.biomethodes.com) that relies on a thermophilic microorganism that allows plate selection of thermostable mutants after overnight cultivation. This methodology is being potentially applicable in metagenome library screening for biofuel enzymes with improved thermal activity and/or stability.

14.5   Protein Engineering of Extremozymes

Regardless of the approach used for the enzyme discovery and development, there is always room for improvement of the properties of a determined enzyme through protein engineering. Even enzymes from extremophiles, which are much more suitable for industrial processes, can benefit from protein engineering. For instance, enzymes such as cellulases, from either mesophilic or extremophilic origin, are known to have low degradation efficiency, and low enzyme activity, which tend to constrain their wider applications in industrial production.

Protein engineering is a powerful tool that allows the development of enzymes with new desirable properties such as thermostability, thermoactivity, specificity, enantioselectivity, pH adaptation, etc. It is based on the use of recombinant DNA technology and a number of different approaches have been developed for modifying proteins by mutagenesis of the parent gene (Antikainen and Martin 2005). The two main approaches are rational design and directed evolution although proteins can also be engineered by a combination of both.

Rational protein design involves site-directed mutagenesis for altering a gene or vector sequence at a selected location. Point mutations, insertions, or deletions are introduced by incorporating primers containing the desired modification(s) in a PCR reaction. Therefore, a vast and detailed knowledge and understanding of the three-dimensional structure, function and even the catalytic mechanism of an enzyme is needed in order to obtain an improvement in protein functionality from a point mutation (Arnold 1993).

There are two well-established methods to achieve site-directed mutagenesis: overlap extension and whole plasmid single round PCR (Antikainen and Martin 2005), and both offer a relatively inexpensive and rapid preparation of mutants with new/improved activities. However, one of the major drawbacks of protein engineering through rational design is that in many cases there is limited amount of structural or functional information of the enzymes, or this is simply unavailable. This situation is even more complex when studying enzymes from extremophilic microorganisms, where the information available is even more limited.

On the other hand, directed evolution works by mimicking evolution by natural selection and involves iterative rounds of random mutagenesis, thus, one of the main advantages is that it requires little prior knowledge of the target protein structure, function or mechanism. With random mutagenesis, point mutations are introduced at random positions in a target gene, usually through PCR employing an error-prone DNA polymerase (error-prone PCR), chemical mutagens, or saturation mutagenesis (Neylon 2004). The randomly mutated sequences are then cloned into an adequate expression vector, and the resulting mutant libraries are then screened with a highly sensitive assay, and pass through a suitable selection process that favors the desired protein properties to identify mutants with altered or improved properties. An additional technique known as DNA shuffling mimic the recombination that occurs naturally during sexual reproduction, mixing and matching pieces of successful variants in order to produce better results (Stemmer 1994; Antikainen and Martin 2005). For best results, both of these processes involve iterative rounds of evolution and selection to engineer a particular enzyme property.

However, one of the disadvantages of directed evolution is that the screening and selection of variants with the desired properties, among the great majority of mutants that are negative or neutral, is time-consuming and in order to automate this process, expensive robotic equipment is needed. Therefore, one of the major challenges of using random methods to engineer proteins is to find or develop an efficient high throughput screening (HTS) method for obtaining desired mutants, and it is also important to take into consideration that not all desired activities can be easily assessed by HTS (Liu et al. 2014).

It should be noted that currently effective pretreatment followed by a high load of industrial hydrolytic enzymes is needed to achieve an efficient, rapid and complete saccharification of biomass. The high cost of the required commercial enzymes constitutes one of the major technical and economical bottlenecks in the overall bioconversion process, hindering the use and commercialization of ligno-cellulosic bioethanol. Therefore, it is suggested that any improvement, either through rational design or directed evolution, in the activity and/or stability of any of the enzymes involved in the process might have the potential to effectively enhance plant biomass degradation and therefore, its utilization and conversion into bioethanol. Some examples of protein engineering of extremophilic carbohydrate-degrading enzymes associated to cellulose and hemicellulose hydrolysis are described below.

14.5.1 Rational Design of Extremophilic Glycoside Hydrolases

The Carbohydrate-Active Enzymes (CAZy) database classifies glycoside hydro-lases in different families according to their specific characteristics such as amino acid sequence, mechanism, and structure (Lombard et al. 2014). This may suggest that if the three-dimensional structure of a glycoside hydrolase within a family has been solved, other members of the same family will probably share the same protein fold, and as protein function is intimately linked to three-dimensional structure, this offers some clues for engineering proteins through rational design.

One example is the hemicellulase α-L-arabinofuranosidase (HiAXHd3) belong-ing to glycoside hydrolase family 43 (GH43). This enzyme from the thermophilic fungus *Humicola insolens* was modified by the mutation of Tyr166 to Ala166, generating a thermophilic hydrolase with both, arabinofuranosidase and xylanase activity. This promising work suggests that any glycoside hydrolase family 43 (all sharing a five-bladed β-propeller fold), provides a structural platform for generating multifunctional enzymes able to hydrolyze complex substrates through different action modes (McKee et al. 2012).

Another example is glycoside hydrolase family 52, where the hemicellulose β-xylosidase (GSxyn) from the thermophilic bacterium *Geobacillus stearothermophilus* 1A05585, has been successfully modified (by the mutation of Tyr509 to Glu509) to introduce a new catalytic xylanase activity, while conserving its β-xylosidase activity (Huang et al. 2014). In this case, rational design was not made based on the three-dimensional structure of the protein, as at the time that Huang and collaborators performed the experiments, there was no structure avail-able for glycoside hydrolase family 52. They did site-directed mutagenesis based on the amino acid sequence alignment of GSxyn and on the information of two other *G. stearothermophilus* xylanases where, in the same position, they have a glutamic acid instead of a tyrosine. GSxyn wild type displayed β-xylosidase activity using pNP-XP ($K_M = 0.48$ mM and $k_{cat} = 36.6$ s^{-1}), but no activity against beechwood xylan; on the other hand, Y509E mutant shows xylanolytic activity whilst retaining β-xylosidase activity ($K_M = 0.51$ mM and kcat $= 20.6$ s^{-1}) (Huang et al. 2014). The introduction of broadened substrate acceptance into GSxyn has verified the possibility for engineering (through a single amino acid substitution) additional catalytic functions, without losing the original enzymatic activity. As most of the residues, including this tyrosine, are strongly conserved across the majority of GH52 (Espina et al. 2014), it is suggested that the members of this family have also the potential to provide a structural scaffold for generating bifunctional enzymes (Huang et al. 2014).

Both examples represent an improvement of thermophilic hemicellulases that might have the potential to diminish the commercial enzyme load required during the enzymatic hydrolysis step of lignocellulosic bioethanol production as one enzyme would be performing the activities of two different enzymes.

To date there are several site-directed mutagenesis kits available on the market (*e.g.* QuickChange II Site-Directed Mutagenesis Kit from Agilent, Q5® Site-Directed Mutagenesis Kit from New England Biolabs, Phusion Site-Directed Mutagenesis Kit from Thermo Fisher) and there are also companies that offer site-directed mutagenesis services. Genewiz Inc., (South Plainfield, New Jersey, US) and GenScript (Piscataway, New Jersey, US) are just two examples of companies that can increase the efficiency of enzymes by performing site-directed mutagenesis, including deletion, insertion, and point mutations to obtain mutant constructs in a fast manner. PEACCEL is another service company specialized in protein engineering and synthetic biology that offer rational design of industrial enzymes. The approaches employed by PEACCEL for rational design are based on evolutionary and functional analysis of protein families, protein structure analysis and modeling, and it employs its unique expertise and robust in-silico methods for guiding design of mutant libraries in site directed mutagenesis, but also offers saturation mutagenesis and directed evolution experiments.

14.5.2 Directed Evolution of Extremophilic Glycoside Hydrolases

One example of successful directed evolution is the highly active Cel5A endoglucanase from *Thermoanaerobacter tengcongensis* MB4, where five variants with improved activities were identified using the Congo Red screening method (Teather and Wood 1982) from 4700 mutants generated after three rounds of error-prone PCR (Liang et al. 2011). When compared with the wild type Cel5A, the two best variants, 3F6 and C3–13, showed 135 and 193% of its specific activity against carboxymethyl cellulose (CMC) substrate (Liang et al. 2011).

Another example is the work of Wang and collaborators (2012) that mutagenized through directed evolution the cellobiohydrolase II (CBHII) encoding gene (*cbh2*) from the thermophilic fungus *Chaetomium thermophilum*. In their work, two mutants, CBHIIX16 and CBHIIX305, showed an optimum temperature of 60 °C and pH level 5 or 6, while the wild-type CBHII, has an optimum reaction temperature of 50 °C and pH 4. Moreover, after 1 h at 80 °C both mutants retained more than 50% of their activities while CBHII lost it all. A possible explanation for this enhanced characteristics of the mutants is given by the sequence analysis that revealed that CBHIIX16 contained five mutated amino acids while CBHIIX305 contained six (Wang et al. 2012).

These examples indicate an improvement of thermophilic cellulases that might also have the potential to diminish the commercial enzyme load required during the enzymatic hydrolysis step of lignocellulosic bioethanol production.

Furthermore, there are several companies dedicated to protein design and the improvement of enzymes through directed evolution. Codexis, Inc. (Redwood City, California, US) is one of the largest companies and has applied its protein engineering platform, CodeEvolver® and their ProSAR™ and MOSAIC® directed evolution technologies to improve biocatalysts such as a carbonic anhydrase (Alvizo et al. 2014). Another interesting company working in protein engineering is Novici Biotech (Vacaville, California, US) that have developed a proprietary gene shuffling technology, called Genetic ReAssortment by MisMatch Resolution (GRAMMR®), that overcomes the limitations of conventional variant gene library construction approaches to produce large (10^3–10^6) and high-quality shuffled gene libraries with exceptional crossover frequencies (10–20 per kb) and extremely granular crossover resolution (Padgett et al. 2010). This technology streamlines directed evolution workflows with rapid construction of initial libraries and production of subsequent libraries of re-shuffled hits at much lower per-gene cost and higher diversity content than can be achieved by other shuffling methods or by gene synthesis.

14.5.3 Semi-Rational Protein Engineering and Design

Even though engineering proteins through the classical directed evolution approach has been proven to be successful, it is widely known that it can also be very challenging due to various reasons, including the intrinsic disadvantages mentioned above, along with the very low coverage of the immense sequence space possible for any average protein, even when generating protein libraries with millions of members (Lutz 2010). In addition, library design is constrained by the degeneracy of the genetic code and the experimental method bias. Therefore, many researchers are now inclined to move towards new strategies for designing more efficient libraries (smaller and of higher quality) through a more rational design rather than larger libraries and more screening and selection methods (Lutz 2010).

Often referred to as semi-rational, smart or knowledge-based library design, these novel approaches utilize information on protein sequence, structure and function and also computational predictive algorithms to preselect promising target sites and restrict amino acid diversity for protein engineering. As a result, the size of the library is dramatically reduced and, when considering evolutionary variability, topological constraints and mechanistic features to weigh-in on amino acid identity libraries with higher functional content are produced (Lutz 2010). Furthermore, as these smaller high-quality libraries require less iteration to identify variants with the desired phenotype, they may possibly soon reduce the need for high-throughput methods (as well as their limitations) during library analysis (Lutz 2010).

A critical component to the success of this emerging engineering strategy is the development and advances in computational tools for the evaluation of protein sequence datasets and the analysis of conformational variations of amino acids in protein (Lutz 2010). For instance, RosettaDesign server identifies low energy amino acid sequences for target protein structures (http://rosettadesign.med.unc.edu). After providing the backbone coordinates of the target structure and specifying which residues to design, the server returns the sequences, coordinates and energies of the designed protein (Liu and Kuhlman 2006). Another server is FoldX, (http://foldx.embl.de/) which is an empirical force field that was developed for the rapid evaluation of the effect of mutations on the stability, folding and dynamics of proteins and nucleic acids through the calculation of the free energy of a macromolecule based on its high-resolution three-dimensional structure (Schymkowitz et al. 2005). Both of these strategies can be also used for computational design of protein thermostability by FRESCO (Framework for Rapid Enzyme Stabilization by Computation), which is a computational protocol that employs molecular dynamics simulations of up to 500 designed variants to eliminate poor designs, which increases the relative abundance of improved variants among the experimentally tested designs (Wijma et al. 2014)

These are promising predictors for altering protein features such as substrate specificity, stereoselectivity and stability by enzyme redesign, as well as the creation of new functions by de novo design (a bottom-up approach that entails designing an entirely new protein, one amino acid at a time).

14.6 Conclusions

The production of bioethanol from plant biomass is a complex task which requires a multi-step process. For the first two steps, pretreatment and enzymatic hydrolysis, the finding of novel biocatalysts adapted to some of the harsh conditions employed is becoming increasingly important.

Extremophiles appear as a rich source of extremozymes, which are better adapted to the process conditions required for biofuel production, and it is suggested that their use can potentially reduce process costs associated to the bioconversion of plant cell wall to biofuels. Extremozymes are superior to their mesophilic counterparts currently used during bioethanol production, their uses have the potential to effectively enhance plant biomass degradation and its utilization and conversion into bioethanol. Among these, thermostable enzymes are playing an important role because of their robustness, thermal stability, and higher specific activity.

In order to accelerate the process of enzymes discovery and development, classical functional approach for the identification of extremozymes along with the use of more sophisticated technologies such as metagenomics and bioinformatics for searching extreme environments is required. As previously discussed, currently only 1% of microorganisms in an environmental sample are culturable. The other vast majority of microbial diversity must be reached either by improving culturing techniques as this certainly holds the key for obtaining better industrial biocatalysts, or through cultivation-independent techniques, although DNA does not always represent a reliable indicator of novel industrial enzymes. In both cases, cloning and recombinant expression is necessary, and appropriate molecular tools, such as novel expression vectors, as well as selection of suitable hosts play a major role on the successful functional expression of genes obtained from extremophiles or DNA isolated from extreme environments. Currently, the majority of the molecular tools available are designed for expression of mesophilic enzymes, so the development of novel cloning and expression tools is one of the biggest challenges faced to advance further the use of extremozymes on industrial settings.

Directed evolution and rational design for protein engineering are currently the most used techniques for improving the performance of new extremozymes allowing the development of improved enzymes with better thermostability, activity, specificity, enantioselectivity, pH adaptation, etc. The most suitable technique to use will mostly depend on the existing structural knowledge of the target enzyme as both methods have been proven to be successful at engineering cellulases and hemicellulases. Semi-rational protein engineering and design offers a good alternative to the most common methods since it does not strictly require all the protein information to rationalize and improve the quality of an enzyme.

The use of novel extreme biocatalysts is certainly required for the generation of new forms of energy based on renewable resources such as lignocellulosic feedstock.

Take Home Message

- Lignocellulosic biomass is the most abundant renewable natural resource and produced from municipal, agricultural and industrial sources. Lignocellulose is chiefly composed of three types of polymers namely cellulose, hemicellulose and lignin. Cellulose is a linear polysaccharide composed of β-1,4 linked glucose units aggregated into microfibrils and it comprises of 35–50% of the plant biomass. Hemicelluloses constitutes 20–35% of plant biomass and is composed of xylan, xyloglucan, mannan and glucomannan. It is a highly-branched heteropolysaccharides composed of pentoses, hexoses and/or uronic acids.
- Lignin is a complex organic aromatic heteropolymer and it constitutes 10–25% of plant biomass. It consists of three methoxylated monolignols incorporated into lignin in the form of guaiacyl, syringyl and p-hydroxyphenyl present in diverse amounts, depending on the source of lignin. In addition to the main

polymers of the lignocellulosic matrix, other minor components are present such as pectin, proteins, lipids, soluble sugars and minerals.

- Lignocellulose is a very complex and recalcitrant material. Lignin is degraded by Lignin-modifying enzymes (LMEs) such as Lignin Peroxidases, Manganese peroxidase, Laccases, Versatile peroxidase, Glucose oxidase, Glyoxal oxidase, Aryl alcohol oxidase, Cellulases (*e.g.* endo-1,4-β-D-glucanases, exo-1,4-β-D-glucanases, exo-1,4-β-D-glucan cellobiohydrolase and β-D-glucosidases) and Hemicellulases (*e.g.* Endo-1,4-β-D-xylanases, β-D-xylosidases, α-L-arabinofuranosidases, α-D-glucuronidases, Acetyl xylan esterases, Feruloyl esterases, p-Coumaroyl esterases, Mannan endo-1,4-β-mannosidase, exo-β-D-mannanase, exo-1,4-β-mannobiohydrolase, Acetyl esterase, Xyloglucan-specific endo-β-1,4-Glucanase and Glucuronoarabinoxylan endo-1,4-β-Xylanase. Lignocellulosic feedstock containing high content of pectin can be treated with the pectinases such as Endo-pectin lyases, Exo-pectin lyases, Endo-pectate lyases, Rhamnogalacturonan lyase, Endo-polygalacturonase, Exo-polygalacturonase, endo-1,4-β-galactanase, β-galactosidase, Arabinan Endo-1,5-α-L-arabinanase, α-L-arabinofuranosidases, pectin esterase, acetyl esterase and rhamnogalacturonan acetyl esterase.
- Extremozymes are the best starting point in search of new biocatalysts for industrial applications and biofuels generation. They can also be improved in their performance through the use of protein engineering leading to the development of new extremozymes with desirable properties as thermostability, thermoactivity, specificity, enantioselectivity, pH adaptation, required by industry.

References

Adrio JL, Demain AL (2014) Microbial enzymes: tools for biotechnological processes. Biomolecules 4:117–139

Alvizo O, Nguyen LJ, Savile CK et al (2014) Directed evolution of an ultrastable carbonic anhydrase for highly efficient carbon capture from flue gas. Proc Natl Acad Sci USA 111:16436–16441

Antikainen NM, Martin SF (2005) Altering protein specificity: techniques and applications. Bioorg Med Chem 13:2701–2716

Arantes V, Saddler JN (2010) Access to cellulose limits the efficiency of enzymatic hydrolysis: the role of amorphogenesis. Biotechnol Biofuels 3:4

Arnold FH (1993) Protein engineering for unusual environments. Curr Opin Biotechnol 4:450–455

Bhalla A, Bansal N, Kumar S et al (2013) Improved lignocellulose conversion to biofuels with thermophilic bacteria and thermostable enzymes. Bioresour Technol 128:751–759

Bhalla A, Bischoff KM, Sani RK (2015) Highly thermostable xylanase production from a thermophilic *Geobacillus* sp. strain WSUCF1 utilizing lignocellulosic biomass. Front Bioeng Biotechnol 3:84

Chen F, Dixon RA (2007) Lignin modification improves fermentable sugar yields for biofuel production. Nat Biotechnol 25:759–761

Christopher LP, Yao B, Ji Y (2014) Lignin biodegradation with laccase-mediator systems. Front Energy Res 2:12

Dewan SS (2014) Global markets for enzymes in industrial applications. BCC Research, Wellesley, MA

Ekkers DM, Cretoiu MS, Kielak AM et al (2012) The great screen anomaly a new frontier in product discovery through functional metagenomics. Appl Microbiol Biotechnol 93:1005–1020

Espina G, Eley K, Pompidor G et al (2014) A novel β-xylosidase structure *from Geobacillus thermoglucosidasius*: the first crystal structure of a glycoside hydrolase family GH52 enzyme reveals unpredicted similarity to other glycoside hydrolase folds. Acta Crystallogr D Biol Crystallogr 70:1366–1374

Ferrer M, Golyshina OV, Chernikova TN et al (2005) Novel hydrolase diversity retrieved from a metagenome library of bovine rumen microflora. Environ Microbiol 7:1996–2010

Gabor EM, Alkema WB, Janssen DB (2004) Quantifying the accessibility of the metagenome by random expression cloning techniques. Environ Microbiol 6:879–886

Graham JE, Clark ME, Nadler DC et al (2011) Identification and characterization of a multidomain hyperthermophilic cellulase from an archaeal enrichment. Nat Commun 2:375

Grant S, Sorokin DY, Grant WD et al (2009) A phylogenetic analysis of Wadi el Natrun soda lake cellulase enrichment cultures and identification of cellulase genes from these cultures. Extremophiles 8:421–429

Healy FG, Ray RM, Aldrich HC et al (1995) Direct isolation of functional genes encoding cellulases from the microbial consortia in a thermophilic, anaerobic digester maintained on lignocellulose. Appl Microbiol Biotechnol 43:667–674

Huang Z, Liu X, Zhang S et al (2014) GH52 xylosidase from *Geobacillus stearothermophilus*: characterization and introduction of xylanase activity by site-directed mutagenesis of Tyr509. J Ind Microbiol Biotechnol 41:65–74

Kim KW, Lee SB (2003) Inhibitory effect of Maillard reaction products on growth of the aerobic marine hyperthermophilic archaeon Aeropyrum pernix. Appl Environ Microbiol 69:4325–4328

Kishishita S, Fujii T, Ishikawa K (2015) Heterologous expression of hyperthermophilic cellulases of archaea *Pyrococcus* sp. by fungus Talaromyces cellulolyticus. J Ind Microbiol Biotechnol 42:137–141

Klippel B, Sahm K, Basner A et al (2014) Carbohydrate-active enzymes identified by metagenomic analysis of deep-sea sediment bacteria. Extremophiles 18:853–863

Li S, Yang X, Yang S (2012) Technology prospecting on enzymes: application, marketing and engineering. Comput Struct Biotechnol J 2:1–11

Liang C, Fioroni M, Rodríguez-Ropero F et al (2011) Directed evolution of a thermophilic endoglucanase (Cel5A) into highly active Cel5A variants with an expanded temperature profile. J Biotechnol 154:46–53

Liu Y, Kuhlman B (2006) RosettaDesign server for protein design. Nucleic Acids Res 34:W235–W238

Liu M, Xie W, Xu H et al (2014) Directed evolution of an exoglucanase facilitated by a co-expressed β-glucosidase and construction of a whole engineered cellulase system in *Escherichia coli*. Biotechnol Lett 36:1801–1807

Lombard V, Ramulu HG, Drula E et al (2014) The carbohydrate-active enzymes database (CAZy) in 2013. Nucleic Acids Res 42:D490–D495

Lutz S (2010) Beyond directed evolution-semi-rational protein engineering and design. Curr Opin Biotechnol 21:734–743

McKee LS, Peña MJ, Rogowski A et al (2012) Introducing endoxylanase activity into an exo-acting arabinofuranosidase that targets side chains. Proc Natl Acad Sci USA 109:6537–6542

Neylon C (2004) Chemical and biochemical strategies for the randomization of protein encoding DNA sequences: library construction methods for directed evolution. Nucleic Acids Res 32:1448–1459

Nichols D, Cahoon N, Trakhtenberg EM et al (2010) Use of ichip for high-throughput in situ cultivation of "uncultivable" microbial species. Appl Environ Microbiol 76:2445–2450

Padgett HS, Lindbo JA, Fitzmaurice WP (2010) Method of increasing complementarity in a heteroduplex. USPTO. United States, Novici Biotech LLC. US7833759 B2

Peng X, Qiao W, Mi S et al (2015) Characterization of hemicellulase and cellulase from the extremely thermophilic bacterium *Caldicellulosiruptor owensensis* and their potential application for bioconversion of lignocellulosic biomass without pretreatment. Biotechnol Biofuels 8:131

Rosano GL, Ceccarelli EA (2014) Recombinant protein expression in *Escherichia coli*: advances and challenges. Front Microbiol 5:172

Scheller HV, Ulvskov P (2010) Hemicelluloses. Annu Rev Plant Biol 61:263–289

Schymkowitz J, Borg J, Stricher F et al (2005) The FoldX web server: an online force field. Nucleic Acids Res 33:W382–W388

Stemmer WP (1994) DNA shuffling by random fragmentation and reassembly: in vitro recombination for molecular evolution. Proc Natl Acad Sci USA 91:10747–10751

Stricker AR, Mach RL, De Graaff LH (2008) Regulation of transcription of cellulases- and hemicellulases-encoding genes in *Aspergillus niger* and *Hypocrea jecorina* (*Trichoderma reesei*). Appl Microbiol Biotechnol 78:211–220

Taupp M, Mewis K, Hallam SJ (2011) The art and design of functional metagenomic screens. Curr Opin Biotechnol 22:465–472

Teather RM, Wood PJ (1982) Use of Congo red-polysaccharide interactions in enumeration and characterization of cellulolytic bacteria from the bovine rumen. Appl Environ Microbiol 43:777–780

Vartoukian SR, Palmer RM, Wade WG (2010) Strategies for culture of 'unculturable' bacteria. FEMS Microbiol Lett 309:1–7

Vester JK, Glaring MA, Stougaard P (2015) Improved cultivation and metagenomics as new tools for bioprospecting in cold environments. Extremophiles 19:17–29

Viikari L, Vehmaanperä J, Koivula A (2012) Lignocellulosic ethanol: from science to industry. Biomass Bioenergy 46:13–24

Wang M, Lai GL, Nie Y et al (2015) Synergistic function of four novel thermostable glycoside hydrolases from a long-term enriched thermophilic methanogenic digester. Front Microbiol 6:509

Wang XJ, Peng YJ, Zhang LQ et al (2012) Directed evolution and structural prediction of cellobiohydrolase II from the thermophilic fungus *Chaetomium thermophilum*. Appl Microbiol Biotechnol 95:1469–1478

Wexler M, Bond PL, Richardson DJ et al (2005) A wide range-host metagenomic library from a waste water treatment plant yields a novel alcohol/aldehyde dehydrogenase. Environ Microbiol 7:1917–1926

Wijma HJ, Floor RJ, Jekel PA et al (2014) Computationally designed libraries for rapid enzyme stabilization. Protein Eng Des Sel 27:49–58

Xing MN, Zhang XZ, Huang H (2012) Application of metagenomic techniques in mining enzymes from microbial communities for biofuel synthesis. Biotechnology Adv 30:920–929

Erratum to: Lytic Polysaccharide Monooxygenases

Madhu Nair Muraleedharan, Ulrika Rova, and Paul Christakopoulos

Erratum to:
Chapter 6 in: R.K. Sani et al. (eds.), *Extremophilic Enzymatic Processing of Lignocellulosic Feedstocks to Bioenergy,*
https://doi.org/10.1007/978-3-319-54684-1_6

The published version of this book (Extremophilic Enzymatic Processing of Lignocellulosic Feedstocks to Bioenergy; Rajesh K. Sani, R. Navanietha Krishnaraj (Eds)) included misspelling in the heading and the first line of the abstract of the chapter Lytic Polysaccharide Monooxygenases (Madhu Nair Muraleedharan, Ulrika Rova, Paul Christakopoulos, pp. 89–98). The word monooxygensases was corrected to monooxygenases.

The updated online version of the original chapter can be found at
https://doi.org/10.1007/978-3-319-54684-1_6

Questions

1. What the factors enhancing the thermostability of enzymes?
2. Describe about the two substrates acted upon by single enzymes and two enzymes acting on a single substrate?
3. All enzymes are not proteins and all proteins are not enzymes. Explain.
4. What are the advantages and disadvantages of feed enzymes for poultry industry? How the limitations can be overcome by the use of extremozymes?
5. What are the protein engineering techniques for improving the structural and catalytic activity of the enzymes?
6. How feather wastes can be treated by enzyme technology? What value added products can be made from them?
7. Why cytochromes are called as dehydrogenases?
8. What are the key criteria to be considered for developing an enzyme for space application?
9. What is a Bionic enzyme? What are its applications?
10. Suggest the best immobilisation strategy for the immobilisation of polyphenol oxidase in a packed bed reactor, for the treatment of effluent containing polyphenol.
11. What is enzymatic electrocatalysis? How can the enzymes immobilised on paper electrodes or screen print electrodes for biosensor applications?
12. What is thermodynamic stability of the enzyme?
13. How many genes coding for the enzymes are there in the human genome of human? What characteristic of the enzyme confers the thermostability?
14. Compare the sequence any one normal enzyme and the thermostable enzyme and analyse what changes could be observed in its sequence.
15. Compare the structure of any one normal enzyme and the thermostable enzyme by superimposing their structures with any one molecular visualisation tool (PYMOL) and analyse what changes could be observed in its structure. What is its RMSD?
16. Microorganisms have feedback mechanisms. Do enzymatic reactions also have feedback mechanisms?

© Springer International Publishing AG 2017
R.K. Sani, R.N. Krishnaraj (eds.), *Extremophilic Enzymatic Processing of Lignocellulosic Feedstocks to Bioenergy*, DOI 10.1007/978-3-319-54684-1

17. How can you find the size of the enzyme?
18. What are constitutive or adaptive enzymes?
19. How to improve the oxygen sensitivity of the enzyme?
20. How can the catalytic activity of the enzyme analysed from the molecular interactions between the substrate and enzyme analysed using bioinformatics approaches?
21. What is consolidated bioprocessing?
22. What is site directed mutagenesis?
23. How can the sequence of amino acids in the active site of the enzyme identified?
24. What is eurythermalism?
25. Which is the rate limiting step in the enzymatic degradation of lignocellulose?
26. What is biomass recacitance?
27. Describe directed evolution, rational design and multifunctional chimeras for improving the enzyme activity.
28. What is the difference in the mechanism of action of endoglucanases, cellobiohydrolases and B-glucosidases?
29. What are the advantages of extremophilic chitinases?
30. What is the application of extremophilic enzymes for leaching of ores?
31. What are the applications of metagenomic techniques in mining enzymes?

Index

A

Absolute substrate specificity, 10, 27
Accessory enzymes, 94, 198, 199, 220
Acetyl esterase, 215, 273, 295
Acetyl mannan esterases (AMEs), 206, 215
Acetyl xylan esterases, 76, 273
Acid hydrolysis, 101, 163
Acid mammalian chitinase (AMCase), 242
Acid-based pretreatments, 35, 38–39, 46, 50
Acidophiles, 2, 4, 26, 100, 117, 118, 157, 158,
    172, 182, 200, 260
Acidophilic enzyme, 26
Activation energy, 6, 16
Active site, 6, 9, 14, 15, 20, 21, 27, 54, 56, 92,
    93, 97, 102, 120, 123, 141–143, 147,
    216, 218, 219, 252, 265, 266
Adsorption, 2, 22, 23, 27, 219
Adverse conditions, 2, 4, 164
Alanine aminotransferase (ALT), 7
Alcohol dehydrogenase, 11
Algae, 2, 22, 117, 159, 227, 243
Alkaline pretreatments, 39
Alkaliphiles, 2, 4, 26, 100, 105, 106, 117, 118,
    157, 158, 182, 259
Alkalophilicity, 81, 82, 267
Alkanine protease (ALP), 7
Allosteric enzymes, 15
Alpha-L-iduronidase, 7
Amino acid modification, 74, 80, 82, 84
Amino acid sequence, 12, 54, 55, 93, 97, 101,
    108, 111, 140, 216, 227, 232, 290, 293
Aminophenols, 138, 276
Ammonia fiber explosion, 36, 39, 40, 42, 44,
    48, 274
Ammonia recycle percolation (ARP), 37, 41

Amylase, 99, 100
α-Amylase, 7, 100–111
Amylopectin, 96, 97, 101, 102, 111
Amylose, 96, 97, 101, 107, 111
Antibiotics, 24, 59, 151, 214
Antibody binding, 159
Apoenzyme, 6, 11, 12
Arabinan Endo-1,5-α-L-arabinanase, 273
Arabinans, 160, 273, 295
Arabinogalactans, 160
Arabinoxylan, 76, 77, 79, 83, 214
Arginase, 10
Aromatic ring cleavage, 144
Arrhenius equation, 16
Aryl alcohol oxidase, 273
Asparaginase, 7
Asparagine, 7, 82

B

Bacteria, 2, 199, 208, 214, 219, 220, 225, 227,
    229, 232, 258, 260, 262, 265, 271, 273,
    274, 281, 288
Bacterial artificial chromosomes (BAC), 59,
    287
Bacterial chitinases, 229–231
Barophiles, 117, 182
Bilirubin oxidase, 7
Biobleaching, 74, 77, 119, 151–153
Biocatalysts, 2, 6, 89, 107, 183–185, 193, 199,
    200, 208, 249, 250, 256, 258, 271, 274,
    275, 279, 282–288, 292–294
Biochemical reaction, 6
Biofuels, 26, 62, 95, 110, 151, 215, 226, 243,
    249, 271–275, 283, 284, 293

R.K. Sani, R.N. Krishnaraj (eds.), *Extremophilic Enzymatic Processing of Lignocellulosic Feedstocks to Bioenergy*, DOI 10.1007/978-3-319-54684-1